MICROBIAL INFECTIONS
Role of Biological Response Modifiers

ADVANCES IN EXPERIMENTAL MEDICINE AND BIOLOGY

Recent Volumes in this Series

A Continuation Order Plan is available for this series. A continuation order will bring delivery of each new volume immediately upon publication. Volumes are billed only upon actual shipment. For further information please contact the publisher.

MICROBIAL INFECTIONS
Role of Biological Response Modifiers

Edited by

**Herman Friedman and
Thomas W. Klein**

University of South Florida
Tampa, Florida

and

Hideyo Yamaguchi

Teikyo University School of Medicine
Tokyo, Japan

Springer Science+Business Media, LLC

Library of Congress Cataloging-in-Publication Data

Microbial infections : role of biological response modifiers / edited
  by Herman Friedman and Thomas W. Klein and Hideyo Yamaguchi.
       p.   cm. -- (Advances in experimental medicine biology ; v.
  319)
     Proceedings of an international conference held May 29-31, 1991,
  in Tampa, Florida.
     Includes bibliographical references and index.
     ISBN 978-1-4613-6519-8   ISBN 978-1-4615-3434-1 (eBook)
     DOI 10.1007/978-1-4615-3434-1

     1. Biological response modifiers--Congresses. 2. Communicable
  diseases--Immunotherapy--Congresses.   I. Friedman, Herman, 1931-
  . II. Klein, Thomas W. III. Yamaguchi, Hideyo, 1934-
  IV. Series.
     [DNLM: 1. Biological Response Modifiers--immunology--congresses.
  2. Communicable Diseases--immunology--congresses. 3. Communicable
  Disaeases--microbiology--congresses.   W1 AD559 v.319 / WC 195 M626
  1991]
  QR185.8.B54M53   1992
  616'.01--dc20
  DNLM/DLC
  for Library of Congress   92-21809
                                                            CIP

Proceedings of an international conference on Microbial Infections:
Role of Biological Response Modifiers, held May 29-31, 1991,
in Tampa, Florida

ISBN 978-1-4615-3434-1

© 1992 Springer Science+Business Media New York
Originally published by Plenum Press, New York in 1992
Softcover reprint of the hardcover 1st edition

PREFACE

This volume is based on the Proceedings of the International Conference on "Microbial Infections: Role of Biological Response Modifiers" held in Tampa, FL, May 29-31, 1991. The major purpose of this conference was to bring together in one forum prominent investigators from around the world studying a variety of microbial pathogens, including bacteria, viruses, and fungi, and the effects of biological response modifiers (BRM) on the immune response to these microorganisms. BRM have been widely utilized in the area of antitumor resistance and include not only experimental tumor cell vaccines, but also biologically active substances such as cytokines, i.e., interferons, tumor necrosis factor, and interleukins, as well as products from bacteria which influence host resistance mechanisms. It is the belief of the organizers of this Conference that it was very timely to discuss in detail BRMs as they impact on microbial infections per se.

It is now widely accepted that immunocompromised individuals, including those exposed to immunosuppressive substances such as antimetabolites used for chemotherapy of malignancies, or infectious agents, such as the human immunodeficiency virus and other viruses which depress the immune response and, in turn, affect a host so as to become highly susceptible to opportunistic microorganisms, benefit from BRM stimulation of their immune system. A wide variety of immunomodulators are now being studied in terms of treating infectious diseases, as well as malignancy and autoimmune diseases. In this regard it was apparent to the organizers as well as participants of this conference, that this was an important topic to discuss and information concerning newer developments in the field of BRM as applied to microbial infection should be presented.

The first review in this proceedings is a discussion by Dr. Evan Hersh, University of Arizona, concerning the value of immunomodulator drugs in management of microbial infection. Dr. John W. Hadden, University of South Florida, then discusses in general the topic of immunotherapy of infectious diseases, including the important contemporary subject of retroviruses which cause acquired immunodeficiency. A series of reviews then follows concerning the utilization and mechanisms of action of BRMs in various infections. Endotoxins derived from gram-negative bacteria are represented first. These have been utilized for nearly half a century for increasing host resistance to not only tumors but also to a wide variety of bacterial, fungal, parasitic, and viral infectious agents. BRM such as cytokines, muramyldipeptides and other microbial products, are considered next along with the importance of these substances in immunomodulation.

The papers on bacterial immunity are followed by reviews of the effects of immune response mechanisms in antifungal immunity, especially *Candida albicans* infection, as well as cryptococcal infections. Several papers then describe the effects of BRMs as antiviral agents, especially in immunocompromised individuals. The topic of

Kampo medicine, or use of herb extracts in disease processes, is an ancient subject and forms the basis of several presentations made at the conference concerning medicinal plants and immune responses. A general review of this subject is presented by the co-organizer of the meeting, Dr. Hideyo Yamaguchi, Japan, who discusses immunomodulation by medicinal plants. Dr. Gerhard Franz, Germany, presents an overview of the effects of polysaccharides from such plants in the immune response system. Papers are then presented concerning effects of herbal products on various pathways of the immune response system, including the effects on hemopoietic cells.

It is the belief of the conference organizers and particiants that this description of newer studies concerning BRMs and the immune response to microorganisms will stimulate increased interest in the biomedical community in general for this rapidly evolving field. The editors wish to thank Ms. Sally Baker, Ms. Judy Flynn and Ms. Ilona Friedman for outstanding editorial assistance in preparing this book.

H. Friedman
T. W. Klein
H. Yamaguchi

Tampa, FL

CONTENTS

IMMUNOMODULATORY DRUGS OF RELEVANCE TO THE

MANAGEMENT OF MICROBIAL INFECTIONS

Evan M. Hersh

Section of Hematology and Oncology
Arizona Cancer Center
Tucson, Arizona   85724

INTRODUCTION

Microbial infections remain a major challenge to the health care system. With the advent of the AIDS epidemic, various infections which cannot be controlled with antimicrobial agents alone have appeared (1). Other immunocompromised populations such as aged subjects and patients receiving immunosuppressive therapy are also increasing. These immunosuppressed populations, some of whom have concomitant myelosuppression may not respond well to antibiotic therapy and therefore experience increased morbidity and mortality from microbial infections.

Clearly, therapy directed at induction and/or restoration of normal immune competence and at correction of myelosuppression is of high priority in these groups. Recently, considerable progress has been made in correcting granulocytopenia through the administration of recombinant DNA produced GM-CSF (2) and G-CSF (3). These agents have both been approved recently by the FDA for the treatment of chemo-therapy-induced granulocytopenia.

It seems to this author that a similar effort should be made to bring immunomod-ulatory drugs into the management of microbial infections in the context of host-defense failure. In this context it is presumed that such drugs would be used in conjunction with antibiotics as well as the bone marrow growth factors. In this review, we will define the nature and characteristics of immunomodulatory drugs and indicate their current status and potential future application to the treatment of microbial infection.

IMMUNOMODULATION

The therapeutic approach of immunomodulation is directed at two distinct circumstances: 1) immunodeficiency, and 2) immune perturbation which can be corrected. Immune perturbation is usually hyperactive function of one or more components of the immune system as in autoimmune or inflammatory diseases. However, it may refer to an imbalance in the components of the immune response. In addition, the term immunomodulation is also applied to non-specific, not strictly

Microbial Infections, Edited by H. Friedman *et al.*
Plenum Press, New York, 1992

"immune" components of the host-defense system, such as alteration of macrophage function, repletion of deficient metabolites or modification of cytokine function (see below).

Immunomodulation is closely related to immunostimulation. That implies the activation of normal mechanisms by active non-specific or specific mechanisms. Obviously, immunomodulation may permit immunostimulation to occur and some agents, such as the cytokines IL-2 and IFN α may be both immunomodulators and immunostimulants (4, 5).

IMMUNOMODULATORS

An immunomodulator is defined as a drug which can normalize impaired, deficient, perturbed or hyperactivated immune responses and host-defense mechanisms. Thus, deficient or suppressed cellular and humoral immune responses are restored while aberrant responses such as an autoimmune response as well as the associated autoimmune disease are reduced or ameliorated. It is of interest, that the same drug can have immunorestorative activity and yet be able to down-regulate immune responses. This may be through a differential effect on the balance between "helper" and "suppressor" cells or factors in the immune system depending upon their status at the onset of therapy.

Immunomodulators are of increasing interest since clinical conditions, in which immune activation and immunodeficiency co-exist are being recognized more frequently (6).

There are drugs for which the only known biological activity is immunomodulation and these agents are properly called immunomodulators. Others, such as corticosteroids, non-steroidal anti-inflammatory agents, cytotoxic or cytostatic cytokines and cytotoxic chemotherapeutic agents have immunomodulatory activity in addition to other biological activity.

CLASSIFICATION

There is no formal classification of immunomodulators which is generally accepted. The classification outlined below has been developed by the author and is felt to be a useful way of examining this class of drugs. Immunomodulator drugs include: 1) thymic hormones, 2) cytokines, 3) chemical immunomodulators, 4) corticosteroids, 4) non-steroidal anti-inflammatory agents, 5) specific T cell immunosuppressants, and 6) cytotoxic chemotherapeutic agents.

Immunomodulatory drugs can also be considered from the point of view of a general classification of immunotherapeutic or biological therapy agents and approaches. These include 1) active non-specific immunotherapy with such agents as muramyl tripeptide (MTP-PE) (7) a macrophage activator, 2) active specific immunotherapy with tumor cell or tumor antigen vaccines (8), 3) cytokine therapy with agents such as interferon α (9), 4) passive serotherapy with monoclonal antibodies (10), 5) adoptive immunotherapy with LAK (11) or TIL (12) cells, and 6) immunomodulatory therapy.

Each of the other categories of immunotherapy has some degree of immunomodulatory or immunorestorative activity. Thus, the active non-specific agents such as the macrophage activators can induce both IL-1 and IFN γ production both of which can be immunorestorative (13). A number of the cytokines including IL-1 (14), IL-2 (15), and IL-4 (16) can drive T cell proliferation and differentiation, and therefore can be immunorestorative. In this review, we will concentrate on those immunotherapeutic agents which are predominantly immunomodulatory.

THYMIC HORMONES

It has been established that the mammalian thymus is the site of T cell maturation. Over the last 25 years, thymic hormones have been identified, purified, characterized and become available for study including for investigation as potential therapeutic agents. The available thymic hormones include: thymosin fraction 5 (17), thymosin alpha-1 (18), thymopoeitin pentapeptide (19) and thymic humoral factor (20). Other thymic hormones have been reported but those mentioned above represent the major thymic hormones receiving clinical attention in the United States.

Thymic hormones induce proliferation and differentiation of T cells and therefore have been considered for immunorestorative therapy in patients with HIV-infection. Several studies has shown that thymic hormone therapy can induce improved host-defense function in patients with HIV-infection (21). However, there have not been any large controlled studies showing clinical benefit of this class of agents. Limited uncontrolled studies suggest a potential benefit in the therapy of opportunistic viral infection in immunocompromised patients (22).

CYTOKINES

Several cytokines have immunorestorative biological activities. These cytokines are usually activators of normal host-defense cells. By inducing proliferation and activation among reduced populations of normal lymphoid cells in immunodeficient subjects these cytokines can restore normal function. The potentially immunorestorative cytokines include IL-2 (23), IFN α (24) and IFN γ (25) which can activate natural killer cells, IFN γ, M-CSF (26) and GM-CSF (27) which can activate monocytes and fixed tissue macrophages, and IL-2 which induces lymphokine activated killer (LAK) cells (28). LAK cell generation requires IFN γ and TNF α as accessary factors. The development and differentiation of T cells, capable of cytotoxic T cell (CTL) activity requires IL-1, IL-2 and other interleukins such as IL-3 and IL-4 (29).

Several cytokines may play an important role in the management of microbial infections. IFN α, via its direct antiviral activity is approved for the therapy of hepatitis C (30) and papovavirus-induced condylomata (31). IFN α also slows the rate of HIV replication (32) and may play an important role in combination with other antiviral agents in patients with HIV-infection. IFN γ, through its ability to activate monocyte phagocytosis and intracellular killing (33) was found to restore antimicrobial function in chronic granulomatous disease (34) and is now approved for use in that condition.

The colony stimulating cytokines GM-CSF (35) and G-CSF (36) by restoring the granulocyte count in neutropenic patients, or by preventing chemotherapy-induced neutropenia can prevent or reduce the incidence of opportunistic bacterial infections. In addition, GM-CSF can activate phagocytes and therefore play an additional role in antimicrobial host-defense (27).

M-CSF, the cytokine responsible for monocyte proliferation and activation can activate monocyte bacteriocidal activity (26) and is being studied as a potential adjunct to antibiotic therapy in severe infections.

CHEMICAL IMMUNOMODULATORS

The chemical immunomodulators form a diverse class of agents which have immunomodulatory activity. Generally, these agents augment both B and T cell functions and result in restoration of immune responses and host-defense mechanisms such as those suppressed by either radiation or chemotherapy. Animals, pretreated with chemical immunomodulators generally show increased resistance to bacterial, fungal and tumor challenges compared to the controls. Chemical immunomodulators of clinical interest include the imidazoles such as levamisole (37), the thiols such as diethyldithio-carbamate (38) and N-acetylcysteine (39), the cyanoaziridine compounds such as azimexon (40) and imexon (41), the nucleic acid analogues such as isoprinosine (42) and inosine methyl ester and tellurium based compounds such as AS101 (43). Certain chemotherapeutic agents such as cyclophosphamide (44) many being immunomodulatory depending on their dose and timing relative to antigen administration. This effect may be related to down-regulation of suppressor cell number or function.

The chemical immunomodulators are still experimental with only a few exceptions. None have been explored extensively in microbial diseases although they have a potential in this regard. Therefore, the clinical data on these agents with relevance to their use in microbial diseases is limited and their potential is speculative.

The imidazole compound have been under study for over 20 years. The sulfur containing imidazole levamisole (37), initially developed as an anthelminthic agent was found to augment delayed hypersensitivity to recall antigens (45). Activity was shown in rheumatoid arthritis, equivalent to other anti-inflammatory drugs in that disease. However, idiosyncratic leukopenia led to its abandonment. Several studies of levamisole as an adjunct to conventional therapy of cancer were done in lung (46), breast (47) and colon cancer (48). Generally results were disappointing with very modest differences in remission rate and/or survival prolongation between treated and control subjects. Because some survival prolongation was seen in metastatic colon cancer, levamisole was tested in the adjuvant therapy of Dukes class B and C colon cancer after potentially curative surgery. Significant prolongation of survival was seen in Dukes C patients treated with 5-fluorouracil (5FU) plus levamisole compared to controls (49). This drug combination is now licensed for this indication in the U.S. The mechanism of action of levamisole in conjunction with 5FU and its potential activity independent of 5FU are not known. Certainly, it should be investigated in conditions of T cell immunodeficiency associated with opportunistic infections.

Thiols exert their immunorestorative action via their sulfur groups. They induce T cell differentiation and augment or restore both T an B cell functions (50). Mechanisms of action of immunomodulation include induction of thymic hormone-like activity in the liver (51), glutathione repletion (52), and free-radical or oxygen scavenging (53). Glutathione has been shown to be essential for immune system function and is known to be depleted by a variety of diseases (54).

Thiols of interest have included diethyldithiocarbamate (38) and more recently N-acetylcystine (39). Diethyldithiocarbamate (DTC) has been shown to reduce bacterial infection in surgical patients (55) and to improve the efficacy of antituberculous therapy (56). It has also shown activity in rheumatoid arthritis (57).

DTC has been studied extensively in animal models of human disease. It is active in adjuvant arthritis in the rat (58) and in the MLR/lpr/lpr murine autoimmune disease model (59). In the retrovirus-induced LP-BM5 murine immune deficiency disease model, DTC prevents or abrogates the development of lymphadenopathy, splenomegaly and hypergammaglobulinemia (60). Furthermore, it prevents mortality. It can be given as late as 10 weeks after virus inoculation, and completely reverse disease manifestations (61).

In the last 5 years DTC has been studied as a therapy for patients with HIV-infection. Five studies including 4 randomized controlled investigations have shown considerable benefit. These have included improved symptoms, reduced lymphadeno-pathy and splenomegaly and most important, a reduced incidence of opportunistic infections (62-66). An increase in CD4+ cells was also noted. The studies were generally short-term. Therefore, only a limited effect was seen on the incidence of

Kaposi's sarcoma and survival. The drug could be administered safely together with zidovudine and benefit was seen in zidovudine treated patients who also received DTC. DTC was essentially without toxicity. It seems clear that thiols, as immunomodulators for patients with immunodeficiency and infection should receive intensive investigation.

The cyanoaziridine dye compounds are a unique class of immunomodulators which also have antitumor activity. Like other immunomodulators, they have broad effects on T and B cell functions (40, 41). In malignant disease models they are immunomodulatory because they reduce the takes and growth of tumors implanted after drug administration. Imexon was utilized in a few small, uncontrolled clinical trials and was well-tolerated and resulted in remissions in breast cancer, colon cancer and melanoma (67). Careful dose-respond, toxicology or pharmacology studies have not been done yet.

One cyanoaziridine dye compound, azimexon was studied in both cancer (68) and AIDS (69). In both conditions it had a modest effect on restoring T cell immunocompetence. In an uncontrolled clinical study, azimexon had some symptom ameliorating and immunomodulatory activity in patients with HIV-infection (70).

We have studied Imexon in the LP-BM5 model (71). The drug was active over a broad dosage range from 50-150 mg/kg, the latter being the MTD. It was active when given, starting 0-2 weeks after virus inoculation to as late as starting 13 weeks after virus inoculation. In the latter experiment lymphadenopathy was abrogated and survival was prolonged. In addition, Imexon was synergistic with AZT and a combination of suboptimal doses of Imexon and AZT not only prolonged survival but restored the disease-suppressed mitogen responses of splenic lymphocytes.

NUCLEIC ACID ANALOGUES

The prototype compound in this class is isoprinosine, a complex of inosine and a para aminobenzoic acid analogue. Isoprinosine restores cell-mediated and humoral immune responses in immune deficient animals and humans (42). Its immunorestorative activity in virus-infected subjects appears to shorten disease duration and improve response to conventional therapy (72). Based on appropriate controlled studies, the drug is licensed in several countries for treatment of viral diseases such as Herpes Simplex.

Early in the course of the HIV epidemic, isoprinosine was reported to ameliorate symptoms and improve immunocompetence (73). Recently, a carefully controlled randomized trial showed that isoprinosine has a major effect on symptoms and signs of disease and survival (74).

Another member of this class of agents is inosine methyl ester. It is immunorestorative in animal models (75) and restores human lymphocyte responses *in vitro*.

CONCLUSIONS AND FUTURE DIRECTIONS

There are a number of immunorestorative and immunomodulatory drugs available to treat human immunodeficiency diseases. Immunorestoration in these diseases has the potential to modulate the response to infection. This has been observed with cytokines, thymic hormones and chemical immunomodulators. This approach to therapeutics has potential, particularly at a time when opportunistic infections are increasing secondary to the drug abuse epidemic, the AIDS epidemic and the increasing use of immunosuppressive therapies for cancers, transplants and autoimmune diseases. Infections in newborns, aged persons, patients receiving immunosuppressive drugs and patients with immunodeficiency diseases should be targeted for study of these drugs. Consideration for their use should be predominantly as prevention prior to the actual development of an opportunistic infection. In addition, they should be considered for use in conjunction with antibiotics.

ACKNOWLEDGEMENT

Supported by grant 5U01 AI 25617-04 from the NIAID, NIH, Bethesda, Maryland.

REFERENCES

1. H. W. Jaffe, K. Choi, P. A. Thomas, H. W. Haverkos, D. M. Auerbach, M. E. Guinan, M. F. Rogers, T. J. Spira, W. W. Darrow, M. A. Kramer, S. M. Friedman, J. M. Monroe, A. E. Friedman-Kien, L. J. Laubenstein, M. Marmor, B. Safai, S. K. Dritz, S. J. Crispi, S. L. Fannin, J. P. Orkwis, A. Kelter, W. R. Rushing, S. B. Thacker, and J. W. Curran, National case-control study of Kaposi's sarcoma and pneumocystis carinii pneumonia in homosexual men: part 1, epidemiologic results, Ann. Intern. Med. 99:145 (1983).

2. J. E. Groopman, J. M. Molina, and D. T. Scadden, Hematopoietic growth factors: biology and clinical applications, N. Engl. J. Med. 321:1449 (1989).

3. S. C. Clark and R. Kamen, The human hematopoietic colony-stimulating factors, Science 236:1229 (1987).

4. A. H. Rook, H. Masur, H. C. Lane, W. Frederick, T. Kasahara, A. M. Macher, J. Y. Djeu, J. F. Manischewitz, L. Jackson, A. S. Fauci, and G. V. Quinnan, Jr., Interleukin-2 enhances the depressed natural killer and cytomegalovirus-specific cytotoxic activities of lymphocytes from patients with the acquired immune deficiency syndrome, J. Clin. Invest. 72:398 (1983).

5. J. R. Neefe, E. A. Phillips, and J. Treat, Augmentation of natural immunity and correlation with tumor response in melanoma patients treated with human lymphoblastoid interferon, Diagnostic Immunol. 4:299 (1986).

6. R. S. Schwartz, Immunoregulation, oncogenic viruses and malignant lymphoma, Lancet I:1266 (1972).

7. E. G. MacEwen, I. D. Kurzman, R. C. Rosenthal, B. W. Smith, P. A. Manley, J. K. Roush, and P. E. Howard, Therapy for osteosarcoma in dogs with intravenous injection of liposome-encapsulated muramyl tripeptide, J. Natl. Cancer Inst. 81:935 (1989).

8. H. C. Hoover, Jr., M. G. Surdyke, R. B. Dangel, L. C. Peters, and M. G. Hanna, Jr., Prospectively randomized trial of adjuvant active-specific immunotherapy for human colorectal cancer, Cancer 55:1236 (1985).

9. G. Sarna, R. Figlin, and J. DeKernion, Interferon in renal cell carcinoma, Cancer 59:610 (1987).

10. R. A. Miller and R. Levy, Response of cutaneous T cell lymphoma to therapy with hybridoma monoclonal antibody, Lancet 8:226 (1981).

11. S. A. Rosenberg, M. T. Lotze, L. M. Muul, A. E. Chang, F. P. Avis, S. Leitman, W. M. Linehan, C. N. Robertson, R. E. Lee, J. T. Rubin, C. A. Seipp, C. G. Simpson, and D. E. White, A progress report on the treatment of 157 patients with advanced cancer using lymphokine-activated killer cells and interleukin-2 or high-dose interleukin-2 alone, N. Engl. J. Med. 316:889 (1987).

12. S. L. Topalian, D. Solomon, F. P. Avis, A. E. Chang, D. L. Freerksen, W. M. Linehan, M. T. Lotze, C. N. Robertson, C. A. Seipp, P. Simon, C. G. Simpson, and S. A. Rosenberg, Immunotherapy of patients with advanced cancer using tumor-infiltrating lymphocytes and recombinant interleukin-2: a pilot study, J. Clin. Oncol. 6:839 (1988).

13. M. Yamada, Y. Sohmura, S. Nakamura, and M. Hashimoto, Interleukin-1α: its possible roles in cancer therapy, Biotherapy 1:327 (1989).

14. K. Muegge, T. M. Williams, J. Kant, M. Karin, R. Chiu, A. Schmidt, U. Siebenlist, H. A. Young, and S. K. Durum, Interleukin-1 costimulatory activity on the interleukin-2 promoter via AP-1, Science 246:249 (1989).

15. V. J. Merluzzi, K. Welte, D. M. Savage, K. Last-Barney, and R. Mertelsmann, Expansion of cyclophosphamide-resistant cytotoxic precursors in vitro and in vivo by purified human interleukin 2, J. Immunol. 131:806 (1983).

16. R. Fernandez-Botran, V. M. Sanders, and E. S. Vitetta, Regulation of the biological effects of IL-4 on murine T and B cells, Cold Spring Harbor Symposia on Quantitative Biology LIV:705 (1989).

17. A. L. Goldstein, G. B. Thurman, T. L. K. Low, J. L. Rossio, and G. E. Trivers, Hormonal influences on the reticuloendothelial system: current status of the role of thymosin in the regulation and modulation of immunity, J. Reticuloendothelial Society 23:253 (1978).

18. R. O. Dillman, J. C. Beauregard, J. Mendelsohn, M. R. Green, S. B. Howell, and I. Royston, Phase I trials of thymosin fraction 5 and thymosin α 1, J. Biol. Response Modifiers 1:35 (1982).

19. S. J. Knox, L. S. Rosenblatt, R. W. Anderson, and B. R. Greenberg, Effect of thymopoietin and interleukin 2 on depressed mitogenic responsiveness and colony formation of lymphocytes from patients with preleukemia, Thymus 8:33 (1986).

20. N. Tranin, M. Pecht, and Z. T. Handzel, Thymic hormones: inducers and regulators of the T-cell system, Immunol. Today, 4:16 (1983).

21. W. Barcellini, P. L. Meroni, D. Frasca, C. Squotti, M. O. Borghi, C. Ubertifoppa, P. Buzzetti, A. Lazzarin, G. Doria, M. Moroni, and C. Zanussi, Effect of subcutaneous thymopentin treatment in drug addicts with persistent generalized lymphadenopathy, Clin. Exp. Immunol. 67:537 (1987).

22. F. Silvestris, A. Gernone, M. Frassanito, and F. Dammacco, Immunologic effect of long-term thymopentin treatment in patients with HIV-induced lymphadenopathy syndrome, J. Lab. Clin. Med. 113:139 (1989).

23. N. Flomenberg, K. Welte, R. Mertelsmann, N. Kernan, N. Ciobanu, S. Venuta, S. Feldman, G. Kruger, D. Kirkpatrick, B. Dupont, and R. O'Reilly, Immunologic effects of interleukin 2 in primary immunodeficiency diseases, J. Immunol. 130:2644 (1983).

24. J. R. Neefe, J. E. Sullivan, M. Ayoob, E. Phillips, and F. P. Smith, Augmented immunity in cancer patients treated with α-interferon, Cancer Res. 45:874 (1985).

25. T. Yamaski, H Handa, J. Yamashita, Y. Watanabe, M. Taguchi, S. Kuwata, Y. Namba, and M. Hanaoka, Immunoregulatory role in gamma interferon production by a T cell growth factor-dependent experimental malignant glioma-specific cytotoxic T lymphocyte clone, J. Neurosurg. 63:763 (1985).

26. E. Wing, A. Ampel, A. Waheed, and M. Shadduck, Macrophage colony-stimulating factor (M-CSF) enhances the capacity of murine macrophages to secrete oxygen reduction products, J. Immunol. 135:2052 (1985).

27. K. H. Grabstein, D. L. Urdal, R. J. Tushinski, D. Y. Mochizuki, V. L. Price, M. A. Cantrell, S. Gillis, and P. J. Conlon, Induction of macrophage tumoricidal activity by granulocyte-macrophage colony-stimulating factor, Science 232:506 (1986).

28. E. A. Grimm, A. Mazumder, H. Z. Zhang, and S. A. Rosenberg, Lymphokine-activated killer cell phenomenon: lysis of natural killer-resistant fresh solid tumor cells by interleukin 2-activated autologous human peripheral blood lymphocytes, J. Exp. Med. 155:1823 (1982).

29. C. A. Dinarello and J. W. Mier, Lymphokines, N. Engl. J. Med. 317:940 (1987).

30. G. L. Davis, L. A. Balart, E. R. Schiff, K. Lindsay, H. C. Bodenheimer, Jr., R. P. Perrillo, W. Carey, I. M. Jacobson, J. Payne, J. L. Dienstag, D. H. VanThiel, C. Tamburro, J. Lefkowitch, J. Albrecht, C. Meschievitz, T. J. Ortego, A. Gibas, and The Hepatitis Interventional Therapy Group, Treatment of chronic hepatitis C with recombinant interferon alfa: a multicenter randomized, controlled trial, N. Engl. J. Med. 321:1501 (1989).

31. L. J. Eron, F. Judson, S. Tucker, S. Prawer, J. Mills, K. Murphy, M. Hickey, M. Rogers, S. Flannigan, N. Hien, H. I. Katz, S. Goldman, A. Gottlieb, K. Adams, P. Burton, D. Tanner, E. Taylor, and E. Peets, Interferon therapy for condylomata acuminata, N. Engl. J. Med. 315:1059 (1986).

32. H. C. Lane, V. Davey, J. A. Kovacs, J. Feinberg, J. A. Metcalf, B. Herpin, R. Walker, L. Deyton, R. T. Davey, Jr., J. Falloon, M. A. Polis, N. P. Salzman, M. Baseler, H. Masur, and A. S. Fauci, Interferon-α in patients with asymptomatic human immunodeficiency virus (HIV) infection, Ann. Intern. Med. 112:805 (1990).

33. G. E. Gifford and M. L. Lohmann-Matthes, γ interferon priming of mouse and human macrophages for induction of tumor necrosis factor production by bacterial lipopolysaccharide, J. Natl. Cancer. Inst. 78:121 (1987).

34. J. I. Gallin, H. L. Malech, R. S. Weening, J. T. Curnutte, P. G. Quie, H. S. H. Jaffe, R. A. B. Ezekowitz, and The International Chronic Granulomatous Disease Cooperative Study Group, A controlled trial of interferon gamma to prevent infection in chronic granulomatous disease, N. Engl. J. Med. 324:509 (1991).

35. A. M. Gianni, M. Bregni, S. Siena, A. Orazi, A. C. Stern, L. Gandola, and G. Bonadonna, Recombinant human granulocyte-macrophage colony-stimulating factor reduces hematologic toxicity and widens clinical applicability of high-dose cyclophosphamide treatment in breast cancer and non-hodgkin's lymphoma, J. Clin. Oncol. 8:768 (1990).

36. J. L. Gabrilove, A. Jakubowski, H. Scher, C. Sternberg, G. Wong, J. Grous, A. Yagoda, K. Fain, M. A. S. Moore, B. Clarkson, H. F. Oettgen, K. Alton, K. Welte, and L. Souza, Effect of granulocyte colony-stimulating factor on neutropenia and associated morbidity due to chemotherapy for transitional-cell carcinoma of the urothelium, N. Engl. J. Med. 318:1414 (1988).

37. W. K. Amery, Overview of levamisole effectiveness in experimental and clinical cancer studies, in: "Immune Modulation and Control of Neoplasia by Adjuvant Therapy," M. A. Chirigos, ed., Raven Press, New York (1978).

38. G. Renoux, M. Renoux, Y. Lebranchu, and P. Bardos, Immunopharmacology of DTC in mice and men, in: "Immunomodulatory Drugs and Modifiers of the Biological Response," B. Serron and U. Rosenfeld, eds., Elsevier Biomedical Press, New York (1982).

39. A. Meister and M. E. Anderson, Glutathione, Ann. Rev. Biochem. 52:711 ((1983).

40. U. Bicker and P. Fuhse, Carcinostatic action of 2-cyanaziridines against a sarcoma in rats, Exp. Pathol. Jena. 10:279 (1975).

41. U. Bicker, BM 06 002: A new immunostimulating compound, in: "Immune Modulation and Control of Neoplasia by Adjuvant Therapy," M. A. Chirigos, ed., Raven Press, New York (1978).

42. D. M. Campoli-Richards, E. M. Sorkin, and R. C. Heel, Inosine pranobex: a preliminary review of its pharmacodynamic and pharmacokinetic properties, and therapeutic efficacy, Drugs 32:383 (1986).

43. B. Sredni, R. R. Caspi, Y. Klein, Y. Kalechman, Y. Danziger, M. BenYa'akov, T. Tamari, F. Shalit, and M. Abeck, A new immunomodulating compound (AS-101) with potential therapeutic application, Nature 330:173 (1987).

44. P. K. Ray and S. Raychaudhuri, Low-dose cyclophosphamide inhibition of transplantable fibrosarcoma growth by augmentation of the host immune response, JNCI 67:1341 (1981).

45. D. Tripodi, L. C. Parks, and J. Brugmans, Drug-induced restoration of cutaneous delayed hypersensitivity in anergic patients with cancer, N. Engl. J. Med. 289:354 (1973).

46. S. Krauss, F. Comas, C. Perez, D. Gordon, G. Philpott, G. Broun, W. Mill, R. Robbins, R. Smalley, O. Mendiondo, P. DeSimone, J. McLaren, J. Keller, J. Durant, R. Birch, and R. Buchanan, Treatment of inoperable non-small cell carcinoma of the lung with radiation therapy, with or without levamisole, Am. J. Clin. Oncol. 7:405 (1984).

47. G. Hortobagyi, H. Y. Yap, G. Blumenschein, J. U. Gutterman, A. U. Buzdar, C. K. Tashima, and E. M. Hersh, Response of disseminated breast cancer to combined modality treatment with chemotherapy and levamisole with or without bacillus calmette guerin, Cancer Treat. Rep. 62:1685 (1978).

48. H. Verhaegen, J. DeCree, W. DeCock, M. L. Verhaegen-Declerq, and F. Verbruggen, Levamisole therapy in patients with colorectal cancer, in: "Immunotherapy of Human Cancer," W. D. Terry and S. A. Rosenberg, eds., Elsevier, New York (1982).

49. C. G. Moertel, T. R. Fleming, J. S. Macdonald, D. G. Haller, J. A. Laurie, P. J. Goodman, J. S. Ungerleider, W. A. Emerson, D. C. Tormey, J. H. Glick, M. H. Veeder, and J. A. Mailliard, Levamisole and fluorouracil for adjuvant therapy of resected colon carcinoma, N. Engl. J. Med. 322:352 (1990).

50. J. M. Lang, F. Oberling, A. Aleksyevic, A. Falkenrodt, and S. Mayer, Immuno-modulation with diethyldithiocarbamate in patients with AIDS-related complex, Lancet 2:1066 (1985).

51. A. Pompidou, M. Renoux, J. M. Guillaumin, B. Mace, P. Michel, F. Coutance, and G. Renoux, Kinetics of the histological changes in lymphoid organs and of the T-cell inducing capacity of serum in mice treated with imuthiol (sodium diethyldithiocarbamate), Int. Arch. Allergy Appl. Immunol. 74:172 (1984).

52. S. M. Deneke and B. L. Fanburg, Involvement of glutathione enzymes in O_2-tolerance development by diethyldithiocarbamate, Chem. Pharmacol. 29:1367 (1980).

53. P. Lacombe, I. Carre, M. Fay, and J. J. Pocidalo, In vitro O_2-induced depression of T and B lymphocyte activation is reversed by diethyldithiocarbamate (DDC) treatment, Immunol. Lett. 18:99 (1988).

54. R. Buhl, K. J. Holroyd, A. Mastrangeli, A. M. Cantin, H. A. Jaffe, F. B. Wells, C. Saltini, and R. G. Crystal, Systemic glutathione deficiency in symptom-free HIV-seropositive individuals, Lancet 12:1294 (1989).

55. G. Champault, G. Biron, P. Boutelier, and M. Defayolle, L'immuno-stimulation en chirurgie viscerale: essai therapeutique controle par diethyl-dithio-carbamate de sodium (124 patients), MCD 12:537 (1983).

56. G. Renoux, M. Renoux, J. Greco, J. Baudoin, M. Lavandier, and E. Lemarie, Intl symposium on new trends, in: "Human Immunology and Cancer Immunothera-py," abstract, Montpellier, France, (1980).

57. C. F. Corke, V. Gul, E. C. Huskisson, E. J. Holborow, and G. Renoux, Immuno-therapy 2:163 (1986).

58. M. Feterah and D. E. Yocum, Inhibition by diethydithiocarbamate of streptococcal cell wall induced arthritis and hepatic granuloma in rats, Clin. Res. 37:124A (1989).

59. M. D. Halpern, E. Hersh, and D. E. Yocum, Diethyldithiocarbamate, a novel immunomodulator, prolongs survival in autoimmune MRL-lpr/lpr mice, Clin. Immunol. and Immunopath. 55:242 (1990).

60. E. M. Hersh, C. Y. Funk, E. A. Petersen, and D. E. Mosier, Effective therapy of the LP-BM5 murine retrovirus induced lymphoproliferative immunodeficiency disease with diethyldithiocarbamate, <u>AIDS Res. and Human Retroviruses</u> (in press).

61. E. M. Hersh, C. Y. Funk, E. A. Petersen, K. L. Ryschon, and D. E. Mosier, Dose response and timing effects in the therapy of the LP-BM5 murine retrovirus induced lymphoproliferative immunodeficiency disease with diethyldithiocarbamate, <u>AIDS Current Science</u> (in press).

62. G. W. Brewton, E. M. Hersh, A. Rios, P. W. A. Mansell, B. Hollinger, and J. M. Reuben, A pilot study of diethyldithiocarbamate in patients with acquired immune deficiency syndrome (AIDS) and the AIDS-related complex, <u>Life Sciences</u> 45:2509 (1989).

63. J. M. Lang, C. Trepo, M. Kirstetter, L. Herviou, G. Retornaz, G. Renoux, M. Musset, J. L. Touraine, P. Choutet, A. Falkenrodt, J. M. Livrozet, F. Touraine, M. Renoux, J. Caraux, and the AIDS Imuthiol French Study Group, Randomized, double-blind, placebo-controlled trial of ditiocarb sodium (imuthiol) in human immunodeficiency virus infection, <u>Lancet</u> 9:702 (1988).

64. E. C. Reisinger, P. Kern, M. Ernst, P. Bock, H. D. Flad, M. Dietrich, and the German DTC Study Group, Inhibition of HIV progression by ditiocarb, <u>Lancet</u> 335:679 (1990).

65. E. M. Hersh, G. Brewton, D. Abrams, J. Bartlett, J. Galpin, P. Gill, R. Gorter, M. Gottlieb, J. Jonikas, S. Landesman, A. Levine, A. Marcel, E. A. Petersen, M. Whiteside, J. Zahradnik, C. Negron, F. Boutitie, J. Caraux, J. M. Dupuy, and L. R. Salmi, Ditiocarb sodium (diethyldithiocarbamate) therapy in patients with symptomatic HIV infection and AIDS: a randomized double-blind placebo-controlled multicenter study, <u>JAMA</u> 265:1538 (1991).

66. C. S. Kaplan, E. A. Petersen, D. Yocum, and E. H. Hersh, A randomized controlled dose response study of intravenous sodium diethyldithiocarbamate in patients with advanced human immunodeficiency virus infection, <u>Life Sciences</u> 45:iii (1989).

67. M. Micksche, E. M. Kokoschka, P. Sagaster, and U. Bicker, Clinical studies of BM 06 002: a new immunostimulant drug, <u>IRCS Med. Sci.</u> 5:192 (1977).

68. Y. Z. Patt, E. M. Hersh, J. M. Reuben, L. Claghorn, and G. M. Mavligit, A phase I study of intravenous azimexon therapy in human cancer, <u>J. Biol. Response Modifiers</u> 5:313 (1986).

69. Y. Z. Patt, G. M. Mavligit, J. M. Reuben, P.W.A. Mansell, S. Li, G. R. Newell, M. Talpaz, and E. M. Hersh, Modulation *in vitro* of immune parameters in homosexual males with the preclinical complex of symptoms related to acquired immune deficiency syndrome by azimexon, <u>J. Biol. Response Modifiers</u> 5:263 (1986).

70. Y. Z. Patt, E. M. Hersh, J. M. Reuben, L. Claghorn, and G. M. Mavligit, A phase I study of intravenous azimexon therapy in human cancer, <u>J. Biol. Response Modifiers</u> 5:313 (1986).

71. E. M. Hersh, C. Funk, C. Gschwind, E. A. Petersen, and D. Mosier, Imexon alters the course of the murine LP-BM5 retrovirus-induced lymphoproliferative immunodeficiency disease, <u>Proceedings AACR</u> 31:302 (1990).

72. C. E. Jones, P. R. Dyken, P. R. Huttenlocher, J. T. Jabbour, and K. W. Maxwell, Inosiplex therapy in subacute sclerosing panencephalitis: a multicentre, non-randomised study in 98 patients, <u>Lancet</u> 1:1034 (1982).

73. M. H. Grieco, M. M. Reddy, D. Manvar, K. K. Ahuja, and M. L. Moriarty, *In-vivo* immunomodulation by isoprinosine in patients with the acquired immunodeficiency syndrome and related complexes, <u>Ann. Intern. Med.</u> 101:206 (1984).

74. C. Pedersen, E. Sandstrom, C. S. Petersen, G. Norkrans, J. Gerstoft, A. Karlsson, K. C. Christensen, C. Hakansson, P. O. Pehrson, J. O. Nielsen, H. J. Jurgensen, and the Scandinavian Isoprinosine Study Group, The efficacy of inosine pranobex in preventing the acquired immunodeficiency syndrome in patients with human immunodeficiency virus infection, N. Engl. J. Med. 322:1757 (1990).

75. J. W. Hadden, E. M. Hadden, Y. Wang, M. Sosa, R. Coffey, and A. Giner-Sorolla, Methyl inosine monophosphate (MIMP) -a new purine immunomodulator, Int. J. Immunopharmac. (in press).

RECENT THOUGHTS ON THE IMMUNOTHERAPY OF INFECTIOUS

DISEASES INCLUDING HIV INFECTION

John W. Hadden

Program of Immunopharmacology
University of South Florida Medical College
Tampa, Florida 33612

The treatment of infections with immunostimulating drugs has been termed by us a "prohost" approach since the therapy is designed to increase host resistance rather than directly kill the pathogen. The origins of this approach go back into history to the use of Chinese traditional herbal medicines to strengthen the host ("stimulate the yin"). At the turn of the century Metchnikoff advocated stimulating the phagocytes to enhance host resistance. More recently, vaccine development has been employed to specifically prepare the host for pathogen challenge.

In recent years a number of reviews (1-6) have summarized the work which supports the concept that nonspecific immunostimulants are effective in both the prevention and the treatment of infectious diseases. No doubt, the great success in the last half century of the development of effective antimicrobial chemotherapy has retarded the development of immunotherapy of infectious diseases.

Infectious scourges, particularly parasitic, still plague the world's population and the epidemic of the acquired immunodeficiency syndrome (AIDS) reminds us of our vulnerability.

The current concepts of the prohost approach derive from the knowledge that when specific host defense mechanisms are impaired, infection results and the type of infection depends on the nature of the immune defect. The definitions of specific and general defects of the immune system have become increasingly clear in recent years and rather than becoming a relatively rare event, cellular immune deficiency is common in the neonatal period, in protein-calorie and vitamin deficiencies, in parasitic, bacterial and viral diseases, in cancer, in aging, and in HIV infection. Large segments of the world's population are, as a result, at risk.

The idea of treating a failing immune system is a relatively new concept; however, in the same way that physicians treat a failing heart, liver or respiratory system, it seems reasonable to treat the defective immune system. It now seems reasonable to do so not only in those who are immunodepressed and infected, but also, immunoprophylactically, in those immunodepressed but not yet infected in order to prevent infection. A large experience has accumulated to show that a variety of crude bacterial and fungal products will enhance defense against pathogens. The increasing repertoire of chemically defined biologicals and drugs having relative specificity for one or another effector cell of the immune system makes the approach to rather specific manipulation possible.

Microbial Infections, Edited by H. Friedman *et al.*
Plenum Press, New York, 1992

Since antibody-mediated mechanisms are well preserved in secondary immunodeficiency and vaccine and adjuvant applications have focused on these mechanisms, cellular immune responses involving thymus-dependent (T) lymphocytes, macrophages and natural killer cells are emphasized here. It is to be noted, however, that recent advances in vaccine development using synthetic adjuvants like MDP coupled to genetically engineered epitopes offer the prospect of conferring specific resistance, cellular or humoral immunity, or both, in the protection against a variety of diseases heretofore unapproached by vaccines (e.g., hepatitis B and malaria) (see Chedid herein for review). Also, the development of monoclonal antibodies, particularly human-human monoclonals, offers great potential in the future for adoptive serotherapy in the treatment of infectious diseases. These two approaches make it likely that vaccination and antibody-mediated mechanisms are also to contribute to future prohost therapies.

Agammaglobulinemia, i.e., the lack of ability to make antibodies, results in infections with high grade pyogenic pathogens (e.g., streptococci, staphylococci and pneumococci) which affect the respiratory tract. Our ability to manage these infections with gammaglobulin therapy and antibiotics make it possible for these patients to live into adulthood.

In contrast, the lack of T cell-mediated immunity, as occurs in AIDS, opens the door to a much larger spectrum of lethal pathogens for which we often lack effective antimicrobial therapy. Central to the therapy of these diseases is the reconstitution of cellular immune function. Bone marrow and thymus transplantation are not generally feasible approaches to treatment. Attempts to replace the humoral role of the thymus have focused on the thymic hormones. A series of proteins have been extracted from the thymus and used in immunoreconstitutive therapy in man and animals with a variety of diseases.

Chemically defined thymic hormones in the form of thymosin α_1, thymopoietin, and thymulin (FTS), and various mixed preparations have been used in large numbers of patients with congenital and acquired cellular immune deficiencies. Rather impressive effects of the administration of the crude thymic extract thymosin fraction V and thymosin α_1, have been observed in chemotherapy-treated immunosuppressed mice to prevent the susceptibility to pathogen challenge (7). Table 1 summarizes some of the infectious disease applications in which these hormones have been employed.

In addition to the thymic hormones, interleukins 1 & 2 (IL-1 & IL-2) regulate T cell function and development. Therapy with IL-2 may be an important adjunct to thymic hormone therapy in T cell deficient individuals.

A number of drugs are thymomimetic in their action (8). Isoprinosine and NPT 15392 are examples of those acting directly to induce prothymocyte differentiation and promote T cell functions including interleukin 2 production. Their immunopharmacologies have been reviewed elsewhere. Isoprinosine has seen the widest clinical experience in viral disease including herpes virus infections, influenza, and measles-associated subacute sclerosing panencephalitis and its effects are likely due to prohost action (see Table 1). Levamisole and other sulphur-containing compounds are indirectly thymomimetic through their ability to promote in vivo the appearance of a thymic hormone-like substance. Diethyldithiocarbamate (DTC-imuthiol) is a sulphur compound sharing a similar immunopharmacology with levamisole presumably through the same mechanisms. Levamisole has had extensive application in a variety of infections in man and animals (see Table 1).

In addition to therapy directed at reconstituting or promoting T cell function, therapy directed at nonspecific components of the system, like macrophages and natural killer cells, has also seen attention. The interferons activate both these populations of cells and have been used in treatment of a number of viral infections in animals and humans (see Table 1).

Table 1.   Immunostimulants and Infections

Thymic Hormones
Acute Encephalitis
CMV
Herpes Keratitis
Recurrent Herpes 1
SSPE
Varicella-Zoster

Isoprinosine

Acute Viral Encephalitis
Hepatitis A & B
Influenza
Measles
Recurrent Herpes 2
Rhinovirus
SSPE
Varicella
Warts

Levamisole
Aphthous Stomatitis
Aspergillosis
Brucellosis
Herpes Zoster
Leprosy
Mucocutaneous candidiasis
Melioidosis
Recurrent Herpes 1 & 2
Tuberculosis

Interferon

Chondyloma Acuminatum
CMV
Hepatitis B
Herpes Simplex
Herpes Zoster
Laryngeal Warts
Rhinovirus

A number of other agents directly activate macrophages for bactericidal activity including lymphokines (MAF's), endotoxins, tuftsin, and muramyl peptides. Treatment of animals, particularly before pathogen challenge, by these agents has been shown to increase host resistance, particularly to facultative intracellular bacterial pathogens and parasites. Other agents like isoprinosine, levamisole, aximexon, and NPT 15392 potentiate their effects and may be effective in promoting macrophage-mediated resistance.

Natural killer cells are thought to be a primary defense against virus challenge and stimulation of these cells with IL-2 or interferon increases their function. This action can be mimicked and/or potentiated by drugs like isoprinosine, NPT 15392, and cyclomunine and by interferon inducers like pyran copolymer (MVE-2) and pyrimidinoles and, in a number of settings enhanced resistance to viral challenge has been demonstrated. One of the relationships which emerges from comparing agents which promote T cell-mediated immunity to those which promote nonspecific resistance at the level of the macrophage or natural killer cell is that, while exceptions exist, the former are generally immunotherapeutic while the latter are immunoprophylactic (Table 2).

Table 2.   Agents with Activity in Animal Models of Bacterial, Viral, and Fungal Challenge

Immunoprophylactic
Interferon
Interferon Inducers
MDP and Analogs
RES Expanders - Glycans

Immunotherapeutic
Interferon
Thymic Hormones
Levamisole
Isoprinosine

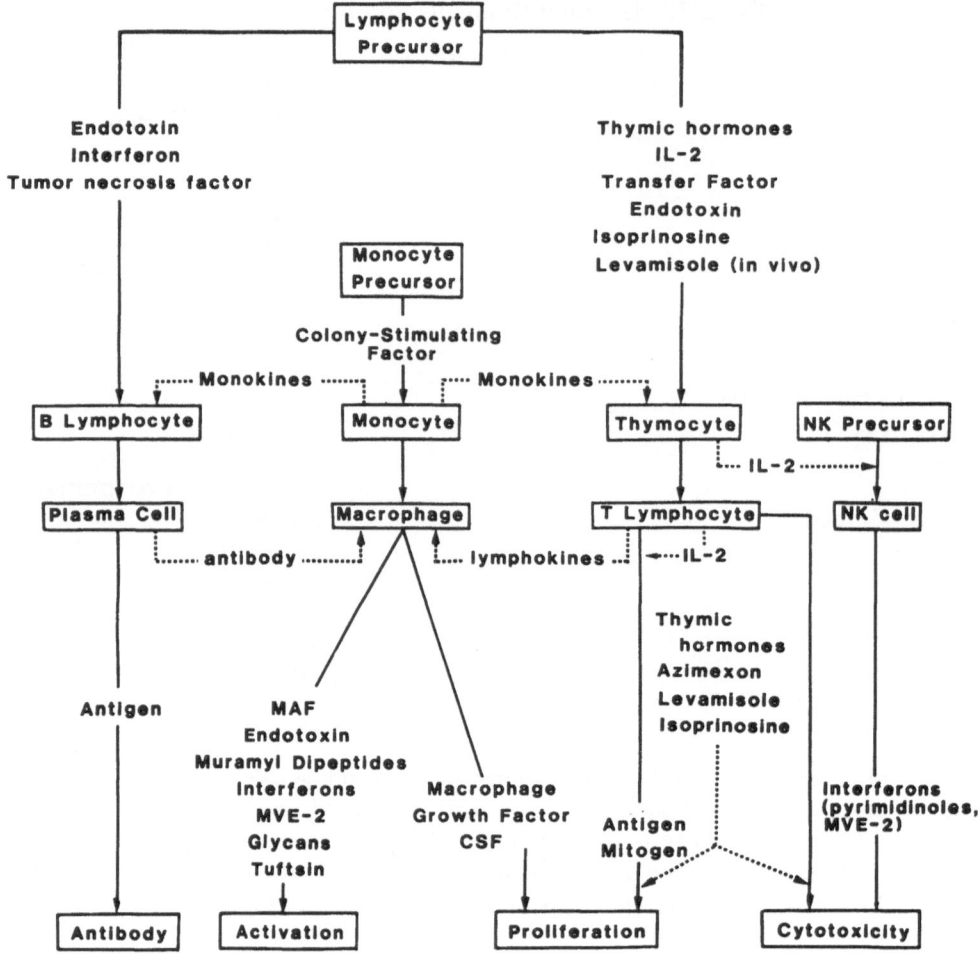

Fig. 1. Immunorestorative and Immunomodulation Therapies.

The number of circumstances in which immunoprophylactic agents may find use in clinical medicine may seem limited since seldom can an infection be diagnosed prior to clinical symptoms let alone prior to exposure. By way of a possible example, it is important to note that a large number of individuals with cancer who have been treated aggressively with chemotherapy are prone to infection. Prophylactic immunorestorative treatment of these patients might significantly reduce their morbidity and mortality. Another population at risk for infection are the elderly. Where individuals in this population can be identified as immunologically compromised and at high risk, immunoprophylactic therapy might be applicable. In other settings, immunologic reconstitution of aged animals has seen benefit in terms of reduced incidence of malignancy and increased longevity (9). In order for us to understand more clearly the therapeutic and immunoprophylactic potential of these biologicals and drugs we need more experience in the clinical setting. We need more specific information about the immune defects which occur in certain infections in order to address our therapies more specifically to correct these defects. Figure 1 summarizes the actions of the various immunomodulators under discussion on their various target cells. Such a schema emphasizes the potential for immunotherapy to be relatively specific in its application depending on the nature of the hosts immune defect and, of course, on the nature of the central mechanism of resistance to a particular pathogen. It also emphasizes the potential for using combinations of agents in therapy and it is notable that a number of experimental examples attest to the synergistic effects of combinations of both drugs and biologicals. It may be that use of immunoprophylactic agents in combination with immunotherapeutic agents will enhance the therapeutic potential of the latter. It also is likely that where antimicrobial therapy exists but is relatively ineffective, e.g., mucocutaneous candidiasis, lepromatous leprosy and drug-resistant tuberculosis, the combination of antimicrobial therapy and immunotherapy may be effective.

THE IMMUNOTHERAPY OF HIV VIRUS INFECTION

Immunotherapy in AIDS as the "end stage" of HIV infection in the absence of concurrent antiviral therapy has not been effective in reversing the immunodeficiency. Interferon-α has been licensed by the Food and Drug Administration (FDA) for treatment of Kaposi's Sarcoma. IFN-α is effective alone (20-40% major response rates) and in conjunction with zidovudine (>60% response rate), but it does not reverse immunodeficiency. Also, unlike the anticancer chemotherapy generally employed in AIDS-dependent Kaposi's sarcoma, it is not immunosuppressive and has not been proven to reduce the incidence of infections.

Intravenous immunoglobulin therapy is generally considered to be useful and nontoxic in the treatment of pediatric AIDS patients for the management of bacterial infections. A multicenter phase III trial with zidovudine is underway to confirm this widely accepted but unproven notion. A number of anecdotal reports indicate that such therapy may improve the course of AIDS and AIDS-related complex (ARC) in adult patients and a multicenter trial has been initiated to test this by using intravenous immunoglobulin therapy alone and in combination with zidovudine.

Initial studies with recombinant IL-2 in AIDS were negative; however, with persistent i.v. infusion of high doses, lymphocytosis without enhanced viral recovery was observed. Toxicity is associated with a "flu-like" syndrome, fluid accumulation, and increased bacterial infections. A number of combined IL-2 and zidovudine phase I protocols are in progress. Preliminary data indicate that the combination is compatible, that the toxicity levels are acceptable and that the effects of IL-2 to increase CD4+ T-cell counts, cellular cytotoxicity and enhanced skin test reactivity were not prevented by zidovudine. No increase in viral antigenemia of viral recovery was observed.

IMMUNORESTORATIVE THERAPY IN ARC AND ASYMPTOMATIC HIV INFECTION

IFN-α modulates various immune responses. Thus, it enhances killer cell activity by lymphocytes, macrophages and natural killer cells while depressing lymphocyte proliferation and secretory responses. Recent reports indicate that IFN-α treatment of pre-AIDS patients reduces HIV virus recovery, p24 antigenemia, and opportunistic infection despite toxic side-effects. Several combined protocols (IFNα with diethyldithi-ocarbamate, IL-2 or zidovudine) are underway.

ImReg-1 is a dialysed leukocyte extract from normal donors in which the oligopeptides Tyr-Gly-Gly and Tyr-Gly are thought to be the nonspecific active components. ImReg-1 modulates lymphokine production *in vitro* and enhances dermal

17

skin test responses *in vivo* when given intracutaneously with tetanus toxoid antigen. In a multicenter trial totalling 141 ARC patients, 93 were treated with ImReg-1 biweekly and showed a significantly reduced tendency to progress to a clinically relevant endpoint (4.3% vs 25%) or to AIDS (3% vs 17%). Marginally significant changes in CD4+ counts and symptom scores were also observed. No toxicity was observed. A larger confirmatory trial is planned.

Thymopentin is the pentapeptide active site of thymopoietin, a chemically defined thymus-derived peptide having thymomimetic activity *in vitro* and *in vivo*. Initial controlled studies with thymopentin showed positive effects. On the basis of these findings, a blind multicenter trial involving 47 thymopentin-treated asymptomatic HIV-positive and ARC patients and 44 placebo-treated controls was set up. Preliminary results indicate that four controls progressed to constitutional symptoms or AIDS, while none of the treated patients progressed ($p < 0.03$). Asymptomatic patients with CD4+ counts greater than $400mm^{-3}$ maintained their levels while controls declined. In symptomatic patients treated with thymopentin, p24 viral antigen levels did not increase and β_2-macroglobulin levels decreased in contrast to controls. No side-effects were observed. Combination trials with zidovudine are planned.

Multicenter controlled studies with diethyldithiocarbamate (DTC) (10) in 143 symptomatic pre-AIDS patients indicate that DTC decreases symptoms and AIDS conversion (13% to 4.5%) and tended to increase CD4+ lymphocytes.

These findings have been extended in a large multicenter study of AIDS and ARC patients treated orally with 400 mm^{-2} DTC weekly, where 389 patients showed a significant (50%) reduction in opportunistic infections in both AIDS and ARC patients (11). Although the number of T cells expressing CD4 increased in the DTC-treated group (255 to 300 mm^{-3}), the levels did not decline, as had been expected, in the control group (unchanged at 270 mm^{-3}); the changes were, therefore, not significant. No significant toxicity was observed. A 1600-patient trial is in progress in Europe to determine the effect of DTC on asymptomatic HIV-positive patients.

A controlled trial with isoprinosine (12) demonstrated significant increases in CD4 counts and natural killer activity and significant decreases in symptoms. Fewer cases of ARC converted to AIDS in patients with progressive generalized lymphadenopathy and mean CD4 counts greater than 500 mm^{-3}.

Several multicenter double-blind trials have been completed. In the USA a trial of 696 HIV-positive symptomatic patients with CD4 counts less than $400mm^{-3}$ did not reveal significant immunological or clinical results (13). By contrast, a Scandinavian trial (14) with 866 HIV-positive symptomatic ARC patients with mean CD4 counts greater than 425 showed a significant reduction in ARC to AIDS conversion (4% to 0.5%) over a six month period, although no significant immune correlates were obtained. In Italy, from a trial with 553 asymptomatic HIV-positive individuals (15), isoprinosine treatment was associated with no new infections; 12 controls developed infections. In this study, decreased symptoms, preservation of CD4/CD8 ratios and an increase in natural killer cells were observed in the treated group. No toxicity was observed. These studies indicate that isoprinosine treatment benefits HIV-positive patients by preventing the development of infection only when CD4+ lymphocyte counts are greater than $400mm^{-3}$.

The molecular strategies to augment T-cell function with thymomimetic agents in pre-AIDS appear to depend on the presence of lymphocytes whose function, when restored, will benefit the host and not result in the activation and spread of virus. Ideally, the system should be replenished with virus-free T lymphocytes produced by the thymus. The thymus involutes and ceases to function relatively early in life and thymic secretory capacity for thymic hormones declines in midlife at a time corresponding to female gonadal menopause. Thus, many conclude that the thymus is nonfunctional. However, studies of radiation injury, cancer chemotherapy, and bone marrow trans-

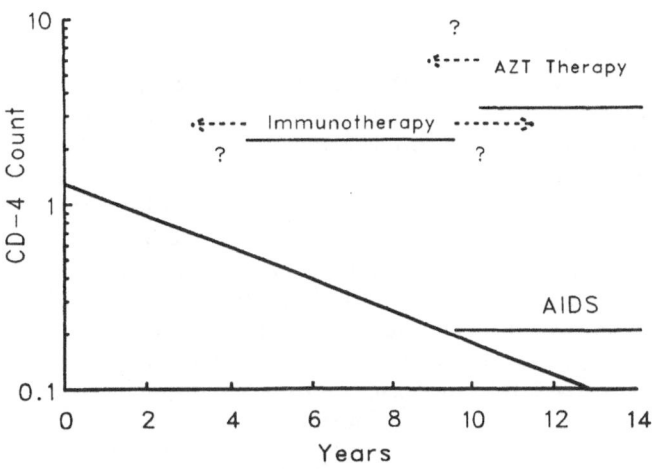

Fig. 2. Therapy Model for HIV Infection.

plantation for leukemia indicate that the thymus can process precursors to restore the T-cell system in the periphery. It appears that when the need arises, signals develop to induce thymic T-cell processing function. These signals have not been identified, but recent evidence implicates both interleukins and pituitary factors.

The thymus apparently is destroyed in AIDS with obliteration of both epithelial and lymphoid components. Once the thymus has been destroyed, T cells cannot be generated anew. It is not clear at what stage of HIV infection the thymus loses its capacity to replenish. Obviously, the administration of thymotrophic substances to enhance T-cell replenishment should logically precede thymus destruction. A useful therapeutic strategy would be to employ thymic and pituitary hormones, and interleukins, to stimulate T-cell development provided that a residual, marginally functional thymus still exists. In the context of HIV infection, this therapy would include antiviral therapy.

The most useful therapeutic approaches oriented toward improving the function of T lymphocytes in HIV disease involve two drugs, DTC and isoprinosine, and two peptides, thymopentin and ImReg for use alone or in combination with antiviral therapy. No significant side-effects with these agents have been noted; thus, they are safe and no incidence of HIV activation has been uncovered. Their combined use with antiviral agents relies on the assumption that the antiviral agents (17) do not inhibit their activity. This remains to be determined.

The total number of reported patients treated with these agents exceeds 2500, which underscores the statistical significance of the findings. The case for immunotherapy is no longer in doubt. The question remains as to how to perfect it.

Figure 2 offers a schematic view of the current use of immunotherapy, which in this context would be more appropriately termed "post-infection immunoprophylaxis". The length of the latency period is variable, presumably related to other concomitant immunosuppressive influences. Zidovudine treatment should be reserved for AIDS itself. Its earlier use may or may not ultimately be fruitful in terms of prolongation of life, because of the risk of development of HIV resistance. Combined use of zidovudine or other antivirals with immunotherapy based upon the experience with isoprinosine, DTC, ImReg-1 and thymopentin appears to be beneficial when CD4 counts are greater than 400.

A NEW DRUG FOR HIV INFECTION

Methyl inosine monophosphate (MIMP) is a new IMP derivative which appears to have distinct advantage over preceding immunomodulatory purines (e.g. isoprinosine or NPT 15392) in showing a more consistent activity. We, therefore, classify it as a "third generation" compound in our effort to produce the most effective purine immunomodulator drug. As far as its immunopharmacology has been analyzed, it has activity parallel to the precursor purines and, therefore, is classified by us as a "thymomimetic" drug. *In vitro*, it augments lymphocyte responses to T lymphocyte mitogens but less so B cell mitogens in both human peripheral blood lymphocytes and murine splenic lymphocytes. The action is apparent at .1 to 1 µg/ml and increases progressively to 100 µg/ml. In more than 50 studies to date, the response is consistent and approaches a 1.5 to 2 fold increase in human and murine lymphocytes. The effect is superior to that observed with isoprinosine, inosine, or IMP itself. MIMP is able to restore depressed lymphoproliferative responses of murine lymphocytes inhibited by glucocorticoids. It restores the responses of human lymphocytes *in vitro* inhibited by interferon α, prostaglandin PGE_2, and also an immunosuppressive synthetic peptide derived from the P41 segment of the intramembranous portion of the GP 160 of the human immunodeficiency virus (HIV). MIMP is effective at reversing immunosuppressive influences when they are mild to moderate; however, in the presence of profound inhibition, the drug effect is lost.

In lymphocytes *in vitro* from patients with pre-AIDS (8 patients), MIMP restored the depressed lymphoproliferative responses to the normal range. MIMP had a negligible effect to restore the PHA response of lymphocytes of patients with AIDS (8 patients). These data indicate that MIMP shows pronounced augmentative and immunorestorative activity on T lymphocytes and suggests its appropriateness to treat individuals with T lymphocyte impairment.

Normal mice were treated with MIMP (1-100 mg/kg) both intraperitoneally and orally. MIMP augmented the PHA responses and ConA responses of T lymphocytes from mouse spleen cells with animals treated by both routes. Mice were immunized with sheep erythrocytes for the assessment of spleen plaque-forming (SPF) antibody producing cells. Treatment with MIMP (1-100 mg/kg) at the time of immunization resulted in augmentation, approximately 2 fold, of the number of plaque-forming cells per spleen. Optimum effects were observed at 50 mg/kg with approximately equivalent effects observed down to 1 mg/kg by both the intraperitoneal and oral route. Mice were immunized with sheep erythrocytes for the assessment of delayed-type hypersensitivity (DTH) to foot pad challenge with SRBC. MIMP administered by both routes augmented the delayed-hypersensitivity reaction and peak response was observed at 50 mg/kg with significant effects at 1 mg/kg.

Mice were challenged with Friend Leukemia Virus (FLV) as a model for retrovirus infection similar to AIDS but more rapidly lethal. Mice infected with FLV were treated with MIMP from day 3 to day 13 following virus infection and survival was monitored. Animals treated with 1, 10 and 50 mg/kg i.p. showed significant prolongation of survival. Animals treated with 1 mg/kg/day every other day and animals treated with 1 mg/kg p.o. daily showed equivalent prolongation of survival (MST = 46 days compared to control of 39 days. P < 0.01) These studies indicate that MIMP is effective to stimulate the proliferative responses of lymphocytes from normal donors and donors suppressed under a variety of influences including retroviral peptide, and that this correlates with enhanced survival in a retrovirus-induced leukemia model. In so far as MIMP has been analyzed from an immunopharmacologic standpoint, the pattern of effects is directly parallel to those observed with other thymomimetic purines; however, the magnitude and consistency of the effect is superior.

In conclusion, the immunotherapy of microbial diseases, particularly viral infections for which no effective chemotherapy exists, and bacterial, fungal or parasitic infections for which antimicrobial therapy is ineffective or poorly effective is at the same time an old and a new approach. The increasing sophistication of the science of immune regulation, i.e. immunopharmacology, implies that such therapies in the future will have unparalleled efficacy and safety. A new era of clinical immunomodulation is clearly emerging.

REFERENCES

1. J. W. Hadden, C. Lopez, R. J. O'Reilly and E. M. Hadden, Levamisole and inosiplex: Antiviral agents with immunopotentiating action, NY Acad. Sci. 284:139 (1976).
2. G. H. Werner, Immunopotentiating substances with antiviral activity, Phar. Ther. 6:235 (1979).
3. H. Friedman and S. Specter, Immunotherapy and immunoregulation, in: "Chemotherapy of Viral Infection," P. E. Came and L. A. Caliguri, eds., Springer-Verlag, Berlin (1982).
4. J. W. Hadden, Immunostimulation therapy in the treatment of infectious diseases, in: "Immunotherapy and Viral Diseases," J. W. Hadden, F. Sorice and J-L. Touraine, eds., EOS Sigma Tau, Rome (1985).
5. M. Lopez-Cepero, S. Specter, and J. W. Hadden, Implications for immunotherapy of viral infections, in: "Virus-Induced Immunosuppression," S. Specter, M. Bendinelli and H. Friedman, eds., Plenum Press, New York (1989).
6. J. W. Hadden, Immunotherapy of human immunodeficiency virus infection, TIPS 12:107 (1991).
7. H. Ishitsuka, Y. Umeda, J. Nakamura and Y. Yagi, Protective activity of thymosin against opportunistic infections in animal models, Cancer Immunol. Immunother. 14:145 (1983).
8. J. W. Hadden, Thymomimetic drugs, in: "Immunopharmacology," P. A. Miescher, L. Bolis and M. Ghione, eds., Serono Symposia Publications, Raven Press, New York (1985).
9. M. Bruley-Rosset, I. Florentin, N. Kiger, J. Shulta and G. Mathe, Restoration of impaired immune functions of aged animals by chronic bestatin treatment, Immunol. 38:75 (1979).
10. J. M. Lang, J-L. Touraine, C. Trepo, P. Choulet, M. Kirkstetter, A. Falkenrodt, L. Herviou, J. M. Livrozet, G. Retornaz, F. Touraine, G. Renoux, M. Renoux, M. Musset, J. Caraux and AIDS-Imuthiol French Study Group, Randomized, double-blind, placebo-controlled trial of ditiocarb sodium ("Imuthiol") in human immunodeficiency virus infection, Lancet 24(2)8613: 702 (1988).
11. E. M. Hersh, G. Brewton, D. Abrams, J. Bartlett, J. Galpin, G. Parkash, R. Gorter, M. Gottlieb, J. J. Jonikas, S. Landesman, A. Levine, A. Marcel, E. A. Petersen, M. Whiteside, J. Zahradnik, C. Negron, F. Boutitie, J. Caroux, J-M. Dupuy and L. R. Salmi, Ditiocarb sodium (Diethyldithiocarbamate) therapy in patients with symptomatic HIV infection and AIDS, JAMA 265:1538 (1991).
12. J. G. Bekesi, P. H. Tsang, J. I. Wallace and J. P. Roboz, Immunorestorative properties of isoprinosine in the treatment of patients at high risk of developing ARC or AIDS, J. Clin. Lab. Immunol. 24:155 (1987).
13. D. Abrams, M. Grieco, M. Gottlieb and M. Speer, eds., AIDS/HIV Experimental Treatment, AMFAR Directory, Vol. 3, No. 1 (1989).

14. C. Pedersen, E. Sandstrom, C. S. Pedersen, G. Norkrans, J. Gersteft, A. Karlsson, K. C. Christiansen, C. Hakansson, P. Pehrson, J. O. Nielsen, H. J. Jurgenssen and the Scandinavian Isoprene Study Group, The efficacy of inosine pranobex in preventing the acquired immunodeficiency syndrome in patients with human immunodeficiency virus infection, New Eng. J. of Med. 322:1757 (1990).

15. C. DeSimone, F. Albertini, M. Almaviva, G. Angarano, F. Chioda, P. Costiglio, S. Delia, A. Ferlini, F. Gritti, G. Mazzarello, F. Milazzo, M. Montioni, P. Marciso, G. Pastore, E. Raise, G. Santini, F. Sorice, A. Terragna, G. Visco and V. Vullo, Clinical and immunological assessment in HIV+ subjects receiving inosine-pranobex: A randomized, multicentric study, Med. Onc. Tum. Pharmacother. 10:299 (1988).

16. J. W. Hadden, "Advances of Immunopharmacology 4," J. W. Hadden, F. Spreafico, Y. Yamamura, K. F. Austen, P. Dukor and K. Masek, eds., Pergamon Press, Oxford (1989).

17. M. I. Luster, D. R. Germolec, K. L. White, B. A. Fuchs, M. M. Fort, J. E. Tomaszewski, M. Thompson, P. C. Blair, J. A. McCay, A. E. Munson and G. J. Rosenthal, A comparison of three nucleoside analogs with anti-retroviral activity on immune and hematopoietic functions in mice: *In vitro* toxicity to precursor cells and microstromal environment, Toxic. Appl. Pharmacol. 101:328 (1989).

18. J. W. Hadden, E. M. Hadden, Y. Wang, M. Sosa, R. Coffey and A. Giner-Sorolla, Methyl inosine monophosphate (MIMP) -A new purine immunomodulator, Fifth Int. Conf. on Immunopharmacol., Tampa, Florida, 1991, abstract #107 (in press Int. J. of Immunopharmacol.).

19. M. Sosa, Y. Wang, A. Saha, T. Wadsworth and J. W. Hadden, Methyl inosine monophosphate (MIMP) promotes immune responses in mice, Ibid, abstract #108.

20. J. -L. Touraine, K. Sanhadji, O. de Bouteiller and J. W. Hadden, *In vitro* effects of inosine monophosphate methyl (Me IMP) on human prothymocyte differentiation, Ibid, abstract #109.

21. M. Sosa, Y. Noso and J. W. Hadden, Methyl inosine monophosphate (MIMP) stimulates macrophage related functions *in vitro* and *in vivo*, Ibid, abstract #110.

22. E. M. Hadden, M. Sosa, M. Strand, R. Coffey and J. W. Hadden, Methyl inosine monophosphate (MIMP) restores depressed lymphoproliferative response of normal human and murine T lymphocytes, Ibid, abstract #111.

23. J. W. Hadden, J. Ongradi, S. Specter, D. Nelson, E. M. Hadden, M. Sosa, C. Monell and M. Strand, Methyl inosine monophosphate (MIMP) restores HIV-associated suppression of the proliferative responses of human lymphocytes *in vitro*, Ibid, abstract #112.

IDENTIFICATION AND CHARACTERIZATION OF MAMMALIAN CELL

MEMBRANE RECEPTORS FOR LPS-ENDOTOXIN

D. C. Morrison, M.-G. Lei, T.-Y. Chen, L. M. Flebbe,
J. Halling, and S. Field

Department of Microbiology, Molecular Genetics and Immunology
University of Kansas Medical Center
39th and Rainbow Boulevard
Kansas City, Kansas 66103

Central to the concept of host defense against potentially lethal invasion by infectious microbes is the capacity of the host to distinguish self from non-self. To this end, the mammalian host has evolved multiple recognition systems by which non-self may be readily distinguished from self. Such discrimination systems exist at many levels, from the relatively simple, yet elegant, mechanisms for polysaccharide-dependent recognition of non-self by factors B, D and C3b of the alternative complement pathway, to the sophisticated major histocompatibility complex-mediated presentation of non-self peptides to membrane-localized T-lymphocyte multicomplement receptors. In other, considerably less complex, recognition systems, antigen binding to surface-localized immunoglobulins on B lymphocytes serves as a relative direct mechanism for host recognition of non-self. In both T and B-lymphocyte-mediated recognition of non-self, however, the receptor repertoire is extensive, with an acknowledged diversity approaching one billion specificities. The existence of such diverse, immunologically-specific, host recognition systems is central to survival of multicellular eukaryotic systems in an environment of abundant prokaryotic life forms.

It would appear, however, that the host has also developed sensitive and highly efficient mechanisms for recognition of non-self by non-immune inflammatory mediator cells. Thus, most microbes and microbial products can serve as potent inflammatory stimuli via their recognized capacity to directly interact with and trigger a response in macrophages and monocytes, polymorphonuclear leukocytes, platelets, endothelial cells and fibroblasts.   In contrast to the immunologically-specific receptor repertoire, however, it is probable that the non-immune receptor repertoire would be relatively restricted, and designed primarily toward non-self recognition of common molecular structures shared by, but unique to, prokaryotes. Thus, whereas the immunologic receptor repertoire encompasses 10^8 to 10^9 different specificities, the potential number of shared molecular structures on prokaryotes would limit the latter repertoire to around 10^1-10^2 specificities. It is noteworthy that, in contrast to immunologically-specific receptors on B and T lymphocytes, knowledge of inflammatory cell receptors for non-self is extremely limited.

Microbial Infections, Edited by H. Friedman *et al.*
Plenum Press, New York, 1992

Table 1. Candidate LPS Receptors Identified on Mammalian Cells

Candidate LPS Receptor	Target Cell	Reference
CD11/18 (Adhesin)	Monocytes, PMN	Wright & Jong, 1986
CD14 (LBP-LPS Complex)	Monocytes, PMN	Wright et al., 1990
95 kDa protein	Macrophage cell line	Hampton et al., 1988
73 kDa protein	Lymphocytes, Macrophages, PMN, Endothelial Cells, Platelets	Lei & Morrison, 1988
70 kDa protein	Macrophages, lymphocytes	Dziarski, 1991
55, 65 kDa protein	Macrophage cell line (J774)	Hara-Kuge et al., 1990
47 kDa protein	Hepatocytes	Parent, 1989
38 kDa protein	Lymphocytes, Macrophages	Lei & Morrison, 1991
18 kDa protein	B cell line (70Z/3)	Kirkland et al., 1990

One of the most potent inflammatory microbial stimuli characterized to date is the complex outer cell membrane constituent lipopolysaccharide (LPS), a common structural component of all gram negative bacteria. While certain components of this complex polysaccharide manifest remarkable antigenic heterogeneity (O-antigen), LPS from diverse gram negative species all possess common, highly conserved, structural features embodied both the inner core oligosaccharide domain as well as the lipid component of this complex polysaccharide. The former structures are characterized by the unique sugars heptose (l-glycero-d-manno-heptose) and KDO (2-keto-3-deoxyoctanoic acid). The latter define the highly conserved lipid A component, a β-1-6-linked diglucosamine with both ester and amide linked long chain fatty acids and substitutions at the 1 and 4' positions with phosphate, phosphorylethanolamine or 4-amino arabinose. Specific recognition of, and response to, both lipid A (and more recently KDO as well) has been well-documented over the past twenty years. Such responses range from those which are highly deleterious to the host, (the endotoxic properties of LPS/lipid A), to those which may be perceived as beneficial, including immunopotentiation and tumor regression.

Efforts to understand the biochemical basis for these diverse host responses to LPS/lipid A have focused upon both a characterization of the *in vitro* responses of isolated cell populations to LPS, and a potential correlation of *in vitro* properties with *in vivo* responses. An impressive list of cellular responses has accumulated in the scientific literature as attributable to the effects of lipid A. More recent efforts have been directed by a number of investigators toward the identification of specific membrane-localized recognition structure(s) responsible for the initial triggering event leading to the diverse cellular responses to LPS. These studies have allowed the identification of a number of candidate LPS receptors and there is experimental evidence that many of these may function in contributing to LPS binding to cells. These candidate LPS receptors have been characterized with respect to both molecular mass and the target cell(s) on which the protein has been identified, and are summarized in Table 1. It is noteworthy, however, that for most of these cell-associated LPS binding proteins, a specific functional role has yet to be unequivocally established.

One of the first LPS binding proteins to be identified was the CD11/18 heterodimeric complex associated with mononuclear phagocytes and neutrophils. This family of cell surface antigens includes several members, each containing a characteristic β (CD11a, b, or c) chain in noncovalent association with an invariant α chain (CD18) of approximately 95 kDa and is now recognized to have an important functional role in adhesion. As shown by Wright and Jong (1), monoclonal antibody to the β chains inhibited the binding of LPS-coated erythrocytes to human monocytes. These results suggested that this heterodimer glycoprotein complex might serve as a potential functional receptor for LPS; however more recent studies have established that monocytes from patients genetically deficient in CD18 expression (leukocyte adhesion defect) nevertheless maintain characteristic *in vitro* cellular responses to LPS (2). Thus, while CD11/18 may well play an important role in cellular recognition of LPS, perhaps most critically in recognition of LPS-containing macroscopic particles (such as LPS coated erythrocytes), it's role in mediating signal transduction would not appear to be essential. It would, nevertheless, be of interest to learn whether LPS bound to erythrocytes does have the capacity to trigger signal transduction in human monocytes.

Highly provocative studies suggesting a functional role for a CD14 phosphatidylin-ositol-linked glycoprotein on mononuclear phagocytes, and to a lesser extent on polymorphonuclear phagocytes, has recently been obtained by Wright et al. (3). These investigators reported the interesting finding that molecular complexes of LPS and a recently described LPS binding protein (LBP) would bind to CD14 on human mononuclear phagocytes. They further provided data to indicate enhanced responses of these cells for production of tumor necrosis factor with LPS-LBP *in vitro*, relative to LPS alone. Whether CD14 actually functions as a specific receptor for LPS-LBP complexes in triggering LPS responses, or whether this membrane glycoprotein simply serves to passively focus LPS at the membrane surfaces where it subsequently interacts with a second LPS receptor remains to be defined. Further, the *in vivo* significance of LBP-LPS complex formation in regulating LPS responses via CD14 is still unresolved. The availability of neutralizing antibody to LBP and/or blocking antibody to murine or rabbit CD14 would undoubtedly provide extremely valuable information toward the resolution of this important question.

Using a somewhat different approach, Hampton et al. (4) have identified a protein of approximately 95 kDa molecular mass associated with a mouse macrophage cell line RAW 264.7. These investigators prepared solubilized cellular extracts which were fractionated by SDS-PAGE, transferred to nitrocellulose and probed for specific binding to a highly radiolabelled lipid A precursor structure. Using this technique, two proteins of 31 and 95 kDa were readily identified by autoradiography. The former protein was suggested by the authors to be a nuclear histone and the latter to potentially function as a candidate LPS receptor. Of some interest in these studies was the fact that, although binding of the radiolabelled lipid A precursor could be readily inhibited by Re-chemotype LPS, an S-form LPS was not capable of inhibiting binding. A more recent reported by Raetz (5) has proposed a potential identity between this 95 kDa protein and the LDL scavenger receptor. In many respects, the capacity of phagocytic cell to take up LPS from the environment for subsequent intracellular degradation and/or processing via a specialized scavenger receptor is attractive conceptually and thus it will be of considerable interest to see how this story unfolds.

The use of photoactivatable crosslinking derivatives of specific ligands to identify specific receptors on target cells has become increasingly popular during the past decade, including efforts to identify LPS specific receptors. Techniques to prepare photoactivatable disulfide-reducible radioiodinated derivatives of LPS were reported by Wollenweber and Morrison (6) in 1985, and have since been employed by several investigators to detect LPS receptors. As described, this photoactivatable probe selectively substitutes phosphorylethanolamine residues associated with the inner core/lipid A region of LPS. The resulting derivatized LPS has been shown by Wollenweber and Morrison to retain full biological activity as assessed by a number of *in vitro* assays.

The initial studies of candidate LPS receptors using these techniques was reported by Lei and Morrison (7) who tentatively identified a protein of approximately 80 kDa molecular mass and pI of approximately 6.5. More recent experiments using one dimensional gels at various concentrations of polyacrylamide have more accurately placed the molecular weight at 73 kDa (8). This LPS-binding protein was identified on mouse B and T lymphocytes as well as mouse B and T cell lines, and was also found on mouse splenic macrophages. It was not detected on erythrocytes or on an undif-

Table 2. Properties of the 73 kDa LPS Binding Protein

Expressed on mouse B lymphocytes, T lymphocytes and macrophages.

Not expressed on erythrocytes or an undifferentiated murine myeloma cell line (SP2/0).

Binding is saturable and both time and temperature dependent.

Inhibitable by native, underivatized, heterologous and homologous LPS, and purified lipid A.

Protein is membrane localized.

Approximately 10^4 molecules per cell on murine splenocytes.

Equivalently expressed on phenotypically LPS normal and LPS non-responder (C3H/HeJ) splenocytes.

---

ferentiated mouse myeloma cell line. The characteristics of this binding interaction with the 73 kDa LPS binding protein are summarized in Table 2, and suggest this protein as a candidate LPS receptor. Of importance, similar type proteins have been detected on human peripheral blood cells, including monocytes, B and T lymphocytes, polymorpho-nuclear phagocytes and platelets. For most of these cells, the 73 kDa protein exists as the dominant LPS binding protein, however in polymorphonuclear leukocytes, several additional LPS binding proteins can be readily detected (9). Finally similar proteins can be found on peripheral blood cells of a variety of mammalian species in addition to mouse and man (10). Further characteristics of this protein will be detailed below.

Other investigators, however, have also employed these photocrosslinking probes to identify LPS binding proteins and, while in several instances, some notable differences have been observed, the reported differences may not be of major significance. One of the more interesting recent reports was published by Hara-Kuge et al. (11) using the J-774 macrophage-like cell line. These investigators described the isolation and characterization of a J774 mutant cell line which was unresponsive to stimulation by LPS. Photocrosslinking of derivatized LPS to both the parent and mutant J774 cells allowed identification of two specific LPS binding proteins of approximately 65 kDa and 55 kDa, neither of which was detectable on the mutant LPS cell line. It is likely that the 65 kDa protein is either identical with, or closely related to, the 73 kDa protein previously identified by Lei and Morrison (7).

Studies reported by Dziarski using similar photocrosslinking techniques have investigated specific binding sites for cell wall peptidoglycan structures on mouse lymphoreticular cells (12, 13). The results of these studies suggested to Dziarski that both peptidoglycan and LPS bind to a specific protein of approximately 70 kDa, most likely the identical binding protein to that previously reported by Lei and Morrison (7). Curiously, these studies suggested that both peptidoglycan and LPS bind to, and compete for, the same 70 kDa binding protein, results of which contrast rather significantly with more recent results of Lei and Morrison (8). The reason for this discrepancy is, at present, not clear. In view of the rather significant structural differences which exist between LPS/lipid A and peptidoglycan, a shared specificity requirement for receptor binding is not readily apparent; nevertheless future studies will undoubtedly unravel these apparent differences. In any case, the capacity of three independent investigators to detect specific binding of LPS to a 65-73 kDa protein on lymphoreticular cells suggests strongly that this protein may have relevance to LPS-mammalian cell interactions.

Other investigators have also reported the identification of specific LPS receptors by photocrosslinking techniques and these studies merit comment. Specifically Parent has recently defined a specific 47 kDa LPS binding protein on hepatocytes with specificity for R-chemotype LPS (14). Further Kirkland et al. (15) have reported an 18 kDa LPS binding protein for Re-LPS on the mouse 70Z/3 pre B cell line. Finally, Lei and Morrison (16) have recently identified a second LPS binding protein of approximately 38 kDa molecular weight on mouse macrophage and lymphocytes. In this latter study, a potential specificity for KDO determinants on the LPS has been suggested. In all of these studies, however, a relationship between LPS binding and a functional cellular response has yet to be established.

At the present time, therefore, we believe that, of all of the candidate LPS receptors listed in Table 1, the most compelling evidence to date would support the hypothesis that the 73 kDa LPS binding protein identified by our laboratory and subsequently confirmed by Hara-Kuge et al. (11) and Dziarski (12, 13) is a true functional receptor for LPS. In the following paragraphs we shall review our own studies over the last three years which have provided experimental evidence in support of this concept. These studies have been predicated upon the observations, summarized in Table 2, that the 73 kDa protein manifests properties essential for membrane-localized receptors. Our more recent studies have, in this respect, documented that monoclonal antibody to this glycoprotein will serve as an agonist for LPS-dependent mouse macrophage responses both *in vitro* and *in vivo*. An understanding of the precise structure of this receptor will provide valuable information on biochemical mechanisms of LPS dependent signal transduction and in a more global sense, relevant perspectives on the nature of non-immunologic mechanisms of host recognition of non-self.

Once our laboratory established that the 73 kDa LPS binding protein is widely distributed on mammalian lymphoreticular and host inflammatory cells, it was imperative to establish unequivocal evidence for a functional role for this glycoprotein. The method selected to achieve this objective was to generate immunologic reagents with binding specificity for the 73 kDa candidate receptor. Strategies were developed for partial purification of the protein using sucrose gradient centrifugation, selective extraction techniques and preparative polyacrylamide gel electrophoresis. The partially purified protein was used to immunize adult Armenian hamsters and splenocytes from immunized animals used to prepare mouse-hamster heterohybridomas. Of more than 2000 hybridoma culture supernatants screened for binding to the partially purified 73 kDa protein, two hybridomas were selected for further study (17). Both inhibited the binding of LPS to the 73 kDa protein, as assessed by photocrosslinking, and both were inhibited in binding to cells by LPS. Of interest, one of these monoclonals, termed MAb5D3, is directed against a protease-sensitive determinant on the 73 kDa protein, whereas the second MAb3D7 is directed against a periodate-sensitive determinant.

We have investigated the functional properties of MAb5D3 in a variety of *in vitro* cell culture assays. We have shown that MAb5D3 will activate bone marrow culture-derived mouse macrophages to become tumoricidal for transformed P815 mastocytoma cells (18). Like LPS, the activity of MAb5D3 is enhanced in the presence of IFN-γ and inactive on C3H/HeJ macrophages. Unlike LPS however, this MAb5D3 antibody activity is insensitive to the addition of polymyxin B and is totally heat-labile. In more recent unpublished studies, we have demonstrated that tumoricidal activity of MAb5D3 correlates precisely with production of nitroxides (S. Green et al. manuscript in preparation). Thus, the *in vitro* macrophage immunostimulatory properties of MAb5D3 correlate well with those of LPS and reinforce the concept that the target of this monoclonal, the 73 kDa LPS binding protein, is in fact a functional receptor for LPS.

Finally we have investigated the *in vivo* biological activities of MAb5D3. Pretreatment of mice with as little as 15 μg of antibody protects these animals against a subsequent lethal dose of LPS in the D-galactosamine model (19). The protective efficacy of MAb5D3, is however, again shown to be heat labile, confirming that the observed activity is not the result of endotoxin contamination of the monoclonal antibody preparation. The protective effects, also similar to LPS, require at least one hour of pretreatment. We have more recently shown that administration of MAb5D3 causes significant increases in hepatic mRNA for IL-1 and, to a lesser extent, TNF, and therefore believe that the protective effects of MAb5D3 may result from low level activation of LPS sensitive macrophages, rendering these cells refractory to subsequent stimulation with LPS. Current efforts are, in part, directed toward the confirmation of this hypothesis.

In summary, therefore, studies from our laboratory during the past five years have focused upon the identification and characterization of specific LPS receptor molecules on macrophages and lymphocytes. We have, to date, defined and characterized a protein of 73 kDa molecular mass, which we have shown to be a prominent LPS binding protein expressed on lymphoreticular cells of a variety of mammalian species, including man. We have reported strong experimental evidence to support the concept that this protein can serve as a functional receptor for LPS on mouse macrophages. At the present time, the precise biochemical identity of this protein has yet to be determined; however preliminary protein sequencing data do not suggest strong homology with any reported protein. Our current research is focused upon the elucidation of the complete primary structure of this protein and it's biochemical mechanism of action in triggering of mammalian lymphoreticular cell responses.

ACKNOWLEDGEMENTS

This work was supported by NIH grants 2R01AI-22943, 2R37AI-23446, the Wesley Foundation and a grant from Centocor Inc. J. Halling and S. Field are scholars of the Wesley Foundation Cancer Research and Training Grant, awarded to the University of Kansas and Kansas State University. The authors acknowledge the expert secretarial assistance of Ms. Janet Hollands.

REFERENCES

1. S. D. Wright and M. T. C. Jong, Adhesion-promotion receptor on human macrophages recognize *Escherichia coli* by binding to lipopolysaccharide, J. Exp. Med. 164:1876 (1986).
2. S. D. Wright, P. A. Detmers, Y. Aida, R. Adamowski, D. C. Anderson, Z .Chad, L. G. Kabbash, and M. J. Pabst, CD18-deficient cells respond to lipopolysaccharide *in vitro*, J. Immunol. 144:2566 (1990).
3. S. D. Wright, R. A. Ramos, P. S. Tobias, F. J. Ulevitch, and J. C. Mathison, CD14, a receptor for complexes of lipopolysaccharide (LPS) and LPS binding protein, Science 249:1431 (1990).
4. R. Y. Hampton, D. T. Golenbock, and C. R. H. Raetz, Lipid A binding sites in membranes of macrophage tumor cells, J. Biol. Chem. 263:14802 (1988).
5. C. R. H. Raetz, K. A. Brozek, T. Clementz, J. D. Coleman, S. M. Galloway, D. T. Golenbock, and R. Y. Hampton, Gram-negative endotoxin: A biologically active lipid, Cold Spring Harbor Symposia on Quantitative Biology, LIII:973 (1988).
6. H. W. Wollenweber, and D. C. Morrison, Synthesis and biochemical characterization of a photoactivatable, iodinatable, cleavable bacterial lipopolysaccharide derivative, J. Biol. Chem. 260:15068 (1985).

7. M. G. Lei, and D. C. Morrison, Specific endotoxin lipopolysaccharide binding proteins on murine splenocytes I. Detection of LPS binding sites on splenocytes and splenocyte subpopulation, J. Immunol. 141:996 (1988).

8. M. G. Lei, S. A. Stimpson, and D. C. Morrison, Specific endotoxic lipopolysaccharide-binding receptors on murine splenocytes III. Binding specificity and characterization, J. Immunol. 147:1925 (1991).

9. J. L. Halling, D. R. Hamill, M. G. Lei, and D. C. Morrison, Identification and characterization of LPS binding proteins on human peripheral blood cell subpopulations. Manuscript submitted for publication.

10. D. Roeder, M. G. Lei, and D. C. Morrison, Specific endotoxic lipopolysaccharide (LPS) binding proteins on lymphoid cells of various animal species-correlation with endotoxin susceptibility, Infect. and Immun. 57:1054 (1989).

11. S. Hara-kuge, F. Amano, M. Nishijima, and Y. Akamatsu, Isolation of a lipopolysaccharide (LPS)-resistant mutant, with defective LPS binding, of cultured macrophage-like cells, J. Biol. Chem. 265:6606 (1990).

12. R. Dziarski, Demonstration of peptidoglycan binding sites on lymphocytes and macrophage by photoaffinity crosslinking, J. Biol. Chem. 266:4713 (1991).

13. R. Dziarski, Peptidoglycan and lipopolysaccharide bind to the same binding site on lymphocytes, J. Biol. Chem. 266:4719 (1991).

14. J. B. Parent, Core specific receptor for lipopolysaccharide on hepatocytes, Circ. Shock 27:341 (1989).

15. T. N. Kirkland, G. D. Virca, T. Kuus-Reichel, F. K. Multzer, S. Y. Kim, R. J. Ulevitch and P. S. Tobias, Identification of lipopolysaccharide-binding proteins in 70Z/3 cells by photoaffinity cross-linking, J. Biol. Chem. 265:9520 (1990).

16. M. G. Lei and D. C. Morrison, Identification of an LPS binding protein with specificity for inner core region (KDO) determinant, FASEB J. 5:A1363 (1991).

17. S. W. Bright, T. Y. Chen, L. M. Flebbe, M. G. Lei, and D. C. Morrison, Generation and characterization of hamster-mouse hybridomas secreting monoclonal antibodies with specificity for lipopolysaccharide receptor, J. Immunol. 145:1 (1990).

18. T. Y. Chen, S. W. Bright, J. L. Pace, S. W. Russell, and D. C. Morrison, Induction of macrophage mediated tumor cytotoxicity by a hamster-monoclonal antibody with specificity for LPS receptor, J. Immunol. 145:8 (1990).

19. D. C. Morrison, R. Silverstein, S. W. Bright, T. Y. Chen, L. M. Flebbe, and M. G. Lei, Monoclonal antibody to mouse lipopolysaccharide receptor protects mice against the lethal effects of endotoxin, J. Infect. Dis. 162:1063 (1990).

BACTERIAL POLYSACCHARIDES, ENDOTOXINS, AND IMMUNOMODULATION

Phillip J. Baker[1], Christopher E. Taylor[1], and Felix S. Ekwunife[2]

Laboratory of Immunogenetics, NIAID, National Institutes of Health, Twinbrook-II Research Facility, 12441 Parklawn Dr., Rockville, Maryland 20852[1], and the Department of Natural Sciences, University of Maryland-Eastern Shore, Princess Anne, Maryland 21853[2]

INTRODUCTION

For many years, the antibody response to the capsular polysaccharide antigen of Type III *Streptococcus pneumoniae* (SSS-III) was considered to be thymus-independent, mainly because athymic nude mice and thymus-bearing mice, make antibody responses of the same magnitude (1-4). However, there is much evidence that the antibody response to SSS-III is indeed regulated in both a positive and a negative manner by two types of thymus-derived (T) lymphocytes with opposing functions (5-7). Such regulatory T cells have been referred to as suppressor T cells (Ts) and amplifier T cells (Ta). Ts limit the extent to which antibody-forming B cells proliferate, whereas Ta drive B cells to multiply further after antigenic stimulation (5-7). Usually, the effects produced by Ts and Ta are counter-balanced, which explains why they have escaped detection for so many years. Consequently, the action of one type of regulatory T cell is not often apparent unless the activity of the other is either altered or eliminated. Fortunately, Ts are CD8+, CD4-, whereas Ta are CD8-, CD4+ so that this can be accomplished by treating cell suspensions with monoclonal anti-CD8 or anti-CD4 antibody plus complement (8-11). For example, if donor cell suspensions containing both Ts and Ta activity are transferred to athymic mice immunized with SSS-III, the magnitude of the resulting antibody response is not changed; however, if Ta activity is eliminated from the same cell suspension before transfer, significant suppression occurs, whereas the antibody response is greatly increased if Ts activity is eliminated before cell transfer (9-12). These findings demonstrate that Ts and Ta act in a competitive manner on B cells to control the magnitude of the antibody response elicited after immunization. All of the experimental data we have obtained thus far are consistent with such a homeostatic model for regulating the antibody response (5-7).

ACTIVATION OF REGULATORY T CELLS

Because the effects produced by Ts and Ta are antigen-specific, one might assume that both types of regulatory T cells are activated directly by antigen; that is not the case and there is no evidence to indicate that Ts and Ta respond directly to SSS-III (5). Instead, both Ts and Ta are activated by immune B cells in the absence of antigen, i.e., even after immune B cells have been treated with an enzyme (polysaccharide depolymerase) to remove any residual antigen that might be bound to their surface (10, 13). Also, the kinetics for the induction of Ts and Ta activity after the infusion of immune B cells are the same as those observed in mice after exposure to antigen (10). Since the effects of Ts and Ta are antigen-specific, this means that the idiotypic determinant (Id) of cell-associated antibody specific for SSS-III, rather than antigen, signals the activation of both types of regulatory T cells (10, 13). This was confirmed by means of double-cell-transfer experiments in which it was shown that the Ts and Ta activity generated in mice exposed to immune B cells could also be transferred by T cells to other mice immunized with SSS-III (10).

It is not known if Ts and Ta are activated in response to Id alone, or Id presented in the context of another molecule, which may be an antigen found only on the surface of activated or immune B cells. If that were not the case, one would expect clones of precursors of antibody-forming cells to activate Ts and Ta spontaneously in unimmunized animals; that rarely -- if ever -- occurs. Other studies indicate that in order for B cells to activate Ts, they must not only express Id but also be able to synthesize protein for either the continued expression of cell-associated antibody (with its appropriate Id) after cell transfer, or for the manufacture of another protein molecule required for the activation of Ts in conjunction with Id (14-15). Although the nature of this molecule -- which appears to be absent in autoimmune NZB mice -- remains to be identified, it does not appear to be Ia antigen (14-15). It may be a B cell differentiation antigen and studies are underway to clarify that issue.

A significant feature of the homeostatic control mechanism just described is the fact that suppression is a natural consequence of an immune response. Consequently, one can not elicit an antibody response in thymus-bearing mice without activating Ts, as well as Ta (7). This implies that in order to increase the magnitude of the antibody to poorly immunogenic antigens, one must either eliminate the inhibitory effects of Ts or increase the activity of Ta. This has important implications since regulatory T cells are not unique for the antibody response to SSS-III and have be shown to regulate the magnitude of the antibody response to other microbial antigens such as *Neisseria meningitidis* group A and C capsular polysaccharides (16-17). *Pseudomonas aeruginosa* lipopolysaccharide (LPS) (17, 18), *Streptococcus mutans* polysaccharide (17), *Serratia marcescens* (LPS) (19, 20), *Escherichia coli* 055 LPS (19, 20) bacterial (*Leuconostoc*) dextran (21) and perhaps many other microbial antigens.

EFFECT OF ENDOTOXIN ON REGULATORY T CELL FUNCTION

Although the endotoxins or lipopolysaccharides (LPS) of gram-negative bacteria are extremely toxic and pyrogenic, they also possess several beneficial properties (22). Indeed, LPS is one of the most potent immunological adjuvants known (22) and the ability of LPS to activate B cells and macrophages with the induction of cytokines no doubt contributes to its adjuvanticity; however, we wished to examine whether some of the adjuvant effects of LPS might be due to its ability to modulate the expression of regulatory T cell function. It has been established that all of the pharmacological and immunological effects of LPS are mediated by the diphospho lipid A region of the molecule which is of course extremely toxic (22); however, if one removes a phosphate group from the reducing end of lipid A, one obtains monophosphoryl lipid A or MPL. MPL retains the adjuvant properties of lipid A and LPS, but is nontoxic, even in large doses (23, 24). The availability of nontoxic MPL enables one to examine whether the adjuvanticity is associated with an alteration of regulatory T cell function.

The prior exposure or priming of mice with a single injection of a subimmunogenic dose (5 ng) of SSS-III results in the development of an antigen-specific form of unresponsiveness termed low-dose immunological paralysis (25); such unresponsiveness persists for several weeks or months after priming with just one injection of SSS-III (25). It is mediated by Ts and can be transferred to other mice with CD8$^+$ T cells (12). If one determines the degree of unresponsiveness generated at different times after priming with 5 ng of SSS-III, the kinetic pattern obtained shows that significant unresponsiveness can be detected as early as 18-24 hr after priming; thereafter, unresponsiveness increases progressively with time, reaching a maximal level, 2-3 days after priming (12). If one treats mice with MPL at the time of priming -- or at different times after priming -- it is possible to determine whether MPL has an effect on the induction or expression of Ts activity. The results of such an experiment showed that treatment with 50 μg of MPL has no effect on the induction of Ts activity; however, it is able to decrease the inhibitory effects produced by Ts, once they have become activated (26). It also was shown that unresponsiveness could be eliminated by giving large doses (100 μg) of MPL or multiple injections of smaller doses MPL --or LPS -- after Ts have been activated (27). The effects produced by MPL appear to be selective for Ts since treatment with the same or larger doses of MPL has no adverse effect on the expression of helper T cell (Th) and Ta activity (26) as well as cytotoxic T lymphocyte (Tc) function (28). There is no evidence to suggest that MPL acts by stimulating the expression of Ta activity (unpublished observations).

C3H/HeN mice are responsive or susceptible to the lethal effects of LPS (LPS^r mice), whereas C3H/HeJ mice are resistant or defective in their sensitivity to LPS (LPS^d mice; 29); however, such mice do not differ in their capacity to make an antibody response to SSS-III, or in the degree of Ts activity generated after prior exposure (priming) to a subimmunogenic dose (5 ng) of SSS-III (30). Therefore, it was of interest to determine if the Ts activity expressed in LPS^r and LPS^d strains of C3H mice differs in sensitivity to MPL. To this end, 6-8 week-old male LPS^r and LPS^d strains of C3H

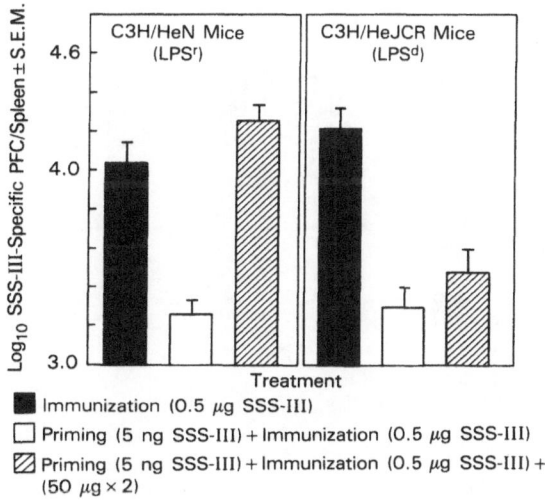

Fig. 1. Effect of *S. minnesota* R595 MPL on the expression of low-dose paralysis to SSS-III in LPS^r and LPS^d strains of C3H mice. Mice were given (i.p.) 50 μg of MPL on days 0 and +1 relative to immunization (day 0) with 0.5 μg of SSS-III.

mice (from the Fort Detrick Cancer Research Center, Frederick, Maryland) were primed with a single injection (i.p.) of 5 ng of SSS-III; 3 days later, they were immunized (i.p.) with an optimally immunogenic dose (0.5 μg) of SSS-III and numbers of plaque-forming cells (PFC) making antibody specific for SSS-III were determined, 5 days after immunization by means of an immune-plaque procedure (31). The results obtained were expressed as the number of SSS-III-specific PFC detected per spleen, which provides a valid measure of the total antibody response elicited (32).

The data of Figure 1 show that priming with 5 ng of SSS-III resulted in the development of significant (P<0.005) immunological unresponsiveness (low-dose paralysis) in both strains of mice. It has been established that such unresponsiveness is antigen-specific and mediated by CD8$^+$ Ts (7, 10, 12). If primed mice are given (i.p.) two injections of 50 μg of *Salmonella minnesota* R595 MPL (Ribi ImmunoChem Research, Inc., Hamilton, Montana), unresponsiveness is eliminated in LPS^r -- but not in LPS^d -- mice. These results are in accord with those of another study using MPL derived from *S. typhimurium* (30), thereby indicating that this is a general property of MPL.

The ability to transfer low-dose paralysis to other mice with CD8$^+$ T cells (10, 12), proves that it is mediated by Ts and permits one to examine directly the ability of MPL to abolish Ts activity (27). A pooled spleen cell suspension was prepared from LPS^r mice, 18-24 hr after priming (i.p.) with 5 ng of SSS-III. It was adjusted with Medium 199 to contain 10^8 nucleated cells/ml and dispensed in 2.5 ml volumes among several tubes. To each tube was added a known amount (5 pg to 5 μg) of MPL or saline (control) in a volume of 50 μl, and the contents were held at 4°C for 30 min after mixing. Then groups of LPS^r mice were given (i.v.) 20 x 10^6 cells in a volume of 0.2 ml at the time of immunization with 0.5 μg of SSS-III. The magnitude of the antibody

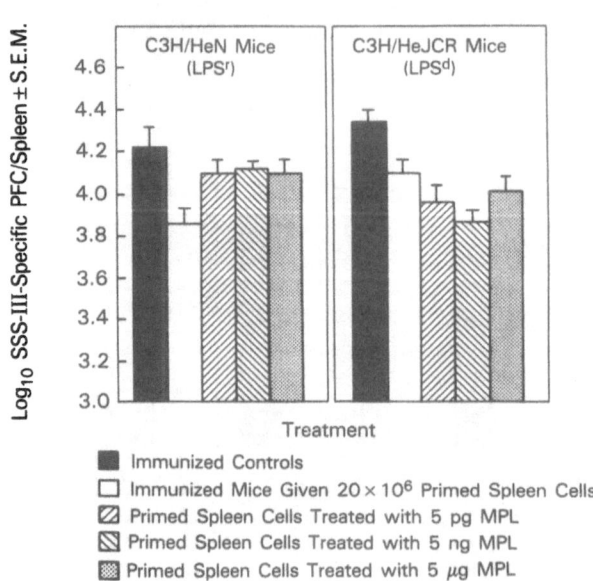

Fig. 2. Effect of *in vitro* treatment with *S. minnesota* R595 MPL on the ability of primed spleen cells to transfer suppression to immunized recipient mice. Primed spleen cells were given (i.v.) to recipients at the time of immunization (i.p.) with 0.5 μg of SSS-III.

(PFC) response produced was determined, 5 days after immunization and compared with that of (a) immunized LPS^r mice not given primed spleen cells, and (b) immunized LPS^r mice given primed spleen cells not treated *in vitro* with MPL.

The transfer of 20 x 10^6 primed spleen cells not treated with MPL suppressed (P < 0.05) the antibody response of LPS^r mice (Figure 2). Prior *in vitro* treatment with 5 pg, 5 ng, or 5 µg of MPL abolished the capacity of primed LPS^r spleen cells to transfer suppression to LPS^r recipient mice. Different results were obtained with LPS^d mice (Figure 2). Although the transfer of 20 x 10^6 primed LPS^d spleen cells likewise resulted in suppression of the antibody response, treatment with the same amounts of MPL failed to abolish the capacity of such cells to transfer suppression. Thus, Ts - mediated unresponsiveness of the same magnitude can be generated in both LPS^r and LPS^d strains of C3H mice after priming with SSS-III, although only the Ts activity of LPS^r mice is sensitive to MPL. This suggests that the activated Ts of LPS^r and LPS^d strains of C3H mice differ with respect to (a) a cell-surface receptor required for the binding and subsequent internalization of MPL, and/or (b) a unique biochemical pathway that is extremely sensitive to inactivation by MPL.

EFFECT PRODUCED BY LPS AND ITS DERIVATIVES

Since the MPL used is extracted from the LPS of *Salmonella minnesota* R595, one might ask whether some of the effects observed might be attributed to the presence of small amounts of contaminating LPS. The data of Figure 3 indicate that is not the case.

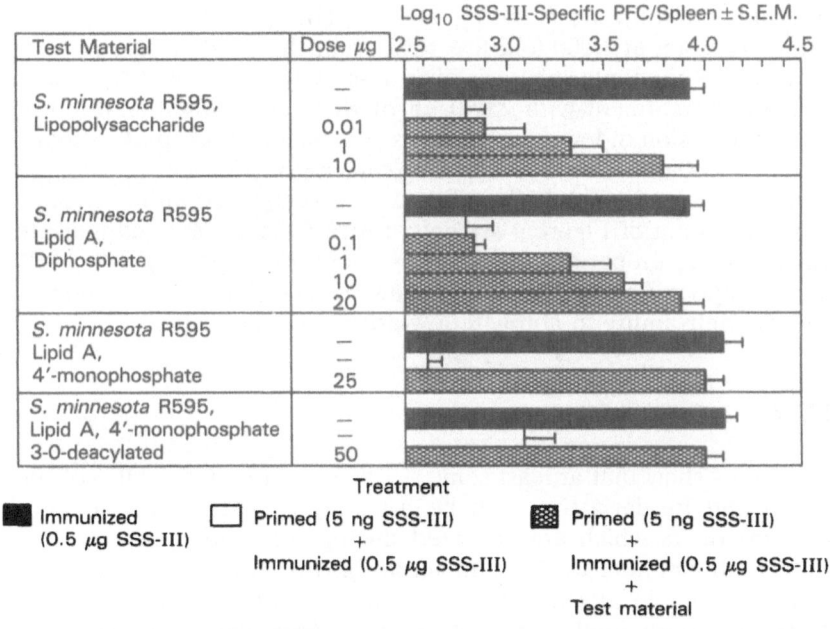

Fig. 3. Effect of treatment with various derivatives of *S. minnesota* R595 LPS on the expression of low-dose paralysis to SSS-III in 8-10 week old female BALB/cByJ mice (Jackson Laboratories, Bar Harbor, Maine). Stated amounts of the preparations used (obtained from Ribi ImmunoChem Research Inc., Hamilton, Montana) were given (i.p.) to primed mice on day 0 and +1 relative to immunization (day 0) with 0.5 µg of SSS-III.

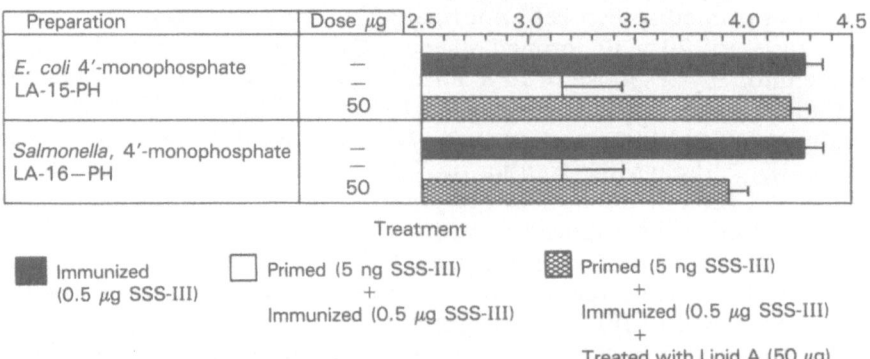

Preparation	Dose µg	Log$_{10}$ SSS-III-Specific PFC/Spleen ± S.E.M.
E. coli 4'-monophosphate LA-15-PH	— — 50	
Salmonella, 4'-monophosphate LA-16—PH	— — 50	

Treatment

■ Immunized (0.5 µg SSS-III) □ Primed (5 ng SSS-III) + Immunized (0.5 µg SSS-III) ▦ Primed (5 ng SSS-III) + Immunized (0.5 µg SSS-III) + Treated with Lipid A (50 µg)

Fig. 4. Effect of treatment with synthetic preparations of lipid A (obtained from ICN Biomedicals Inc., Costa Mesa, California) on the expression of low-dose paralysis to SSS-III in BALB/cByJ mice. Primed mice were given (i.p.) 50 µg of each preparation used on days 0 and +1 relative to immunization (day 0) with 0.5 µg of SSS-III.

Although the administration of 1-10 µg of LPS results in either a significant reduction or abrogation of low-dose paralysis, similar results were obtained under the same experimental conditions using the same amounts of lipid A diphosphate; however, both of these materials are quite toxic and pyrogenic. The MPL (4'-monophosphate) of S. minnesota lipid A is about 1,000-fold less toxic than the lipid A diphosphate, whereas the 3-0-deacylated 4'-monophosphate is even less toxic - perhaps by a factor of 8 or 10 - than MPL. Treatment with 25-50 µg of these relatively nontoxic preparations eliminated the expression of low-dose paralysis. Thus, it is most unlikely that the effects of MPL and its 3-0-deacylated derivative can be attributed to the presence of contaminating LPS, even though both LPS and lipid A can also influence the expression of Ts activity. The data of Figure 4 strongly affirm such a view and indicate that two injections of synthetic preparations of the 4'-monophosphoryl lipid A of E. coli and Salmonella, which do not contain LPS, can likewise abolish the expression of Ts activity as evidenced by their ability to abrogate low-dose paralysis.

SUMMARY AND CONCLUSIONS

These studies show that at least some --though certainly not all -- of the adjuvant effects of LPS and its derivatives can be attributed to its ability to eliminate the inhibitory effects of Ts which are activated during the course of a normal immune response. The ability of nontoxic MPL to act in this fashion suggests that it can be used as a safe and acceptable alternative to Freund's complete adjuvant to increase the immunogenicity of poorly immunogenic antigens. More important, the ability of MPL to eliminate the expression of Ts activity, without adversely influencing other T cell functions (e.g., Th, Ta, and Tc activity) makes its use as an adjuvant even more promising since it can then permit those T cell functions to be expressed in a much more efficient manner. Obviously, this would have great significance for the development of tumor immunity.

REFERENCES

1. J. H. Humphrey, D. M. V. Parrot, and J. East, Studies on globulin and antibody production in mice at birth, Immunology 7:419 (1964).

2. J. G. Howard, G. H. Christie, B. M. Courtney, E. Leuchars, and A. J. S. Davies, Studies on immunological paralysis. VI. Thymic-independence of tolerance and immunity to Type III pneumococcal polysaccharide, Cell. Immunol. 3:614 (1971).

3. J. K. Manning, N. D. Reed, and J. W. Jutila, Antibody response to *Escherichia coli* lipopolysaccharide and Type III pneumococcal polysaccharide by congenitally thymus-less (nude) mice, J. Immunol. 108:1470 (1972).

4. P. J. Baker, N. D. Reed, P. W. Stashak, D. F. Amsbaugh, and B. Prescott, Regulation of the antibody response to Type III pneumococcal polysaccharide, I. Nature of regulatory cells, J. Exp. Med. 137:1431 (1973).

5. P. J. Baker, Homeostatic control of antibody responses: a model based on the recognition of cell-associated antibody by regulatory T cell, Transplant. Rev. 26:1 (1975).

6. P. J. Baker and B. Prescott, Regulation of the antibody response to pneumococcal polysaccharides by thymus-derived (T) cells: mode of action of suppressor and amplifier T cells, p. 67-105, In: "Immunology of Bacterial Polysaccharides," J. A. Rudbach and P. J. Baker, eds., Elsevier/North-Holland Publishing Co., New York (1979).

7. P. J. Baker, Regulation of magnitude of antibody response to bacterial polysaccharide antigens by thymus-derived lymphocytes, Infect. Immun. 58:3465 (1990).

8. K. L. McCoy, P. J. Baker, P. W. Stashak, and T. M. Chused, Two defects of old New Zealand black mice are involved in the loss of low-dose paralysis to Type III pneumococcal polysaccharide, J. Immunol. 135:2438 (1985).

9. C. E. Taylor, D. F. Amsbaugh, P. W. Stashak, G. Caldes, B. Prescott, and P. J. Baker, Cell surface antigens and other characteristics of T cells regulating the antibody response to Type III pneumococcal polysaccharide, J. Immunol. 130:19 (1983).

10. C. E. Taylor, P. W. Stashak, G. Caldes, B. Prescott, T. E. Chused, A. Brooks, and P. J. Baker, Activation of antigen-specific suppressor T cells by B cells from mice immunized with Type III pneumococcal polysaccharide, J. Exp. Med. 158:703 (1983).

11. C. E. Taylor, P. W. Stashak, J. Chiang, W. M. Leiserson, G. Caldes, B. Prescott, and P. J. Baker, Characteristics of amplifier T cells involved in the antibody response to the capsular polysaccharide of Type III *Streptococcus pneumoniae*, J. Immunol. 132:3103 (1984).

12. P. J. Baker, D. F. Amsbaugh, P. W. Stashak, G. Caldes, and B. Prescott, Direct evidence for the involvement of thymus-derived (T) suppressor cells in the expression of low-dose paralysis to Type III pneumococcal polysaccharide, J. Immunol. 128:1059 (1982).

13. P. J. Baker, C. E. Taylor, M. B. Fauntleroy, P. W. Stashak, and B. Prescott, The role of antigen in the activation of regulatory T cells by immune B cells, Cell. Immunol. 96:376 (1985).

14. K. L. Elkins, P. W. Stashak, and P. J. Baker, Transferred B cells from auto-immune NZB/N mice fail to activate T suppressor cells, Cell. Immunol. 110:14 (1987).

15. K. L. Elkins, P. W. Stashak, and P. J. Baker, Metabolic activity is necessary for activation of T suppressor cells by B cells, J. Immunol. 56:259 (1988).

16. E. Muller and M. A. Apicella, T-cell regulation of the murine antibody response to *Neisseria meningitides* group A capsular polysaccharide, Infect. Immun. 56:259 (1988).

17. C. E. Taylor and R. Bright, T-cell modulation of the antibody response to bacterial polysaccharide antigens, Infect. Immun. 57:180 (1989).

18. W. G. Powderly, G. B. Pier, and R. B. Markham, *In vitro* T cell-mediated killing of *Pseudomonas aeruginosa*, IV. Nonresponsiveness in polysaccharide-immunized BALB/c mice is attributable to vinblastine-sensitive suppressor T cells, J. Immunol. 137:2025 (1986).

19. K. L. Elkins, P. W. Stashak, and P. J. Baker, Prior exposure to subimmunogenic amounts of some bacterial lipopolysaccharides induces specific immunological unresponsiveness, Infect. Immun. 55:3085 (1987).

20. K. L. Elkins, P. W. Stashak, and P. J. Baker, Mechanisms of specific immunological unresponsiveness to bacterial lipopolysaccharides, Infect. Immun. 55:3093 (1987).

21. K. A. Haslov, M. B. Fauntleroy, P. W. Stashak, C. E. Taylor, and P. J. Baker, T cells regulate the IgM antibody response of BALB/c mice to dextran B1355, Immunobiology 182:100 (1990).

22. M. A. Freudenberg, C. Galanos, Bacterial lipopolysaccharides: structure, metabolism and mechanisms of action, Intern. Rev., Immunol. 6:207 (1990).

23. E. Ribi, Beneficial modification of the endotoxin molecule, J. Biol. Response Modifiers 3:1 (1984).

24. E. Ribi, J. L. Cantrell, K. Takayama, N. Qureshi, J. Peterson, and H. O. Ribi, Lipid A and immunotherapy, Rev. Infect. Dis. 6:567 (1984).

25. P. J. Baker, P. W. Stashak, D. F. Amsbaugh, and B. Prescott, Characterization of the antibody response to Type III pneumococcal polysaccharide at the cellular level. I. Dose-response studies and the effect of prior immunization on the magnitude of the antibody response, Immunology 20:469 (1971).

26. P. J. Baker, J. R. Hiernaux, M. B. Fauntleroy, B. Prescott, J. L. Cantrell, and J. A. Rudbach, Inactivation of suppressor T-cell activity by nontoxic monophosphoryl lipid A, Infect. Immun. 56:1076 (1988).

27. P. J. Baker, C. E. Taylor, P. W. Stashak, M. B. Fauntleroy, K. Haslov, N. Qureshi, and K. Takayama, Inactivation of suppressor T cell activity by the nontoxic lipopolysaccharide (LPS) of *Rhodopseudomonas sphaeroides*, Infect. Immun. 58:2862 (1990).

28. F. Esquivel, C. E. Taylor, and P. J. Baker, Differential sensitivity of CD8$^+$ suppressor T cells and cytotoxic T lymphocytes to inactivation by bacterial monophosphoryl lipid A, Infect. Immun. 59:2994 (1991).

29. D. C. Morrison, C3H/HeJ mouse strain: role in elucidation of host responses to bacterial endotoxin, Microbiology pp. 23-23 (1986).

30. J. R. Hiernaux, P. W. Stashak, J. L. Cantrell, J. A. Rudbach, and P. J. Baker, Immunomodulatory activity of monophosphoryl lipid A in C3H/HeJ and C3H/HeSnJ mice, Infect. Immun. 57:1483 (1989).

31. P. J. Baker, P. W. Stashak, and B. Prescott, The use of erythrocytes sensitized with purified pneumococcal polysaccharides for the assay of antibody and antibody-forming cells, Appl. Microbiol. 17:422 (1969).

32. P. J. Baker and P. W. Stashak, Quantitative and qualitative studies on the primary antibody response to pneumococcal polysaccharide at the cellular level, J. Immunol. 103:1342 (1969).

IMMUNOMODULATION OF C3H/HeJ CELLS BY ENDOTOXIN

ASSOCIATED PROTEIN AND LIPOPOLYSACCHARIDE ENDOTOXIN

Barnet M. Sultzer, Jayant Bandekar,
Raymond Castagna and Khaled Abu-Lawi

State University of New York
Health Science Center at Brooklyn
Brooklyn, New York 11203

INTRODUCTION

Selected outer membrane polypeptides from Gram-negative bacteria are associated with lipopolysaccharide endotoxin (LPS) upon extraction of the organism by the trichloracetic acid method (1). This endotoxin associated protein (EP) when purified is known to be a potent polyclonal activator of B lymphocytes from mice and humans and can act as a potent adjuvant as well (2-4). However, the mechanism by which EP acts as an immunomodulator has not been elucidated. On the other hand, some of the early events initiated by the cross-linking of surface immunoglobulin on B cells have been described which are believed to be important for inducing DNA synthesis. Anti-sIg antibodies stimulate the turnover of phosphatidyl inositol (PI) leading to an increase in diacylglycerol (DAG) and inositol triphosphate (IP$_3$) which, in turn, induce the activation of protein kinase C (PKC) and the release of intracellular Ca^{++}, respectively (5). In contrast, LPS acts like a DAG analogue directly causing the translocation and activation of PKC without changes in PI or Ca^{++} ions (5).

In previous experiments designed to elucidate the immunoregulatory properties of EP, we found that when LPS was used in conjunction with EP, activation of C3H/HeJ splenic B lymphocytes was suppressed by LPS (6). Maximal inhibition occurred 9 to 12 hr into the culture period suggesting some step(s) during the G1 phase of the cell cycle was affected. No evidence was found that suppressor cells were activated by LPS so that our working hypothesis has been that the observed suppression was the result of a biochemical event(s) in C3H/HeJ B cells that acted as a negative signal (6).

Based on these findings, we chose to examine the early biochemical events in EP activation of C3H/HeJ B cells and to further characterize the LPS suppression phenomenon in these terms. In this manner, we hope to learn more about the mechanism of immunomodulation by these biologically potent molecules.

Microbial Infections, Edited by H. Friedman *et al*.
Plenum Press, New York, 1992

MATERIALS AND METHODS

Animals

C3H/HeJ mice were obtained from the National Cancer Institute (Frederick, MD.). Mice of either sex were used separately at 4 to 6 months of age and maintained on water and Purina Chow *ad lib*.

Materials

EP from *Salmonella typhi* 0-901 was prepared as described previously (1). LPS was prepared from *S. typhi* 0-901 by the phenol-water method (7). Radioiodinated LPS was prepared from the p-hydroxy-methylbenzimidate derivative of LPS by the chloramine T method (8). 1-(5-isoquinoline-sulfonyl)-2-methylpiperazine dihydrochloride (H-7), staurosporine and phorbol myristate acetate (PMA) were obtained from Calbiochem (San Diego, CA). Acridine orange was obtained from Polyscience Inc. (Warrington, PA). Pertussis toxin (PT) and the B-oligomer of PT (PTB) were obtained from List Biological Labs (Campbell, CA). DNase I, RNase, ionomycin, p-OH methylbenzimidate and Triton-X-100 were from Sigma Chemical Co. (St. Louis, MO). Excellulose GF-5 desalting columns were from Pierce (Rockford, IL), the mouse T cell recovery kit from Biotex Labs., Inc. (Edmonton, Canada) and the PGE_2 ELISA kit was obtained from Cayman Chemical (Ann Arbor, MI).

Purification of B Cells

Mice were sacrificed by cervical dislocation and single-cell suspensions were prepared from the spleens as previously described (1). The red blood cells (RBCs) were lysed by tris-ammonium chloride treatment (9) and T cells were depleted by treatment with anti-Thy 1.2 monoclonal antibodies (Sigma Chemical Co.) and rabbit Low Tox Complement (Cedarlane Laboratories, Hornby, Canada). Dense resting B cells were obtained by Percoll (Pharmacia, Uppsala, Sweden) density gradient separation. Cells in the band showing 1.076 to 1.08 density were selected and washed twice with RPMI 1640 medium supplemented with 5% fetal calf serum (Gibco Laboratories, Grand Island, NY), 100U of penicillin and 100 µg of streptomycin.

Purification of T Cells

Single-cell suspensions were prepared from spleens. RBCs were lysed by treatment with tris-ammonium chloride and T cells were obtained by passing through Biotex mouse T cell recovery columns. Cells were washed twice with RPMI 1640 culture medium containing 5% FCS before culturing in microtiter plates.

Macrophage Stimulation

Resident peritoneal macrophages were obtained by washing the peritoneal cavity with RPMI 1640 culture medium containing 10% FCS. The cells (4×10^6/ml) were incubated in 24 well flat bottom plates for 2 hr, washed free of nonadhering cells and cultured with the appropriate stimulant for 24 hr. The supernatants were then recovered, centrifuged to remove any cells, and stored at -20°C until assayed.

DNA Synthesis

Purified resting B cells, and T cells were cultured in RPMI supplemented with 5% FCS at 2.5×10^5 cells per well in 96 - U-bottom well microtiter plate (Falcon Labware, Lincoln Park, NJ). EP or other mitogens in the presence or absence of the inhibitors, H7, staurosporine, PT or PTB were added at various times to the cultures as indicated in the experiments. The cells were incubated at 37°C in 10% CO_2 for 48 hr and DNA synthesis was measured by (^{3}H) thymidine incorporation by adding 1 μCi of (^{3}H)thymidine (ICN Biochemicals, Costa Mesa, CA) in the last 18 hr of incubation. The cultures were harvested onto glass fiber filters using a cell harvester (Brandel, Gaithersburg, MD) and (^{3}H)thymidine incorporation was measured by a Beckman liquid scintillation counter, model # 6000 IC (Beckman Instruments, Fullerton, CA).

To examine the effect of PKC depletion on DNA synthesis, 2.5×10^6 cells/ml were incubated in tubes with or without 5 μg PMA for 16 hr at 37°C. Then 1 μg EP or 100 g PMA plus 1 μM ionomycin were added and the cells were cultured for another 48 hr. (^{3}H)thymidine was added 18 hr before harvesting as described above. All cultures in microtiter plate experiments were done in six replicates. Variation about the mean did not exceed ± 10%. Unless otherwise noted, experiments were repeated at least 3 times.

Cell Cycle Analysis by Cytofluorometry

The cell cycle analysis was done by measuring cellular RNA and DNA content using acridine orange dye by the method of Darzynkiewickz et al. (10). Acridine orange binding to double stranded DNA results in green (530 nm) fluorescence while binding to single stranded RNA gives red (> 600 nm) fluorescence. Purified resting B cells (2.5 $\times 10^6$/ml) from C3H/HeJ mice were cultured in RPMI-1640 medium supplemented with 5% FCS with various mitogens and with or without LPS. At specific time intervals between zero and 66h, 1 ml FCS was added to the culture and then the cells were fixed with 6 ml of 70% cold ethanol and kept at 4°C until analyzed. The flow cytometric analysis was done using an argon ion laser (388 nm) for excitation and measuring green and red fluorescence by Cytofluorograph IIS (Ortho Instruments, MA) linked to a 2140 Computer (Ortho Instruments). The fluorescence from 10,000 cells was measured.

Assay for Prostaglandin E2 (PGE$_2$)

PGE$_2$ in the culture supernatant was estimated using an ELISA kit (Cayman Chemical, Ann Arbor, MI). In brief, a 96-well microtiter plate was coated with mouse

Table 1. Stimulation of DNA Synthesis by EP in C3H/HeJ Resting B Lymphocytes

EP (ng)	(^{3}H) Thymidine Incorporation Mean cpm (S.I.)[a]	
–	1,121	
10	8,239	(8.3)
20	16,244	(15.5)
50	28,398	(26.3)
100	39,398	(36.1)

[a]S.I. = Stimulation index or the total mean cpm of stimulated culture divided by the total mean cpm of the background control.

antiglobulin. Wells were then washed and to each well 50 µl of each of the following was added: dilutions of the culture supernatants, acetylcholinesterase-labeled PGE_2 and anti-PGE_2. Plates were incubated overnight at room temperature, washed, and then a constant amount of acetylcholine derivative was added, and the color was allowed to develop for 3 hr. The color change was read at 420 nM wave length using a microtiter reader (ARTEK Sys. Corp. NY) and PGE_2 was estimated using a standard curve that was prepared using known amounts of pure PGE_2.

RESULTS

The Role of Protein Kinase C in EP-Stimulated DNA Synthesis in C3H/HeJ Resting B Cells

As previously reported, EP stimulates splenic B cells from LPS nonresponder C3H/HeJ mice to synthesize DNA and produce polyclonal antibodies (1). Likewise, EP induces purified resting B cells to proliferate as shown in Table 1. When these cells were stimulated in the presence of various concentrations of H-7 or staurosporine (St.) added at the initiation of the culture, DNA synthesis measured after 48 hr was inhibited in a dose dependent manner (Table 2). B cells stimulated with concentrations of PMA and ionomycin which synergize to activate PKC directly were also inhibited as expected.

When B cells are exposed to a high concentration of PMA alone for a prolonged period, degradation of PKC occurs so that cells are unable to respond to various mitogens such as LPS (11, 12). When C3H/HeJ resting B cells were incubated with PMA (5 µg) for 16 hr at 37°C and then stimulated with EP or PMA and ionomycin, they were unable to respond to either stimulus (Table 3).

To examine the role of PKC in later stages of activation, H7 was added 12 hr after the culture was initiated and DNA synthesis was measured after 48 hr. When B cells were exposed to either EP or the combination of PMA and ionomycin under these conditions, 77% and 83% inhibition was achieved respectively.

Taken together, the results described above clearly indicate that PKC activity is needed for B-cells to progress through the G1 phase of the cell cycle into the S phages of DNA synthesis when stimulated by EP and that a continuum of PKC activity is needed past the first minutes of activation.

Table 2. Inhibition of DNA Synthesis by PKC Inhibitors in Activated C3H/ HeJ B Lymphocytes

Mitogen	Inhibitor	(^{3}H)Thymidine Incorporation Stimulation Index	% Inhibition
EP (100 ng)	-	29.9	-
"	H-7 (40 µM)	1.8	97.2
PMA (50 ng + Ionomycin (1µM)	-	121.0	-
"	H-7 (50 µM)	3.6	98.0
EP (10 ng)	-	14.4	-
"	St. (0.1 µM)	0.9	100.0
"	" (0.01 µM)	8.6	43.8
"	" (0.001 µM)	14.5	0

Table 3. Effect of PKC Depletion on EP Induced Proliferation of C3H/HeJ B Lymphocytes

B Cell Preexposure	Mitogen[a]	(^{3}H) Thymidine Incorporation Stimulation Index
Culture medium	EP	29.4
"	PMA + Ionomycin	26.6
PMA	EP	2.0
"	PMA + Ionomycin	0.1

[a]EP (1 μg) or PMA (100 ng) and Ionomycin (1μM) were added to the cultures and (^{3}H)thymidine incorporation was measured after 48 hr.

The Role of G Proteins in EP Stimulated Resting B Cells

Evidence that guanosine 5'trisphosphate binding (G) regulatory proteins may also be involved in B cell activation by LPS has recently appeared (13). Pertussis toxin (PT), which inactivates the alpha subunit of the Gi and other G proteins by ADP ribosylation (14), has been shown to inhibit LPS-induced murine B cell proliferation, as well as LPS-induced signalling in WEHI-231 B lymphoma cells (15).

Predicated on these findings, we examined the involvement of G proteins in B cell activation by EP using the inhibitor PT and the B-subunit of PT which does not inactivate the Gi protein (16). As shown in Table 4, PT blocks EP induced DNA synthesis when added at the initiation of the culture of resting B cells in a dose dependent manner. However, at the same time, the B oligomer (PTB) which serves to bind the holotoxin to the cell surface receptor, was able to inhibit DNA synthesis at concentrations equivalent to the doses of the holotoxin. The level of inhibition of both

Table 4. Inhibition of DNA Synthesis by Pertussis Toxin in C3H/HeJ Activated B Lymphocytes

Mitogen	Inhibitor (ng)	(^{3}H) Thymidine Incorporation Stimulation Index	% Inhibition
EP (100 ng)	–	35.8	–
"	PT (100)	20.8	42.2
"	" (400)	15.3	58.2
PMA (100 ng) + Ionomycin (1μM)	–	21.1	–
"	PT (100)	18.8	11.7
"	" (400)	19.1	9.9
EP (100 ng)	–	32.1	–
"	PTB (75)	21.6	38.6
"	" (300)	16.0	55.6
PMA (100 ng) + Ionomycin (1μM)	–	21.2	–
"	PTB (75)	25.4	–
"	" (300)	19.0	10.8

Table 5. Suppression of EP Stimulated DNA Synthesis in C3H/HeJ Activated B Lymphocytes by LPS

EP Mitogen (100 ng)	LPS (μg)	(^{3}H)Thymidine Incorporation Mean cpm (S.I.)	% Inhibition
+	–	39,398 (36.1)	–
+	2	19,946 (18.8)	49.8
+	10	16,750 (15.9)	57.5
+	20	11,143 (10.9)	71.7
+	50	5,147 (5.6)	86.9

the holotoxin and the B oligomer were strikingly similar and, furthermore, the combination of PMA and ionomycin stimulation of DNA synthesis was not affected. Therefore, the inhibition observed with the pertussis holotoxin most likely is due to the B-oligomer rather than the inactivation of the G protein by the PT - A subunit.

LPS Suppression of C3H/HeJ B Cells Stimulated by EP

As previously noted, we have found that LPS could inhibit C3H/HeJ lymphocytes when activated by other mitogens. A representative example of this suppression is seen in Table 5 wherein the DNA synthesis of resting B cells stimulated with EP is blocked in a dose responsive manner by LPS. To determine when in the cell cycle this inhibition took place, we employed cytofluorometry to assess the relative amounts of RNA and DNA content per cell. In this manner, we could establish the stage of the cell cycle through which the cells were progressing in the presence of the stimulant EP and where the inhibitor LPS blocked the cells. As shown in Table 6, by 12 hr, 35% of the B cells were inhibited from progressing through the G1 phase of the cell cycle which accumulated to about 65% by 24 to 36 hr. Consequently by 48 hr about 50% of the cells had not synthesized DNA during the S phase. Similar results were obtained when PMA and Ionomycin were used with LPS (data not shown).

Extending these studies to other cell types, we wanted to determine whether the LPS suppression would affect T cells and macrophages if they were activated in a

Table 6. Inhibition by LPS of the Progression of Activated C3H/HeJ B Cells through the Cell Cycle

Time	% Inhibition[a] G1	S
12 hr	35.4	–
24 hr	59.5	42.7
36 hr	62.6	44.6
48 hr	64.8	48.8

[a]B cells were stimulated with 5 μg EP in the presence or absence of 250 μg LPS. Average value from two trials.

Table 7. Suppression of DNA Synthesis by LPS in C3H/HeJ Activated T Lymphocytes

Mitogen[a]	LPS[b]	(^{3}H)Thymidine Incorporation Stimulation Index	% Inhibition
Con A	−	42.6	−
"	+	21.0	51.9
PMA + Ionomycin	−	161.0	−
	+	98.3	39.2

[a]Concanavalin A was used at 100 ng per culture. The final concentration of PMA was 10 ng per culture and Ionomycin 0.5 μM.
[b]The concentration of LPS was 50 μg per culture.

similar manner. When purified T cells from C3H/HeJ mice were cultured with the mitogen Concanavalin A in the presence and absence of LPS, DNA synthesis was found to be inhibited to a significant degree (Table 7). Likewise, T cells initiated by PMA and ionomycin were also inhibited, although the maximal inhibition obtained in both instances (52% and 39% respectively) was less than that seen with activated B cells.

When resident peritoneal macrophages were cultured in the presence of EP, substantial PGE_2 synthesis was obtained again in a dose responsive manner. With optimal concentrations of LPS added to this system, significant inhibition of C3H/HeJ macrophage production of PGE_2 was obtained (Table 8). Consequently, not only are proliferation signals suppressed by LPS as in the C3H/HeJ B and T lymphocytes but also the metabolic chain of events involved in the conversion of arachidonic acid to prostaglandin E2.

DISCUSSION

Resting, noncycling (G0) B lymphocytes can be activated by a variety of stimulants. The process is recognized to be complex and multiple pathways of signal transduction may be involved. Whereas the stimulation of murine B cells by anti-sIg

Table 8. Suppression of PGE_2 by LPS in C3H/HeJ Macrophages Stimulated by EP

EP (ng)	LPS (μg)	Stimulation Index[a]	% Inhibition
−	−	1.0	−
10	−	9.7	−
10	50	1.0	100
50	−	26.0	−
50	100	4.6	83

[a]Stimulation Index = $\dfrac{PGE_2 \text{ from activated macrophages}}{PGE_2 \text{ from control macrophages}}$.

leads to the activation of phospholipase C and the subsequent activation of PKC by the second messenger DAG, LPS appears to act directly as an analogue of DAG (5, 17). The pivotal role of PKC in the eventual proliferation of B cells is emphasized by the inhibition of DNA synthesis by H-7 of both LPS and anti-sIg stimulation (11, 18).

In the experiments reported here, C3H/HeJ resting B cells stimulated by EP are likewise inhibited by two major inhibitors of PKC. Furthermore, by depleting PKC by prolonged exposure to PMA, the cells become unresponsive to EP adding more evidence to suggest that PKC activation is needed for B cells to enter the S phase of DNA synthesis.

The role of G proteins in the activation of B cells also has received attention. For example, Dziarski concluded that G proteins are involved in B cell activation since ribosylation of PT substrates correlated with the inhibition of LPS induced DNA synthesis (13). Also, PT blocks LPS stimulation of IL-1 production at doses consistent with G protein modification (13). However, in our experiments, the B-oligomer of PT which does not modify G proteins is as good as the PT in inhibiting EP stimulated DNA synthesis. It appears, therefore, that with at least EP the inhibition observed with PT involves the receptors for the binding unit of the holotoxin independent of G protein in a manner yet to be determined.

It should be noted that in preliminary experiments (data not shown), we have found that the specific inhibitor of protein tyrosine kinase (PTK), Genisteine, also blocks EP stimulated DNA synthesis so that it is likely that multiple pathways of early signal transduction are turned on by EP. How they are interrelated remains to be explained.

That EP activation of C3H/HeJ immunocompetent cells can be inhibited by LPS is an intriguing problem. This novel phenomenon is applicable to B cells, T cells and macrophages, although the suppression of T cell proliferation is relatively limited. In binding experiments, we have found that the level of binding of ^{125}I-LPS to purified B cells of C3H/HeJ mice between 0 to 12 hr was unaffected by the presence of the stimulant EP and did not differ from that obtained with LPS responder B cells from C3H/OuJ mice. In addition, from wash-out experiments, optimal inhibition by LPS obtained 9 to 12 hr after C3H/HeJ splenic B cells are cultured with a mitogen also suggested that some step(s) after the cell was triggered was affected by LPS (6). This was confirmed by the analysis of the cell cycle phases reported here. LPS blocked the cells from progressing through the G1 phase resulting in few cells capable of synthesizing DNA. In other experiments wherein we measured RNA synthesis, by (^{3}H)-uridine incorporation, significant inhibition by LPS was obtained in B cells from 0 to 12 hr after the culture was initiated confirming the cell cycle results.

Since we have shown PKC is critical to the activation of C3H/HeJ B cells by EP, LPS was used to inhibit the proliferation stimulated by the synergistic combination of PMA and Ionomycin which activates PKC directly, bypassing the PI pathway. The results clearly show that PKC activation, as measured by DNA synthesis, was suppressed by LPS in C3H/HeJ cells at concentrations of LPS which are mitogenic for normal responder B cells of C3H/OuJ mice or other strains. This suggests EP activation of B cells to synthesize DNA is suppressed because PKC activity is shut down. Furthermore, this activity appears necessary at least as long as 12 hr into the G1 phase and is not confined to the early minutes after cells are exposed to the mitogen. In this regard, direct measurement of the phosphorylation resulting from PKC and PTK activity are needed in the presence and absence of LPS to verify the inhibitor results obtained so far.

Nevertheless, this present study suggests one or more protein kinases are activated by EP and possibly suppressed by LPS. In addition, investigation of other secondary events including changes in ion release and transfer (Ca^{++} and K^+), as well as alterations in the membrane potential of B cells is necessary before a more complete picture can be drawn of the events leading to gene activation and the eventual proliferation of C3H/HeJ B cells.

Altogether, the results reported here confirm and extend the suppressive effect of LPS on the so-called LPS nonresponder cells of the C3H/HeJ mouse. That is not to say that the failure of these cells to respond to low but activating amounts of LPS for responder cells is due to this suppressive effect. Rather it is suggested that a negative signal can be turned on by relatively high concentrations of LPS which contributes to the lack of responsiveness of C3H/HeJ cells to endotoxin.

SUMMARY

Protein kinase C plays a vital role in the activation of C3H/HeJ B lymphocytes by endotoxin associated protein; however, it is unlikely that G proteins are involved in the early signals stimulated by EP. On the other hand, LPS suppresses C3H/HeJ B cell DNA synthesis induced by EP which may be the result of PKC down regulation. LPS inhibits C3H/HeJ B cells from progressing through the G1 phase of the cell cycle blocking RNA synthesis within the first 12 hr after the cells are stimulated. Finally, this inhibition extends to activation of the arachidonic acid metabolism in C3H/HeJ macrophages and T cell proliferation to a limited extent.

ACKNOWLEDGEMENTS

This work was supported in part by NIH-NIAID Grant 28526. We thank Dr. Allen J. Norin and Ms. Mino Sadeghian for assistance with the cytofluorometry. Also we thank Ms. Janice Howard for her excellent preparation of the manuscript.

REFERENCES

1. B. M. Sultzer and G. W. Goodman, Endotoxin protein: A B-cell mitogen and polyclonal activator of C3H/HeJ lymphocytes, J. Exp. Med. 144:821 (1976).
2. G. W. Goodman and B. M. Sultzer, Further studies on the activation of lymphocytes by endotoxin protein, J. Immunol. 122:1329 (1979).
3. G. W. Goodman and B. M. Sultzer, Endotoxin protein is a mitogen and a polyclonal activator of human B lymphocytes, J. Exp. Med. 149:713 (1979).
4. B. M. Sultzer, J. P. Craig, and R. Castagna, Endotoxin associated proteins and their polyclonal and adjuvant activities, in: "Immunobiology and Immunophar-macology of bacterial endotoxins," A. Szentivanyi and H. Friedman, ed., Plenum Publishing Corp., New York (1986).
5. J. C. Cambier and J. T. Ransom, Molecular mechanisms of transmembrane signaling in B lymphocytes, Ann. Rev. Immunol. 5:175 (1987).
6. B. M. Sultzer and R. Castagna, Inhibition of activated nonresponder C3H/HeJ lymphocytes by lipopolysaccharide endotoxin, Inf. Immun. 56:3040 (1988).
7. O. Westphal, O. Luderitz, and F. Bister, Uber die extraktion von bakterien mit phenol-wasser, Z. Naturforsch. Teil B 7:148 (1952).
8. R. J. Ulevitch, The preparation and characterization of a radioiodinated bacterial lipopolysaccharide, Immunochem. 15:157 (1978).
9. W. Boyle, An extension of the ^{51}Cr-release assay for the estimation of mouse cytotoxins, Transplant. 6:761 (1968).

10. Z. Darzynkiewickz, F. Traganos, T. Sharpless, and M. R. Melamed, Lymphocyte stimulation: a rapid multiparameter analysis, Proc. Natl. Acad. Sci. USA, 73:2881 (1976).

11. S. Gupta, S. Gollapudi, and B. Vayuvegula, Lipopolysaccharide-induced murine B cell proliferation: A role of protein kinase C, Cell. Immunol. 117:425 (1988).

12. A. Rodriguez-Pena and E. Rozenhurt, Disappearance of Ca^{2+} sensitive phospholipid-dependent protein kinase activity in phorbol ester-treated 3T3 cells, Biochem. Biophys. Res. Commun. 120:1053 (1984).

13. R. Dziarski, Correlation between ribosylation of pertussis toxin substrates and inhibition of peptidoglycan-, muramyl dipeptide- and lipopolysaccharide-induced mitogenic stimulation in B lymphocytes, Eur. J. Immunol. 19:125 (1989).

14. J. Codina, J. Hildebrandt, R. Iyengar, L. Birnbaumer, R. D. Sekura, and C. R. Manclark, Pertussis toxin substrate, the putative Ni component of adenylyl cyclase, is an ab heterodimer regulated by guanosine nucleotide and magnesium, Proc. Natl. Acad. Sci. USA 80:4276 (1983).

15. J. P. Jakway and A. L. DeFranco, Pertussis toxin inhibition of B cell and macrophage response to bacterial lipopolysaccharide, Science 234:743 (1986).

16. M. Tamura, K. Nogimori, S. Murai, M. Yajima, K. Ito, T. Katada, M. Ui, and S. Ishii, Subunit structure of islet-activating protein, pertussis toxin, in conformity with the A-B model, Biochem. 21:5516 (1982).

17. S. A. Grupp and J. A. K. Harmony, Increased phosphatidylinositol metabolism is an important but not an obligatory early event in B lymphocyte activation, J. Immunol. 134:4087 (1985).

18. H. Kikutani, R. Kimura, H. Nakamura, R. Sato, A. Muraguchi, N. Kawamura, R. R. Hardy, and T. Kishimoto, Expression and function of an early activation marker restricted to human B cells, J. Immunol. 136:4019 (1986).

BACTERIAL PROTEIN-LPS COMPLEXES AND IMMUNOMODULATION

Kathryn Nixdorff, Gabriele Weber, Kerstin Kaniecki,
Waltraud Ruiner and Sigrid Schell

Institut für Mikrobiologie, Technical University Darmstadt
Federal Republic of Germany

INTRODUCTION

The cell surface or outer membrane of Gram-negative bacteria contains several components that can exert a variety of effects on cells of the immune system. Lipopolysaccharide (LPS) is a classic example of such immunologically active agents. LPS is tightly bound to other components of the native outer membrane, particularly to proteins. The immune system of a host is therefore most likely confronted with complexes of immunologically active components in addition to single constituents.

For this reason we have been investigating the immunomodulating activities of defined complexes of outer membrane components, particularly the effects that proteins have on the specific responses to LPS. In previous studies it was shown that a major protein isolated from purified cell walls of *Proteus mirabilis* (39 kDa protein) is a strong immunogen in mice (1). In addition, this protein is a potent modulator of the specific immune responses to LPS. When complexed with LPS, the 39 kDa protein greatly enhances IgG antibody-producing cell responses strictly specific for the serotype of LPS used for immunization (2,3). In the present report, we show that IgG2a responses are predominantly enhanced, although all IgG subclasses of antibody-producing cells are augmented.

Our results further indicate that T cells play a decisive role in strengthening the LPS-specific, IgG responses. Since the participation of T cells in immune reactions is regulated to a great extent by signals from antigen-presenting cells, we have also examined possible immunomodulating effects of the 39 kDa protein on LPS-macrophage interactions. It was determined that the 39 kDa protein is indeed a strong modulator in this system, exerting its effects in a very differential manner with respect to various parameters of LPS-induced activation of macrophages.

MATERIALS AND METHODS

Bacteria

Proteus mirabilis strains 19 and VI and *Escherichia coli* K12 were from the collection of H.H. Martin, Institut für Mikrobiologie, Technical University of Darmstadt.

Microbial Infections, Edited by H. Friedman *et al.*
Plenum Press, New York, 1992

Lipopolysaccharides

LPS I from *P. mirabilis* 19 and *P. mirabilis* VI were extracted from whole cells by the phenol/water method of Westphal et al. (4). LPS I was purified further according to Gmeiner (5) by treatment with ribonuclease and deoxyribonuclease, followed by repeated washing in the ultracentrifuge. In some cases, LPS I was reextracted with phenol/water to remove contaminating protein. LPS from *E. coli* K12 was also extracted by the phenol/water method and purified in a manner similar to the isolation of *P. mirabilis* LPS I. LPS from *Salmonella abortus equi* was obtained from Calbiochem (Frankfurt am Main, FRG). Usually, a concentration of 1 mg LPS/ml was prepared in Gey's balanced salt solution (GBSS) (Gibco BRL GmbH, Eggenstein, FRG). A homogeneous suspension was obtained by mixing the LPS solution in a sonication bath for 2 min. Finally, the solution was diluted in the appropriate medium to the concentration to be used in the test system.

Isolation of the 39 kDa Protein

The 39 kDa protein was extracted from cell walls of *P. mirabilis* 19 free of cytoplamic membranes and cell contents with 1% (w/v) sodium deoxycholate and purified according to a method previously described (2). To ensure solubilization, 1 mg of the protein was first dissolved in one part NaOH at pH 9.0 and immediately neutralized with 1 part phosphate buffered saline (two-fold concentrated) and 2 parts of GBSS. The protein was subsequently sonicated for 2 min, and diluted to the desired concentration in the appropriate medium.

Mixtures of LPS and Protein

LPS and the 39 kDa protein were mixed together in the ratios to be used in the test system. The protein was first dissolved as described above, and LPS was subsequently added. The mixture was then sonicated for 2 min in the sonication bath and diluted in the appropriate medium to the desired concentration.

Commercial Proteins and Sera

Bovine serum albumin (BSA) was obtained from Serva (Heidelberg, FRG). Methylated BSA (m-BSA) was purchased from Sigma Chemical Co. (Munich, FRG). Solutions of the proteins in GBSS (either alone or with LPS) were sonicated before use. Fetal calf serum (FCS) and horse serum were purchased from Gibco KG (Eggenstein, FRG). They were heat-inactivated for 30 min at 56°C before use.

Cell Lines

Murine lymphoma EL4 cells were cultivated in culture medium I plus 5% FCS. CTLL-2 cells, a murine cytotoxic T lymphocyte cell line which has lost the ability to lyse target cells, is interleukin 2 (IL-2) dependent for growth. L 929 S cells, a fibroblast cell line from a C3H/An mouse (H-2^k) was used as a source of macrophage colony stimulating factor (M-CSF or CSF-1) (6) and as target cells for measuring macrophage cytotoxic activity. The CTLL-2 and L 929 S cells were cultivated in culture medium I supplemented with 10% (v/v) FCS, 1% (v/v) of non-essential amino acids (NEAA, 100 x concentrated; Gibco KG, Eggenstein, FRG), 1% (v/v) of 100 mM sodium pyruvate (Biochrom KG, Berlin, FRG) and 5 x 10^{-5} M 2-mercaptoethanol (2-ME; Sigma Chemical Co., Munich, FRG). All cell lines were kindly provided by P.G. Munder, Max-Planck-Institut für Immunbiologie, Freiburg i. Brsg., FRG.

Basic Medium

Click-RPMI 1640 medium containing L-glutamin and 1.5 g/l glucose, but no NaHC0$_3$ (Biochrom KG, Berlin), was used as basic medium. An amount of dehydrated medium designated for 10 l was reconstituted in 9 l with distilled water, sterilized by filtration and kept at 4°C.

Culture Medium I

19 ml of 7.5% (w/v) NaHC0$_3$ (Gibco KG, Eggenstein, FRG), 10 ml of 1 M N-2-hydroxyethylpiperazine-N'-2-ethanesulfonic acid (HEPES) (Gibco BRL, Eggenstein, FRG), and 10 ml of 200 mM L-glutamin were added to 900 ml of the basic medium.

Culture Medium II

19 ml of 7.5% (w/v) NaHC0$_3$, 10 ml of 1 M HEPES, 20 ml of 200 mM L-glutamin and 3 g of glucose were added to 900 ml of the basic medium.

Macrophage Differentiation Medium

30% (v/v) L-cell conditioned medium as a source of M-CSF (6), 10% (v/v) FCS, 5% (v/v) horse serum, 1 mM sodium pyruvate (Gibco BRL GmbH, Eggenstein, FRG) and 5 x 10^{-5} M 2-mercaptoethanol (2-ME) were added to culture medium II.

Preparation of L-cell Conditioned Medium

L 929 S cells were cultivated in culture medium I plus 10% FCS at a density of 1 x 10^5 cells/ml for 7 days at 37°C under a humidified atmosphere with 5% C0$_2$ in teflon bags (Biofolie 25, Heraeus Holding GmbH, Hanau, FRG) on the hydrophilic side of the foil as described by Flesch et al. (6). The supernatant was collected and cell debris was removed by centrifugation (350 x g, 15 min, 10°C). Finally, the supernatant was sterilized by filtration, divided into aliquots and stored at -20°C.

Macrophage Culture Medium

10% (v/v) FCS and 1% (v/v) NEAA (100 x concentrated) were added to culture medium II.

Animals

NMRI mice were obtained from Zentralinstitut für Versuchstierzucht, Hannover, FRG. C57BL/6 mice were from Charles River Wiga, Sulzfeld, FRG. All animals were specific pathogen-free, female mice 6-10 wks old at the time of the experiment.

Differentiation of Macrophages Derived from Bone Marrow

Murine bone marrow stem cells were isolated from C57BL/6 mice as described by Flesch et al. (6). 2 x 10^6 bone marrow stem cells in 30 ml of macrophage differentiation medium were cultivated in teflon foil bags (Biofolie 25, Heraeus Holding GmbH, Hanau, FRG) with the hydrophobic surface facing inward (6). Bags measured 10 x 30 cm and were sterilized by autoclaving at 120°C. Bone marrow stem cells were incubated in the bags at 37°C under a humidified atmosphere with 5% C0$_2$ for 7-9 days.

At the end of this incubation period, 98% of the cells were macrophages according to the butyrate-esterase (7) and the Giemsa staining procedures. At the time of harvesting, the cells were washed twice in macrophage culture medium and counted. About 1-2 x 10^7 macrophages were obtained from a culture of 2 x 10^6 stem cells.

Chemiluminescence Assay

Macrophages obtained as described above were adjusted to a density of 2.5 x 10^6 cells/ml in GBSS. 100 µl of these cells were placed in polystyrene test tubes (Abimed Analysen-Technik GmbH, Düsseldorf, FRG) together with 700 µl GBSS, 100 µl of 1 mM N,N'dimethyl-9,9'-biacridinium dinitrate (lucigenin) (Sigma Chemie GmbH, Deisenhofen, FRG), and 100 µl of LPS, protein, or mixtures of LPS with protein contained in GBSS. Controls included macrophages and lucigenin without stimulants. All reagents and cell suspensions were brought to 37°C before starting the experiment. Measurements (cpm) were made at 2 min intervals in a Picolite 6500 luminometer (Packard Instruments, Frankfurt am Main, FRG) for 30-60 min total measuring time, and the peak cpm are reported for each time interval.

Interleukin 1 (IL-1) Assay

An IL-2 dependent IL-1 assay was performed according to Simon et al. (8). Briefly, 1 x 10^6/ml macrophages derived from bone marrow of C57BL/6 mice were stimulated to IL-1 production in macrophage culture medium with LPS I from *P. mirabilis* 19 or with mixtures of LPS and 39 kDa protein for 24 h. 100 µl of the supernatants were added to 1 x 10^5 EL4 cells in the presence of 5 x 10^{-7} M Ca^{++}-ionophore A23187 in culture medium I plus 5 % FCS (final volume 200 µl per well) to induce IL-2 production. After 24 hr incubation, 100 µl of the supernatants containing IL-2 were transferred to 1 x 10^4 IL-2 dependent CTLL-2 cells in a final volume of 200 µl per well. After 40 hr incubation the cultures were pulsed with 0.25 µCi of ^{3}H-thymidine and incubated for an additional 8 hr. Cell proliferation was determined by the amount of ^{3}H-thymidine incorporated. The stimulation index was calculated by setting the values for unstimulated cultures at 1.0.

Determination of Macrophage Cytotoxic Activity

The cytotoxic activity of macrophage culture supernatants was measured according to Drysdale et al. (9). Briefly, 8 x 10^5/ml macrophages derived from bone marrow of C57BL/6 mice were stimulated in macrophage culture medium with LPS I from *P. mirabilis* 19, 39 kDa protein or mixtures of LPS and 39 kDa protein for 1 hr. Supernatants were collected and assayed in a cytotoxicity test with L 929 S cells as targets. 5 x 10^5/ml L 929 S cells were incubated 1 hr with 2 mg/ml actinomycin D in the medium used to culture CTLL-2 cells. 100 µl of these cells were added to 100 µl of dilutions of the macrophage supernatants. After 18 hr incubation the remaining viable cells were stained with crystal violett. A detergent solution was used to solubilize the stained cells and the absorbance at 570 nm was measured in an ELISA Processor II Reader (Behringwerke AG, Frankfurt/Main, FRG). A unit of cytotoxicity (U) is defined as the reciprocal dilution of the sample required to reduce the absorbance by 50% (9).

Immunization of Mice

For characterization of the secondary antibody-producing cell responses *in vivo*, NMRI mice 6-9 wks old were given 25 µg LPS from *P. mirabilis* VI or a mixture of 25 µg LPS and 12.5 µg 39 kDa protein per injection. A primary i.p. injection was given on day 0 and a secondary on day 14. Responses were measured on day 18 in the hemolytic plaque assay.

Stimulation of Spleen Cells

Secondary antibody-producing cell responses were measured in spleen cells of C57BL/6 mice given a primary i.p. injection of 0.625 µg LPS from *P. mirabilis* VI or a mixture of 0.625 µg LPS and 1.25 µg 39 kDa protein 14 days earlier. Spleen cells from these animals were cultivated and stimulated to antibody production in the Mishell-Dutton system (10). Briefly, 3×10^6 spleen cells in 500 µl RPMI 1640 supplemented medium (10,11) containing 5% FCS were placed in wells of cell culture dishes (well size 16 mm). Antigen (0.625 ng LPS alone or mixed with 1.25 ng 39 kDa protein) contained in 100 µl medium was added to each culture and the dishes were incubated at 37°C under a humidified atmosphere with 5% CO_2 for 4 days. Two drops of nutrient cocktail (10) were added to each well daily. Cells were collected for use in hemolytic plaque assays by gently scraping the bottom of wells with a rubber policeman and aspirating with a Pasteur pipette. Cells pooled from 4 wells were centrifuged ($350 \times g$) for 10 min at 4°C and resuspended in Hank's balanced salt solution (10) in an appropriate volume (usually 4-8 ml).

Treatment of Spleen Cells with Anti-Thy 1.2 Plus Complement

Spleen cells were washed 3 times and resuspended to a density of 5×10^7/ml in minimum essential medium (Eagle) (MEM; Biochrom KG, Berlin, FRG) without FCS. 4.0 ml of these cells were mixed with 4.0 ml of a 1:20 dilution of anti-Thy 1.2 (ICN Biomedicals GmbH, Eschwege, FRG) and 4.0 ml of a 1:20 dilution of guinea pig serum (Behringwerke AG, Marburg, FRG). Controls containing no antibody were included. These mixtures were incubated at 37°C in a shaker water bath for 30 min. The mixtures were then cooled in an ice bath for 10 min and washed 2 times in MEM with a cushion of FCS and resuspended in RPMI medium to be used for stimulation *in vitro* as described above.

Assay of Antibody-producing Cells

The IgM and the IgG antibody-producing cell responses to LPS were measured in the hemolytic plaque assay using a modification of the microscope slide assay (10) with sheep red blood cells (SRBC) sensitized with alkali-treated LPS as previously described (2). For the measurement of IgG subclass responses, indirect plaques were developed with specific rabbit anti-mouse IgG1, IgG2a, IgG2b, IgG2ab or IgG3 sera (Nordic Laboratories, Tilburg, NL). The specificity of these IgG subclass antisera has been documented (2).

RESULTS

Modulation of the IgG Subclass Responses to LPS by the 39 kDa Protein from *P. mirabilis*

Table 1 presents the effects of the 39 kDa protein in modulating the antibody-producing cell responses of NMRI mice to LPS I from *P. mirabilis* VI. We have shown previously that modulating effects occur at the time of primary immunization, but can be detected most clearly during the secondary response, when memory cells develop into antibody-producing cells (3). The 39 kDa protein did not modulate IgM responses

Table 1. Comparison of the secondary *in vivo* responses to LPS alone and to a mixture of LPS and 39 kDa protein from *P. mirabilis*

Immunogen[a]	Antibody Type	PFC/10^6 Spleen Cells on Day 18[b]	% IgG Subclass of Total IgG	Adjuvant Factor[c]
LPS	IgM	704 ± 31	-	1
	IgG1	92 ± 10	37	1
	IgG2a	11 ± 5	4	1
	IgG2b	133 ± 38	54	1
	IgG3	10 ± 4	4	1
LPS	IgM	1183 ± 69	-	2
+	IgG1	915 ± 14	22	10
39 kDa	IgG2a	1132 ± 86	27	103
Protein	IgG2b	1286 ± 138	31	10
	IgG3	841 ± 54	20	84

[a]Dosages per injection were 25 µg LPS and 12.5 µg 39 kDa protein. Mice received a primary injection on day 0 and a secondary injection on day 14.
[b]Geometric means ± standard errors of the numbers of plaque-forming cells (PFC) from 3 separate experiments. Responses were measured against *P. mirabilis* LPS coupled to SRBC.
[c]Adjuvant factor was calculated by dividing the PFC obtained with LPS alone as immunogen (taken as factor of 1) into the PFC of the respective Ig isotype obtained with LPS + 39 kDa protein as immunogen.

appreciably, but the numbers of IgG-producing cells specific for LPS were greatly enhanced when the protein was mixed with LPS before injection. Although all subclasses of IgG-producing cells were substantially augmented, IgG2a responses were predominantly enhanced (adjuvant factor of 103). The strict specificity of the IgG responses for the serotype of LPS used for stimulation has been documented previously (2,3).

IgG Responses Enhanced by the 39 kDa Protein Are T Cell Dependent

Treatment of spleen cells with anti-Thy 1 plus complement before secondary stimulation *in vitro* reduced all IgG subclass responses induced by the LPS-protein complex dramatically, compared to the same spleen cell population treated with complement alone (Table 2). Spleen cells stimulated with LPS alone were unaffected by treatment with antibody and complement. It should be noted that treatment of spleen cells with anti-Thy 1 plus complement was very effective in eliminating T cells, while leaving B cells intact; cells treated in this way did not respond to stimulation with Concanavalin A, but did respond well to LPS (data not shown). Therefore, the residual numbers of PFC after treatment with anti-Thy 1 plus complement most probably represent T cell independent responses.

Table 2. Secondary *in vitro* responses to LPS alone and to a mixture of LPS and 39 kDa protein from *P. mirabilis* before and after treatment of spleen cells from C57BL/6 mice with anti-Thy 1.2 plus complement

| Treatment of Spleen Cells | Antibody Type | PFC/Culture on Day 18[a] after Secondary Stimulation of Spleen Cells with | |
		LPS[b]	LPS + 39 kDa Protein[c]
	IgM	574 ± 25	2918 ± 1155
	IgG1	196 ± 40	1885 ± 194
Complement	IgG2ab	79 ± 10	2072 ± 187
	IgG3	0	915 ± 122
	IgM	499 ± 20	881 ± 241
anti-Thy 1.2	IgG1	114 ± 15	365 ± 85
+	IgG2ab	75 ± 5	334 ± 79
Complement	IgG3	0	148 ± 61

[a]Responses were measured against *P. mirabilis* LPS coupled to SRBC. The numbers of plaque-forming cells (PFC) are reported as geometric means ± standard errors from 2-3 separate experiments.
[b]Mice were given a primary injection of 0.625 µg LPS on day 0. Spleen cells were stimulated with 0.625 ng LPS on day 14.
[c]Mice were given a primary injection of 0.625 µg LPS mixed with 1.25 µg 39 kDa protein on day 0. Spleen cells were stimulated with 0.625 ng LPS mixed with 1.25 ng 39 kDa protein on day 14.

Modulation of the Interaction of LPS with Macrophages by The 39 kDa Protein

The effects of the 39 kDa protein in modulating the interaction of LPS with macrophages were tested using various parameters of macrophage activation. The production of oxygen radicals by macrophages was measured in the chemiluminescence reaction with lucigenin as amplifier (12). At a concentration of 50 µg/ml, LPS I from *P. mirabilis* 19 was a strong inducer of oxygen radical production in macrophages derived from bone marrow of C57BL/6 mice (Fig. 1). A peak response of 3.48×10^7 cpm in the chemiluminescence reaction was reached after 14 min reaction time. As little as 1.56 µg/ml of the 39 kDa protein mixed with the LPS was effective in reducing the response significantly, while 12.5 µg/ml of the protein reduced the reaction 10-fold. Higher amounts of the protein had no apparent additional effect. In order to test the specificity of the action of the 39 kDa protein in this system, LPS was also mixed with two model proteins, bovine serum albumin (BSA), which is relatively hydrophilic, and methylated BSA (m-BSA), which is more hydrophobic. The results presented in Fig. 2 show that the 39 kDa protein again reduced the chemiluminescence drastically. In contrast, both BSA and m-BSA dramatically enhanced the activity of LPS. It should be

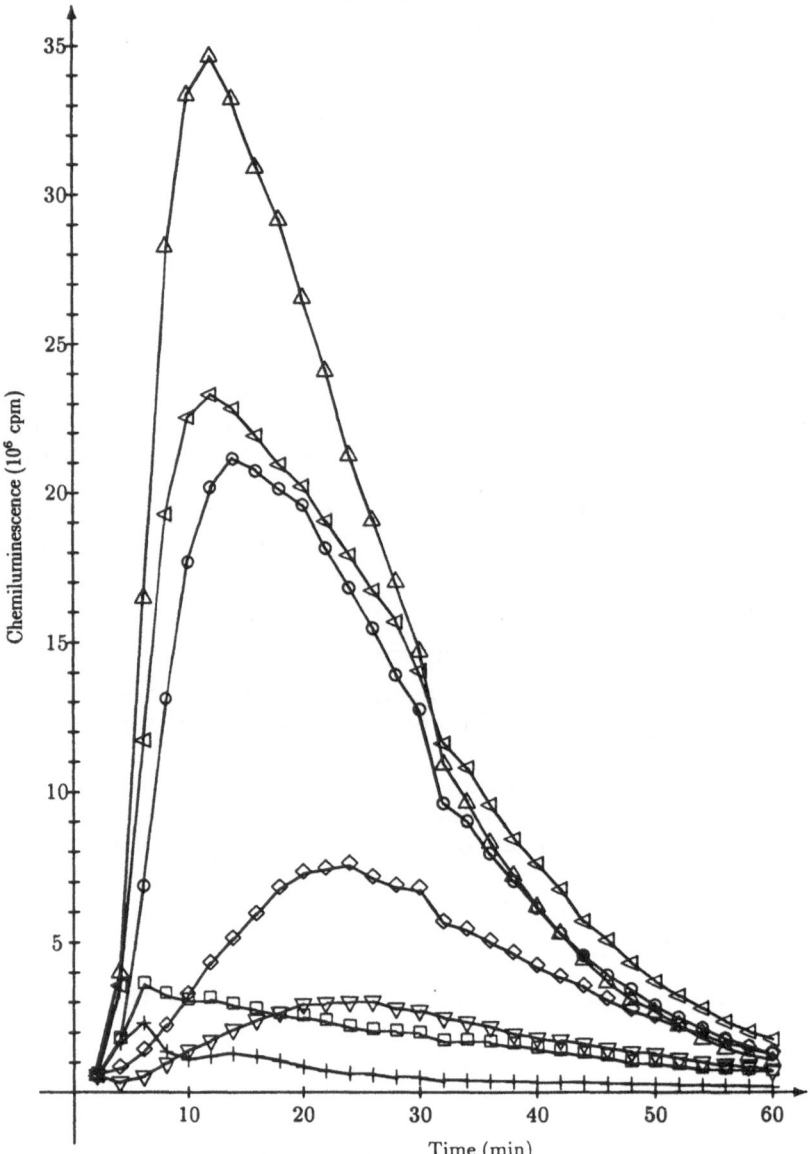

Fig. 1.   Effect of the 39 kDa protein from *P. mirabilis* 19 on the activity of LPS-stimulated macrophages derived from bone marrow of C57BL/6 mice in the chemiluminescence reaction with lucigenin as amplifier.   The 39 kDa protein was mixed with LPS I from *P. mirabilis* 19 for 2 min in a sonication bath.   Cpm are reported as geometric means of values taken from 2-3 separate experiments.   (+) 0 µg/ml LPS, 0 µg/ml protein; (△) 50 µg/ml LPS; (◁) 50 µg/ml LPS + 1.56 µg/ml protein; (○) 50 µg/ml LPS + 3.12 µg/ml protein; (◇) 50 µg/ml LPS + 6.25 µg/ml protein; (▽) 50 µg/ml LPS + 12.5 µg/ml protein; (□) 50 µg/ml LPS + 25 µg/ml protein.

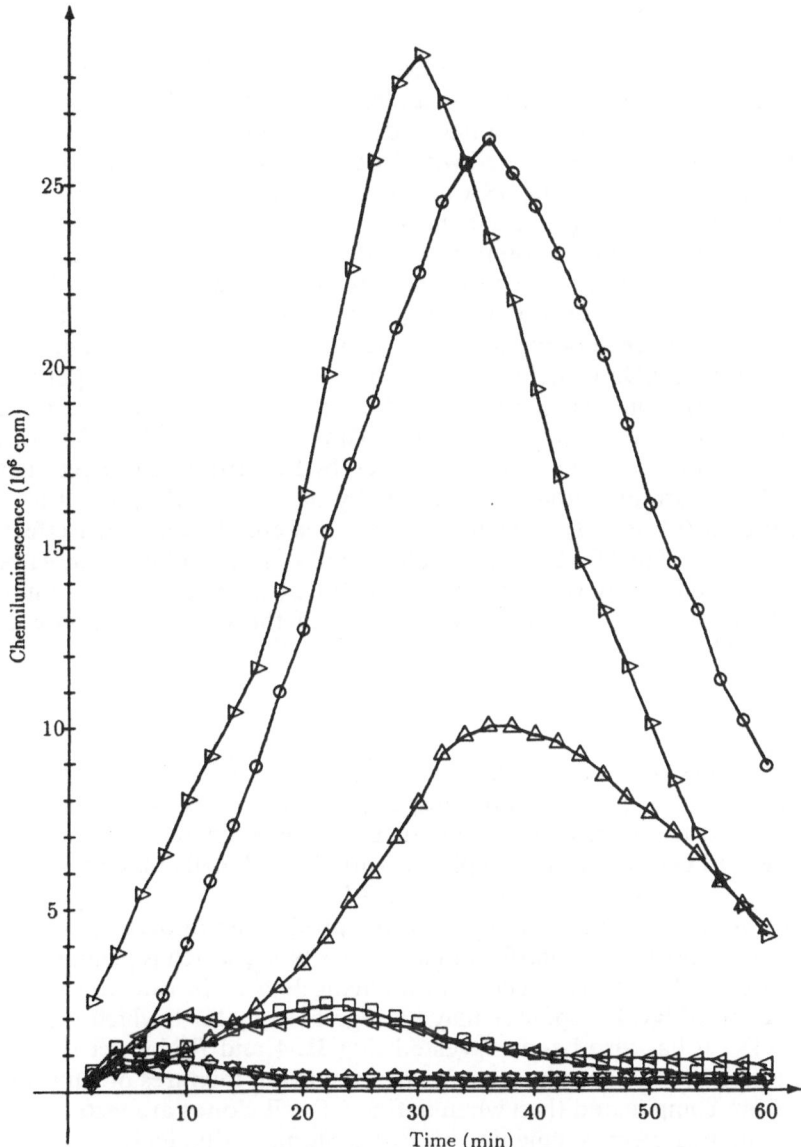

Fig. 2.   Chemiluminescence with lucigenin as amplifier after
stimulation of macrophages derived from bone marrow of
C57BL/6 mice with LPS I from *P. mirabilis* 19 and with
mixtures of LPS I and 39 kDa protein from *P. mirabilis*
19, LPS I and bovine serum albumin (BSA), or LPS I and
methylated BSA (m-BSA). Cpm are reported as geometric means of values from 2-3 separate experiments. (+)
0 μg/ml LPS, 0 μg/ml protein; (△) 50 μg/ml LPS; (▽)
6.25 μg/ml 39 kDa protein; (□) 50 μg/ml LPS + 6.25
μg/ml 39 kDa protein; (◇) 6.25 μg/ml BSA; (○) 50 μg/ml
LPS + 6.25 μg/ml BSA; (◁) 6.25 μg/ml m-BSA; (▷) 50
μg/ml LPS + 6.25 μg/ml m-BSA.

noted that the 39 kDa protein alone had no effect on macrophages in the chemilumines-cence reaction; responses of macrophages to the protein were similar to those of unstimulated cultures.

We also tested the effects of the 39 kDa protein in modulating LPS-induced IL-1 production by macrophages (Fig. 3). 10 µg/ml LPS I from *P. mirabilis* 19 induced substantial IL-1 production in macrophages derived from bone marrow of C57BL/6 mice. The 39 kDa protein, when mixed with LPS, inhibited IL-1 production in a dose-dependent manner. The effects of BSA and m-BSA were very similar to those of the 39 kDa protein in this system (data not shown).

The ability of the 39 kDa protein to modulate LPS-induced cytotoxic activity of macrophages was next examined. The results are presented in Fig. 4. 10 µg/ml LPS I from *P. mirabilis* 19 (A) induced macrophages to produce appreciable amounts of cytotoxic activity in culture supernatants (800 U/ml). When this amount of LPS was mixed with 5 µg/ml 39 kDa protein (B), the cytotoxic activity of the supernatants was greatly enhanced. In contrast, 5 µg/ml BSA (C) or m-BSA (D) had no effect in augmenting the activity of LPS in this system. The ability of the 39 kDa protein to modulate the cytotoxicity of macrophages induced by LPS from other bacteria was also tested. The 39 kDa protein enhanced the activity of *E. coli* LPS (E and F) and LPS from *S. abortus equi* (G and H) considerably, but it seemed to be most effective with LPS from *P. mirabilis*. The 39 kDa protein alone had only a minimal stimulatory effect on macrophages in this assay (I). Preliminary results (G. Weber) using antibodies to tumor necrosis factor (TNF) suggest that the bulk of the cytotoxic activity measured is indeed due to TNF.

DISCUSSION

Our results show that the 39 kDa protein is a strong modulator of the interaction of LPS with lymphocytes and macrophages, exerting its effects in a very differential manner with respect to various parameters of LPS-induced activation.

When the 39 kDa protein is complexed with LPS, T cells play an active role in changing the strength and the character of LPS-specific, antibody-producing cell responses. This may be due at least in part to the production of particular cytokines by T cells. It has been shown that interferon-gamma (IFN-γ), which is primarily produced by the Th1 subset of T cells, is effective in inducing B cells *in vitro* to secrete IgG2a, while enhancement of IgG1 responses may be regulated by IL-4, which is produced by Th2 cells (13,14). It has also been suggested that IL-4 and IFN-γ act reciprocally in regulating Ig isotype (14). The situation *in vivo*, involving mixtures of Th subsets, is of course much more complicated than when defined T cell clones are tested. Indeed, all IgG isotype responses were augmented in our system. Still, IgG2a responses were predominantly enhanced by the presence of the 39 kDa protein, which may well indicate a pronounced participation of Th1 cells. The T cell dependence of the IgG3 responses in our system is more difficult to explain, since production of this Ig isotype is normally considered to be T cell independent. Furthermore, IL-4 and IFN-γ can apparently inhibit production of IgG3 (14). Nevertheless, our results indicate that the T cell dependence of the IgG3 responses specific for LPS was substantial.

The 39 kDa protein also acts differentially to modulate the interaction of LPS with macrophages. When mixed with LPS, this protein inhibited oxygen radical production as well as IL-1 secretion by macrophages, but greatly enhanced the LPS-induced cytotoxicity of macrophages. The differential effects of the protein are accentuated by the action of BSA and m-BSA in the system. In contrast to the 39 kDa protein, BSA and the more hydrophobic m-BSA increased the LPS-induced oxygen radical production by macrophages. This may be due to a dispersing effect of BSA, as this protein has been used to solubilize lipid A (15). However, both BSA and m-BSA failed to enhance the LPS-induced cytotoxic activity of macrophages, at least not when applied in amounts that were very effective for the 39 kDa protein.

LPS can apparently interact with several targets or structures on the surface of macrophages; depending upon the path taken by LPS in this interaction, different responses are manifested (16). For example, it has been shown that LPS-induced TNF production by macrophages is greatly enhanced by the action of an LPS-binding protein, an acute phase protein formed in the liver (17,18). In complex with LPS, this protein interacts with CD14 on the surface of macrophages (19), thus targeting LPS to that particular structure.

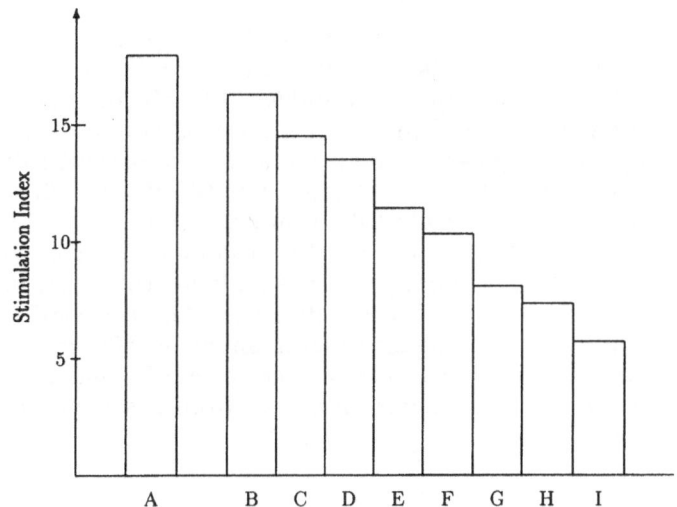

Fig. 3.　　Effect of the 39 kDa protein on IL-1 production by LPS-stimulated macrophages derived from bone marrow of C57BL/6 mice measured in an IL-2 dependent IL-1 assay. Macrophages were treated with 10 μg/ml LPS I from *P. mirabilis* 19 (A) or mixtures of 10 μg/ml LPS I and 0.01 μg/ml (B), 0.05 μg/ml (C), 0.1 μg/ml (D), 0.5 μg/ml (E), 1.0 μg/ml (F), 5 μg/ml (G), 10 μg/ml (H) or 20 μg/ml (I) 39 kDa protein. Stimulation index was calculated by setting cpm of unstimulated control cultures at 1.0. Results represent geometrical means of values from 2 separate experiments.

It therefore seems reasonable to suggest that the 39 kDa protein, in binding to LPS, changes the properties of LPS-macrophage interaction, not only by preventing LPS from reaching a target on the macrophage surface, but also by directing it to other structures, possibly in a manner similar to the action of the acute phase LPS-binding protein. Consequently, the type of macrophage response produced (IL-1, TNF) may be regulated at least in part by targeting LPS to different receptors on the cell surface.

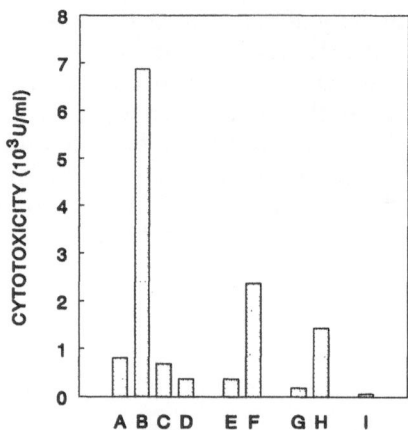

Fig. 4.   Enhancement of the LPS-induced cytotoxicity of macrophages derived from bone marrow of C57BL/6 mice by the 39
kDa protein from *P. mirabilis* 19. Macrophages were
stimulated with 10 µg/ml LPS alone or with mixtures of 10
µg/ml LPS plus 5 µg/ml protein. (A) *P. mirabilis* LPS; (B)
P. mirabilis LPS + 39 kDa protein; (C) *P. mirabilis* LPS +
BSA; (D) *P. mirabilis* LPS + m-BSA; (E) *E. coli* K12 LPS;
(F) *E. coli* K12 LPS + 39 kDa protein; (G) *S. abortus equi*
LPS; (H) *S. abortus equi* LPS + 39 kDa protein; (I) 39 kDa
protein. A unit (U) is defined as the reciprocal of the
dilution of macrophage supernatant giving 50% cytotoxicity
measured against L 929 S target cells. Results represent
the geometrical means of values from 2 separate experiments.

ACKNOWLEDGEMENTS

        We are grateful to E. Ferber and P. G. Munder, Max-Planck-Institut für
Immunbiologie, Freiburg i. Brsg, FRG, for their kind help in setting up the cultivation
of macrophages derived from bone marrow in teflon film bags. We are also indebted

to W. Bessler, Institut für Immunbiologie, University of Freiburg, for discussions
concerning immunomodulation and macrophages, and for suggesting the use of bone
marrow macrophages.

        This work was supported by the Deutsche Forschungsgemeinschaft.

REFERENCES

1. H. Karch and K. Nixdorff, Antibody-producing cell responses to an isolated outer
    membrane protein and to complexes of this antigen with lipopolysaccharide or
    with vesicles of phospholipids from *Proteus mirabilis*, Infect. Immun. 31:862
    (1981).

2. H. Karch, J. Gmeiner and K. Nixdorff, Alteration of the immunoglobulin G subclass responses in mice to lipopolysaccharide: effects of nonbacterial proteins and bacterial membrane phospholipids or outer membrane proteins of *Proteus mirabilis*, Infect. Immun. 40:157 (1983).

3. H. Karch and K. Nixdorff, Modulation of the IgG subclass responses to lipopolysaccharide by bacterial membrane components: differential adjuvant effects produced by primary and secondary stimulation, J. Immunol. 131:6 (1983).

4. O. Westphal, O. Lüderitz and F. Bister, Über die Extraktion von Bakterien mit Phenol/Wasser. Z. Naturforsch. 7b:147 (1952).

5. J. Gmeiner, The isolation of two different lipopolysaccharides from various *Proteus mirabilis* strains, Eur. J. Biochem. 58:621 (1975).

6. I. Flesch, B. Ecker and E. Ferber, Acyltransferase-catalysed cleavage of arachidonic acid from phospholipids and transfer to lysophosphatides in macrophages derived from bone marrow. Comparison of different donor- and acceptor substrate combinations, Eur. J. Biochem. 139:431 (1984).

7. G. A. Miller and P. S. Morahan, 1981. Use of nonspecific esterase stain, in: "Methods for studying mononuclear phagocytes", D. O. Adams, P. J. Edelson and H. L. Koren, eds., Academic Press, New York (1981).

8. P. L. Simon, J. T. Laydon and J. C. Lee, A modified assay for interleukin 1 (IL-1), J. Immunol. Methods 84:85 (1985).

9. B. -E. Drysdale, L. M. Zacharchuk and H. S. Shin, Mechanism of macrophage-mediated cytotoxicity: production of a soluble cytotoxic factor, J. Immunol. 131:2362 (1983).

10. R. I. Mishell and D. W. Dutton, Immunization of dissociated spleen cell cultures from normal mice, J. Exp. Med. 126:423 (1967).

11. B. B. Mishell and R. I. Mishell, Primary immunization in suspension cultures, in: "Selected Methods in Cellular Immunology", B. B. Mishell and S. M. Shiigi, eds., W. H. Freeman and Company, San Francisco (1980).

12. R. Müller-Peddinghaus, *In vitro* determination of phagocyte activity by luminol- and lucigenin-amplified chemiluminescence, Int. J. Immunopharmacol. 6:455 (1984).

13. T. R. Mosmann, H. Cherwinski, M.W. Bond, M.A. Giedlin and R. L. Coffman, Two types of murine helper T cell clone. I. Definition according to profiles of lymphokine activities and secreted proteins, J. Immunol. 136:2348 (1986).

14. C. M. Snapper and W. E. Paul, Interferon-γ and B cell stimulatory factor-1 reciprocally regulate Ig isotype production, Science 236:944 (1987).

15. C. Galanos, E. T. Rietschel, O. Lüderitz, O. Westphal, Y. B. Kim and D. W. Watson, Biological activities of lipid A complexed with bovine serum albumin, Eur. J. Biochem. 31:230 (1972).

16. S. D. Wright, Multiple receptors for endotoxin. Current Opinion Immunol. 3:83 (1991).

17. P. S. Tobias and R. J. Ulevitch, Structure and function of lipopolysaccharide binding protein, Science 249:1429 (1990).

18. R. R. Schumann, S. R. Leong, G. W. Flaggs, P. W. Gray, S. D. Wright, J. C. Mathison, P. S. Tobias, K. Soldau and R. J. Ulevitch, Isolation of a lipopolysaccharide-binding acute phase reactant from rabbit serum, J. Exp. Med. 164:777 (1986).

19. S. D. Wright, R. A. Ramos, P. S. Tobias, R. J. Ulevitch and J. C. Mathison, CD14, a receptor for complexes of lipopolysaccharide (LPS) and LPS binding protein, Science 249:1431 (1990).

SUPPRESSION OF ANTIBODY FORMING CELLS BY LIPID A ANALOGS

Arthur G. Johnson, Marilyn J. Odean and Akira Hasegawa

Department of Medical Microbiology/Immunology
University of Minnesota
Duluth, Minnesota
Department of Agriculture Chemistry
Gifu University
Gifu 501-11, Japan

Endotoxic lipopolysaccharides of Gram negative bacteria are profound immuno-modulating agents (1). They affect immunity in three major ways: (a) they are powerful adjuvants to the immune response when given with antigen (2) (or several days after antigen in the case of pneumococcal antigens [3]); (b) contrariwise when given 1-2 days before antigen a profound suppression of antibody ensues (4) and (c) 1-2 days after LPS a marked increase in resistance to challenge with a variety of unrelated microbial infectious agents occurs, including bacteria, viruses and parasites (5). Paradoxically this is the same time period when antigen stimulus with respect to antibody formation is negated.

All three of these phenomenon were found initially to be properties of whole gram negative bacteria (5) and later of the lipopolysaccharides (LPS) isolated from the outer membrane (1). Subsequently, further study ascribed the immunomodulating activity to the lipid A moiety of the LPS (1, 6). Exploitation of these properties for human medicine were dampened by the considerable toxicity associated with LPS and Lipid A. However, in 1981, Takayama et al. (7) reported the isolation of a monophos-phoryl lipid A (MPL) with markedly diminished toxicity as compared to the diphospho-ryl compound (cf. Figure 1). Biological characterization of this analog documented the retention of the adjuvant activity (8, 9), suppressive activity (9), and non-specific increase in resistance (10). The actual synthesis of lipid A followed shortly (11) and revealed the adjuvant action to be associated with a phosphorylated glucosamine disaccharide with beta hydroxy myristic acids at C_2, C_3, and C'_2 and C'_3. In addition, recently both reducing (12) and non-reducing (13) monosaccharide subunits of lipid A have been synthesized along with multiple analogs differing in the number and position of the fatty acids. Their biological properties are now being assessed. The reducing monosaccharide (lipid X) when highly purified has been shown to be of reduced toxicity (at least 100 fold less lethal in galactosamine sensitized mice and non-pyrogenic). Nevertheless, it is of considerable interest, that it served to block a number of the harmful responses to toxic LPS. Contrary to previous reports, the purified lipid X was devoid of immunostimulatory effects (14).

Microbial Infections, Edited by H. Friedman *et al.*
Plenum Press, New York, 1992

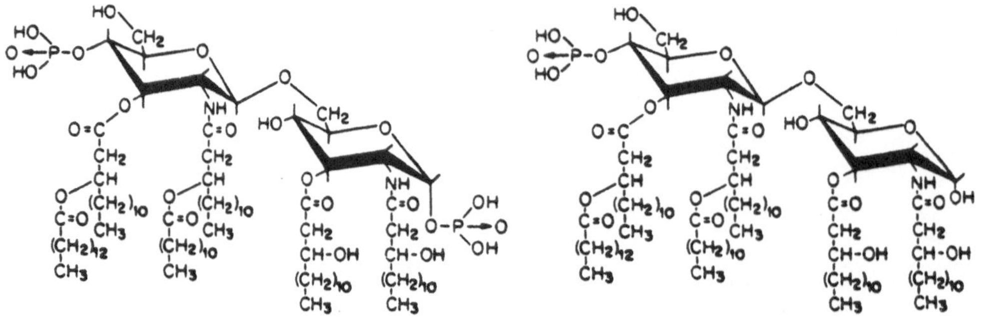

Diphosphoryl lipid A   Monophosphoryl lipid A

Fig. 1. Chemical structure of diphosphoryl (DPL) and monophosphoryl (MPL) lipid A. Note the absence of the phosphate on the reducing end of MPL.

On the other hand, certain of the non-reducing monosaccharides have been found to be active as adjuvants (15) and in enhancing non-specific resistance against infectious agents (16) as well as tumors (16, 17). Matsuura et al., (18) have shown at least a hundred fold reduction in pyrogenicity and lethality for galactosamine sensitized mice of several glycolipid A preparations (GLA) representing the non-reducing half of Lipid A. Nevertheless, the 4-0-monophosphoryl-glucosamine derivative possessing 2-N-3-tetradecanoyl-oxy-tetradecanoyl and 3-0-tetradecanoyl groups (termed GLA 27) exhibited limulus activity, interferon and TNF inducement. In a later paper, its proficiency at mitogenicity, polyclonal B cell and adjuvant activity were documented (19). A second analog (GLA 47) similar to GLA 27 excepting for substitution of an acyl-oxy-acyl (C_{14}-0-C_{14}) at C_3 also was found much less capable of exerting these properties with an exception being B cell mitogenicity.

The activities of the (R) and (S) forms of GLA 27 and a 1-deoxy compound of GLA 27 (GLA 40) were also measured (20). The (S) isomers of both were stronger than that of the (R) isomers in inducing TNF, while the opposite occurred when measuring B cell mitogenicity and polyclonal B cell activation. Phagocytic activity was greater with the R form of GLA 27, but both the R and S forms of GLA 40 were equally active in this function. Of importance, both forms of each compound protected mice against vaccinia virus when given a day before challenge (20). Thus, 2 of the 3 major immunomodulating activities of lipid A have been shown to lie in the non-reducing half of the molecule.

Since the third activity, the ability of the GLA compounds to elicit suppression of the immune response when given prior to antigen was unknown, our laboratory recently tested nine of the synthetic analogs of these non-reducing acylated monosaccharides for their capacity to induce this parameter of lipid A activity. Their structures are depicted in Figure 2. The analogs were injected ip 18 hr before BALB/c mice were given 1 x 10^8 sheep red blood cells; splenic antibody forming cells were quantitated four days later. As may be seen in Figure 2, five of the nine (GLA 27, 60, 78, 115, 147) were found capable of suppressing the antibody response, while four others (GLA 47, 57, 58, 69) did not. Consequently, these results established that the non-reducing acylated monosaccharides represent the minimal unit of LPS capable of inducing suppression.

When comparisons were made of the structures of those inducing suppression with the inactive compounds, some influences relative to the structural requirements for eliciting this phenomenon were revealed. Thus, the number of fatty acids attached to both C_2 and C_3 appears important as attested by examination of the structures of the non-suppressant GLA 47 carrying four myristoyl groups (14 carbon chain) with that of

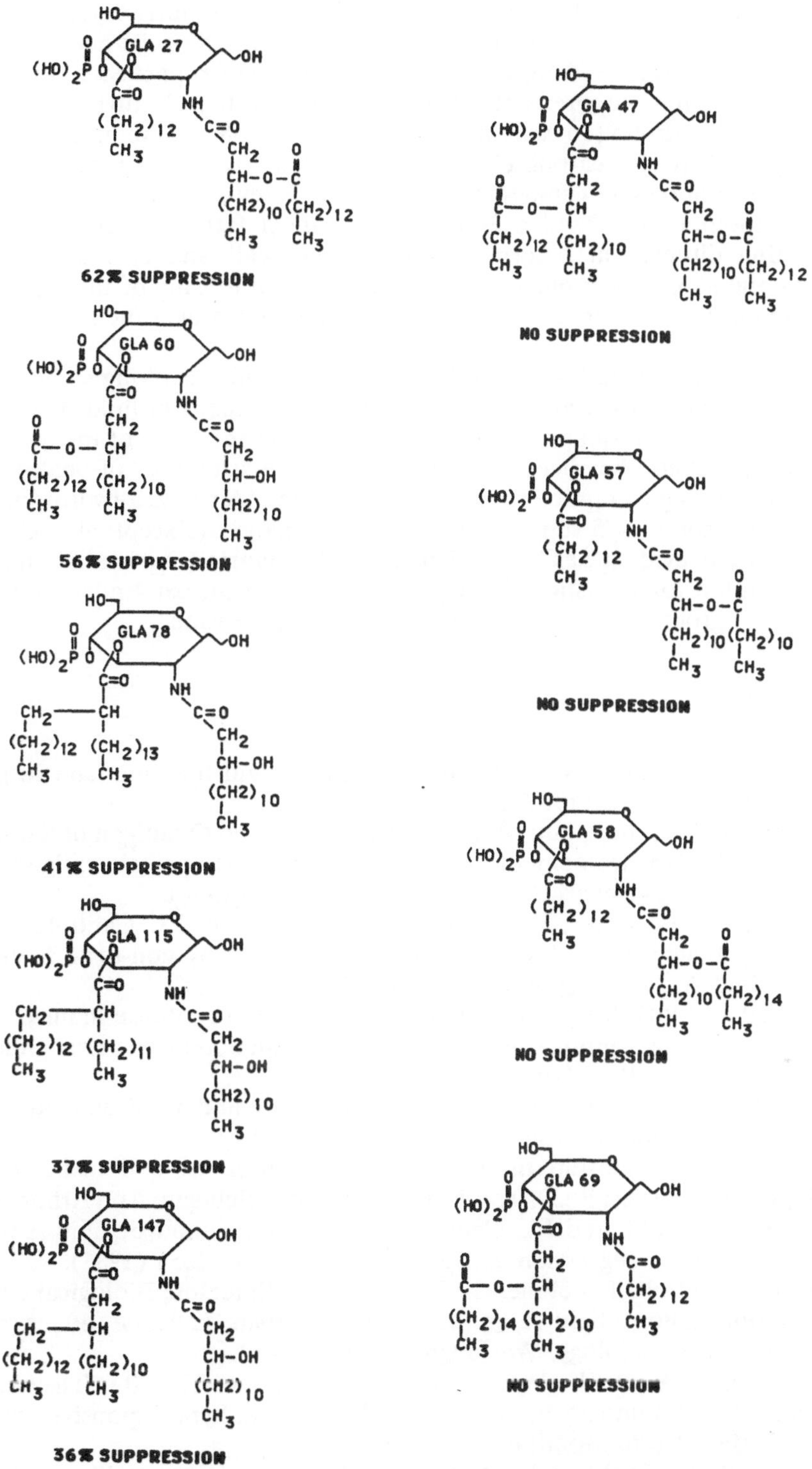

Fig. 2. Chemical structure and suppressive characteristics of the GLA compounds tested. These compounds are synthetic monosaccharide subunits of the non-reducing end of lipid A.

its suppressant counterparts, GLA 27 and 60 carrying one less myristoyl group. Also, a surprising specificity was revealed when one notes the length of the acyloxyacyl fatty acid on C_2 of compounds 57 and 58, which were incapable of suppression, with that of GLA 27, a strong immunosuppressant. Thus, lengthening the C_{14} myristoyl group of GLA 27 to a palmitoyl group (16 carbon chain), as in GLA 58, or diminishing the chain to a lauryl fatty acid (twelve carbon chain) as in GLA 57, rendered these compounds devoid of immunosuppressing competence. It is also noteworthy that four of the five suppressing compounds possessed a 3-OH-tetradecanoyl fatty acid at C_2. Ongoing studies are being directed at determining the target cell and cytokine mediators responsible for the suppressive phenomenon, and the relationship of the latter to the non-specific increase in resistance to microbial agents induced over the same period as suppression.

In summary, non-reducing glucosamine monosaccharides carrying a phosphate at C'_4 and acyl substituents at C'_2 and C'_3 are the smallest analogs of lipid A capable of eliciting nonspecific immunosuppression. A model system for study of the mechanism of nonspecific suppression has been established with the structures of analogs which induce suppression being compared to those which do not. A single 3-hydroxymyristoyl group at C'_2 is common to 4/5 suppression inducing analogs. (Exception = GLA 27). Extreme specificity is evident at C'_2 of GLA 27 in that lauric (C_{12}) and palmitic (16) cannot be substituted for myristic acid (C_{14}). Addition of an oxetradecanoyl to the 3-hydroxymyristoyl group at either C'_2 or C'_3 negates suppression.

REFERENCES

1. D. E. Morrison and J.L. Ryan, Bacterial entoxins and host immune responses, Adv. Immunol. 28:293 (1979).
2. A. G. Johnson, S. Gaines, and M. Landy, Studies on the O antigen of Salmonella typhosa. V. Enhancement of the antibody response to protein antigens by the purified lipopolysaccharide, J. Exp. Med. 103:225 (1956).
3. P. J. Baker, J.R. Hiernaux, M.B. Faunterloy, B. Prescott, J.L. Cantrell, and J.A. Rudbach, Inactivation of suppressor T cell activity by non-toxic monophosphoryl lipid A, Infect. Immun. 56:1076 (1988).
4. R. E. Franzl and P.D. McMaster, The primary immune response in mice. I. The enhancement and suppression of hemolysin production by bacterial endotoxin, J. Exp. Med. 127:1087 (1968).
5. Whole book cited, "Bacterial Endotoxins". M. Landy and W. Braun, ed., New Brunswick, Institute of Microbiology Press (1964).
6. A. Abdelnoor, Relationship of chemical composition of endotoxins to their biological properties, Ph.D. Thesis, University of Michigan, Ann Arbor (1969).
7. K. Takayama, E. Ribi, and J.L. Cantrell, Isolation of a nontoxic lipid A fraction containing tumor regression activity, Cancer Res. 41:2654 (1981).
8. E. Ribi, J.L. Cantrell, T. Feldner, K. Myers, and J. Peterson, Biological activities of monophosphoryl lipid A, in: Microbiology 1986, L. Levie, ed., American Society for Microbiology, Washington, D.C. (1986).
9. M. A. Tomai, M.A., L. E. Solem, A.G. Johnson and E. Ribi, The adjuvant properties of a nontoxic monophosphoryl lipid A in hyporesponsive and aging mice, J. Biol. Resp. Modif. 6:99 (1987).
10. J. T.Ulrich, K. N. Masihi, and W.Lange, Mechanisms of nonspecific resistance to microbial infections induced by trehalose mycolate and monophosphoryl lipid A, Adv. Biosciences 68:167 (1988).
11. M. Imoto, H. Yoshimura, N. Sakaguchi, S. Kusomoto, and T. Shiba, Total synthesis of Escherichia coli lipid A, Tetrahedron Lett. 26:1545 (1985).

12. H. Stuetz, Aschauer, J. Hildebrandt, C. Lam, H. Loibner, I. Macher, D. Scholz, E. Schuetze and H. Vypiel, Chemical synthesis of endotoxin analogues and some structure activity relationships, in: Cellular and Molecular Aspects of Endotoxin Reactions. A. Nowotny, J.J. Spitzer and E.J. Ziegler, ed., Elsevier Science Publishers, B.V. (1990).

13. M. Kiso, H. Ishida,and A. Hasegawa, Synthesis of biologically active, novel monosaccharide analogs of lipid A, Agric. Biol. Chem. 48:251 (1984).

14. C. Lam, J. Hildebrandt, E. Schutze, B. Rosenwirth, R.A. Proctor, E. Liehl, and P. Stutz, Immunostimulatory, but not anti-endotoxin, activity of lipid X is due to small amounts of contaminating N, 0-acylated disaccharide-l-phosphate: *In vitro* and *in vivo* reevaluation of the biological activity of synthetic lipid X, Infect. Immun. 59:2351 (1991).

15. Y. Kumazawa, M. Matsuura, J.Y. Homma, Y. Nakatsuru, M. Kiso, and A. Hasegawa, B cell activation and adjuvant activities of chemically synthesized analogues of the non-reducing sugar moiety of lipid A, Eur. J. Immunol. 15:199 (1985).

16. M. Nakatsuka, Y. Kumazawa, M. Matsuura, J.Y. Homma, M. Kiso, and A. Hasegawa, Enhancement of non-specific resistance to bacterial infections and tumor regressions by treatment with synthetic lipid A-subunit analogs. Critical role of N- and 3-0-linked acyl groups in 4-0-phosphono-D-glucosamine derivatives, Int. J. Immunopharm. 11:349 (1989).

17. M. Nakatsuka, Y. Kumazawa, J.Y. Homma, M. Kiso and A. Hasegawa, Inhibition in mice of experimental metastasis of B16 melanoma by the synthetic lipid A-subunit analogue GLA-60, Int. J. Immunopharmac. 13:11 (1991).

18. M. Matsuura, A. Yamamoto, Y. Kojima, J.Y. Homma, M. Kiso, and A. Hasegawa, Biological activities of chemically synthesized partial structure analogues of lipid A, J. Biochem. 98:1229 (1985).

19. Y. Kumazawa, M. Nakatsuka, H. Takimoto, T. Furuya, T. Nagumo, A. Yama-moto, J.Y. Homma, K. Knada, M. Yoshida, M. Kiso, and A. Hasegawa, Importance of fatty acid substituents of chemically synthesized lipid A-subunit analogs in the expression of immunopharmacological activity, Infect. Immun. 56:149 (1988).

20. Y. Kumazawa, Y., S. Ikida, H. Takimoto, C. Nishimura, M. Nakatsuka, J.Y. Homma, A. Yamamoto, M. Kiso, and A. Hasegawa, Effect of stereospecificity of chemically synthesized lipid A-subunit analogues GLA 27 and GLA 40 on the expression of immunopharmacological activities, Eur. J. Immunol. 17:663 (1987).

BRM ACTIVITIES OF LOW-TOXIC *BORDETELLA PERTUSSIS*

LIPOPOLYSACCHARIDES

Shigeru Abe[1], Megumi Ohnishi[1], Sadao Kimura[1],
Masatoshi Yamazaki[2], Haruyuki Oshima[3], Denichi Mizuno[3]
and Hideyo Yamaguchi[1]

Teikyo University School of Medicine[1], Faculty of Pharmaceutical
Sciences[2] and Biotechnology Research Center, Tokyo, Japan[3]

ABSTRACT

A low-toxic lipopolysaccharide (BP-LPS) was isolated from killed *Bordetella pertussis* (Tohama strain). LD_{50} of BP-LPS was about 0.8 mg/mouse which was about 10-fold higher than the LD_{50} of *E. coli*-LPS(80 μg/mouse). Toxicity measured by decrease in body weight of BP-LPS-injected mice was similarly low. BP-LPS had strong antitumor activities against various murine syngeneic tumors, and its systemic administration caused clear regression of such as MM46 mammary carcinoma and Meth A fibrosarcoma. It is noteworthy that a tolerable dosage of BP-LPS (375 μg/mouse) showed clear antitumor activity against MH134 hepatoma, which is known to be insusceptible to usual types of BRM including bacterial LPS. These findings suggest that BP-LPS is a promising candidate as an antitumor agent for clinical use. Biological activities of BP-LPS were examined and compared with those of toxic LPS extracted from *Escherichia coli* and other enterobacteria. Activation or stimulation of macrophages and lymphocytes by these LPS, including TNF induction, was found to be similar. However, activation of human or murine neutrophils, as estimated by neutrophil-adherence assay *in vitro*, though induced by all other toxic LPS tested, was not induced by BP-LPS. This inability of BP-LPS to activate neutrophils is assumed to be related to its low toxicity.

INTRODUCTION

Bacterial lipopolysaccharides (LPS) are known to have antitumor activity against experimental animal tumors. Though the benefit of LPS has been demonstrated in tumor regression even in human patients, its severe toxicity and even lethality has prohibited its development as an antitumor therapeutic agent. Trials to find a low toxic LPS with antitumor activity and to elucidate its biological characteristics are therefore crucial to break through this barrier. Among these trials are those on detoxification of toxic LPS (1), and a search for active nontoxic derivatives of LPS or Lipid A. Our approach to this problem is to pursue less toxic LPS with antitumor activity from bacteria. We previously found that systemic administration of killed cells of *Bordetella pertussis* (Tohama strain) significantly inhibited the growth of tumors but induced no

Microbial Infections, Edited by H. Friedman *et al.*
Plenum Press, New York, 1992

toxic reaction in tumor-bearing mice (2,3). This encouraged us to attempt to isolate and identify a low-toxic LPS from this microorganism. Here, we report that LPS isolated from *Bordetella pertussis* Tohama-strain (BP-LPS) is less toxic than LPS from other Gram-negative rods such as *E. coli* and *Salmonella typhimurium* and it possesses potent antitumor activities against murine tumors. Comparative studies on immunological properties of BP-LPS suggested that its low toxicity might be related to its characteristic action on neutrophils.

MATERIALS and METHODS

Animals and tumors were maintained as previously described (4). BP-LPS, which was extracted from killed *B. pertussis* (Tohama strain) by hot phenol method (5) and purified by alcohol-precipitation, was provided by the Biotechnology Research Center, Teikyo University. *E. coli*-LPS(0127:B8) was purchased from Difco Lab(Detroit, Mich.), *S. typhimurium*-LPS (S,Ra,Rc,) was from Sigma Chemical Co.(St. Louis, Mo.), both BP-LPS (165 strain) and detoxified endotoxin from Ribi Immunochem. Res. Inc.(Hamilton, MT). OK432 and Lentinan were kindly provided by Chugai Pharmaceutical Co.(Tokyo) and Ajinomoto Co.(Kawasaki), respectively. Human recombinant tumor necrosis factor alpha (TNF) was donated by Asahi Chemical Industries (Tokyo). Blastogenesis of spleen cells and TNF induction was tested as previously described (6,7). Neutrophil adherence test was performed by the method established by Sendo and his colleagues (8).

RESULTS

Toxicities of BP-LPS and other LPS were compared in several systems. Lethal toxicity of intravenous injection of BP-LPS was tested for normal C3H/HeN mice. LD_{50} of BP-LPS, which was calculated by the method of Behrens and Kärber (9), was about 800 µg/mouse and LD_{50} of *E. coli*-LPS was less than 80 µg/mouse. Lethal toxicity of BP-LPS to normal mice was about one tenth that of *E. coli*-LPS. Similar difference in lethal toxicity between BP-LPS and *E. coli*-LPS was observed in galactosamine-loaded C57BL/6 mice. As a toxic response parameter, decrease in body weight of LPS-treated mice was examined. ICR male mice 4 weeks-old were injected intraperitoneally with LPS preparations and 24 hrs later changes in body weight were checked. Doses less than or equal to 15 µg of BP-LPS did not induce significant change in body weight. Table 1 shows loss in body weight of the mice injected with 3 µg of various LPS preparations. Most LPS preparations tested induced 5-10% body weight loss but BP-LPS (Tohama strain) and detoxified endotoxin, a monophosphoryl Lipid A, did not. From these results we concluded that BP-LPS (Tohama strain) is a relatively low-toxic LPS.

In studies on the antitumor activity of BP-LPS, the therapeutic effects against biological response modifier(BRM)-susceptible tumors such as MM46 mammary carcinoma and Meth A fibrosarcoma were initially tested. We have reported that MM46 mammary carcinoma was highly susceptible to BRM including bacterial LPS, especially when administered 1-2 wks after tumor inoculation (10). Table 2 shows that a single intravenous injection of BP-LPS caused complete tumor regression of all mice tested. Mizuno et al., reported that intradermal administration was one of the best routes for LPS to achieve antitumor efficacy without toxic response (11). Table 2 also shows that intradermal injection of 1 µg of BP-LPS significantly inhibited the growth of MM46 carcinoma. This indicates that even at a dose of 1/800 of LD_{50} (800 µg), BP-LPS demonstrates its antitumor activity.

Table 1. Toxicity of LPS *In vivo* and Neutrophil-adherence Induced by LPS *In vitro*

LPS	Loss in body weight[a] (%)	Relative adherence[b] Neutrophil adherence induced by LPS (%)
Control	0 ± 1.3	0
E. coli	-7.4 ± 3.3	138 ± 5
S. typhimurium (S)	-6.6 ± 3.8	121 ± 21
S. typhimurium (Ra)	-8.8 ± 4.6	130 ± 5
S. typhimurium (Rc)	-11.3 ± 2.8	167 ± 8
B. pertussis (165)	-5.7 ± 3.3	90 ± 8
B. pertussis (Tohama)	-1.7 ± 1.8	16 ± 6
Detoxified endotoxin	-0.3 ± 2.9	24 ± 10

[a] ICR male mice (5 weeks old) were injected ip with 3 μg of each LPS. Loss in body weight at 24 hrs after injection is shown as value relative to total body weight.

[b] See footnote to Fig.2. Neutrophil adherence was measured in the presence of each LPS (1 μg/ml) in vitro. Relative adherence was calculated as follows : Relative adherence = [Adherence(LPS)-Adherence(Medium)] / [Adherence (TNF1u/ml) - Adherence(Medium)] x 100%

Meth A fibrosarcoma in BALB/C mice is known to be susceptible to LPS. A preliminary experiment indicated that intravenous administration of a tolerable dose of *E. coli*-LPS (15 μg) at 9 and 16 days after tumor inoculation caused only a slight retardation of growth of Meth A sarcomas (data not shown). This weak effectiveness is thought to due to late timing of LPS administration, because the timing optimal for antitumor efficacy is limited to 5-7 days after tumor inoculation (12). As shown in Fig.1, a higher dose, 75 μg of BP-LPS or *E. coli*-LPS, clearly inhibited the growth of the tumors. As expected from the described LD_{50} of these LPS, 3 of 5 mice treated with *E. coli*-LPS were killed by its toxicity but all mice treated with BP-LPS survived 25 days after tumor inoculation. This suggests that BP-LPS may expand the limit of LPS therapeutic effectiveness which had been restricted up to now by the toxicity.

Table 2. Antitumor Activity of *B. pertussis* LPS against MM46 Mammary Carcinoma

		Tumor diameter (mm)	
		2 weeks	4 weeks
Exp.1 — — iv	(n=8)	9.6 ± 0.6	13.8 ± 1.1
LPS 75μg iv	(n=6)	4.7 ± 1.3**	0**
Exp.2 — — id	(n=8)	10.7 ± 1.5	13.1 ± 2.4
LPS 1μg id	(n=7)	6.4 ± 2.3**	10.1 ± 3.9
3μg id	(n=7)	7.0 ± 2.3**	11.7 ± 4.5

C3H/He mice received intradermally inoculation of 1.0×10^6 MM46 mammary carcinoma cells on day 0. In Exp.1, they were treated intravenously once with 75 μg BP-LPS on day 7, and in Exp.2 intradermally with BP-LPS on days 1,2,3 and 4.
* P < 0.05, ** P < 0.01

BP-LPS and *E. coli*-LPS were also compared in antitumor activities against MH134 hepatoma which is known to be insusceptible to LPS and other BRM and to easily develop metastatic nodules in lymph nodes (13). Treatment by both LPS in the dose range of 15-75 µg/mouse transiently inhibited the growth of tumors, but then permitted them to regrow for 26 days after tumor inoculation. BP-LPS by an increased dose of 375 µg clearly caused the tumors to regress and all mice survived. On the other hand, 375 µg of *E. coli*-LPS killed all six experimental mice by its toxicity. This result suggests that BP-LPS was able to provide therapeutic benefits in the mice with tumors in a wider spectrum than the other LPS. We therefore concluded that low-toxic BP-LPS has powerful antitumor activity.

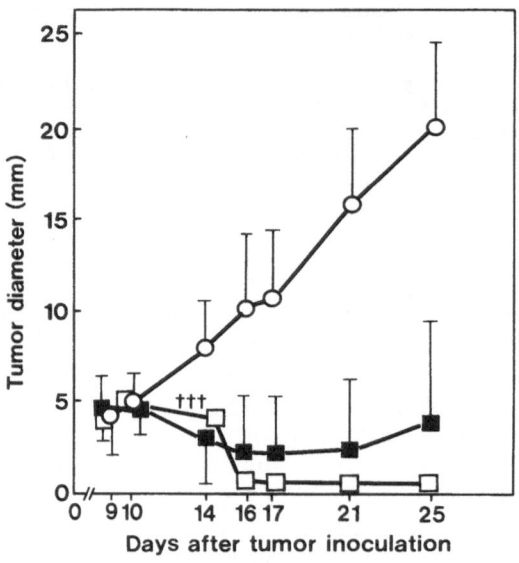

Fig. 1.   Antitumor activity of BP-LPS against Meth A fibrosarcoma.
BALB/C mice were intradermally inoculated with 2.0×10^5 cells of Meth A fibrosarcoma on day 0 and intravenously treated with 75 µg/mouse of BP-LPS (■) or *E. coli*-LPS (□) at days 9 and 16. Saline (○).   + represents death of a mouse tested.
** $p < 0.01$.

In order to answer why antitumor BP-LPS is less toxic than other LPS, we investigated its immunological properties in comparison with these of *E. coli*-LPS. Blastogenic activity of these LPS on spleen cells of C3H/He mice was examined by the method previously described (6). Spleen cells, cultured for 3 days in the presence of more than 2 µg/ml of BP-LPS or *E. coli*-LPS, were similarly stimulated; their stimulation index as shown by uptake of ^{3}H-thymidine was 6 to 10 fold that of unstimulated control. Macrophage activation by LPS was judged by *in vitro* TNF production by resident peritoneal macrophages of C3H/He mice as described by Okutomi and Yamazaki (14). More than 10 ng/ml of BP-LPS or *E. coli*-LPS similarly induced a significant level of TNF (1 to 3 u/ml) in culture supernatant within 2 hrs.

TNF-inducing activity in vivo , which was reported to be at least partially responsible for the toxic activity of LPS (15), was assessed by measuring the concentration of TNF in mouse serum after iv injection of each LPS. At 1 to 2 hrs after iv-injection of BP-LPS or *E. coli*-LPS (3 μg) into C3H/He normal mice, the level of serum TNF reached maximum (about 100 u/ml) similarly and then decreased within 4 hrs. This suggested that the difference in toxicity of BP-LPS and *E. coli*-LPS does not depend on the levels of inducible TNF.

Table 3. Antitumor Activity of BP-LPS against MH134 Hepatoma

	Dose (5μg/mouse)	Tumor diameter (mm) 2 weeks	4 weeks
——	——	13.7 ± 2.2	23.4 ± 1.5
BP-LPS	15	11.2 ± 2.4	22.3 ± 2.1
	75	10.3 ± 1.6*	20.3 ± 4.0
	375	4.8 ± 3.8**	7.2 ± 7.7**
E. coli-LPS	15	10.5 ± 0.8*	22.8 ± 0.9
	75	8.8 ± 0.4**	23.8 ± 0.4

C3H/He mice were intradermally inoculated with 2.0 x 10^5 cells of MH134 hepatoma on day 0. They received LPS intravenously at days 9 and 16.
* P < 0.05,   ** P < 0.01

Effects of BP-LPS and *E. coli*-LPS on neutrophil function was compared by neutrophil adherent assay, a simple method recently developed to detect neutrophil-activation (8). As shown in Fig. 2, more than 100 ng/ml of *E. coli*-LPS induced adherence of human peripheral blood neutrophils *in vitro*, while BP-LPS hardly induced this reaction. Using murine peritoneal neutrophils induced by casein sodium, similar results were obtained. To examine whether or not these qualitatively different reactions to neutrophils correspond to the toxicities of various LPS, activities of other LPS were also tested. As shown in Table 1, other than BP-LPS (Tohama-strain) and detoxified LPS, all LPS preparations shown to be toxic in Table 1, induced neutrophil adherence. From these results, we can assume that the capacity of LPS to induce neutrophil-adherence is correlated with their toxic properties and that lack of the capacity to induce neutrophil adherence is a characteristic immunological property of low-toxic BP-LPS.

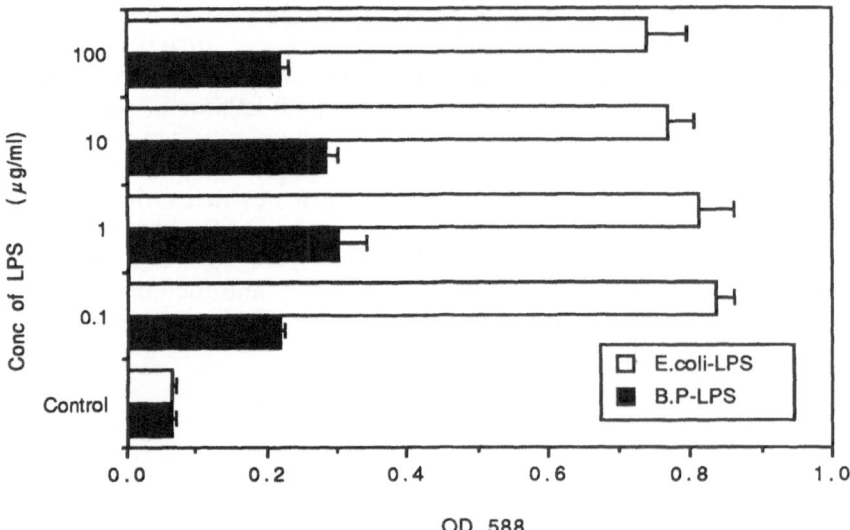

Fig. 2. Adherence of neutrophils induced by LPS *in vitro*.
Neutrophils prepared from human blood were incubated with
various concentrations of BP-LPS (closed bar) or *E. coli*-LPS
(open bar) for 60 min at 37°C. The cells adhering to plastic
plates were stained by crystal violet. Degree of adherence
was estimated photometrically by dye concentration.
PMN adherence induced by LPSs.

DISCUSSION

BP-LPS, LPS isolated from *B. pertussis* (Tohama strain) has low toxicity and has
potent antitumor activity. Especially, BP-LPS was effective against tumors usually
insusceptible to BRM, and even in a small dose (1/800 LD_{50}) demonstrated efficacy
against some tumors. As far as we know, this is the first report showing that a MH134
hepatoma, established *in situ* and more than 3 mm in diameter, was clearly inhibited in
growth by a systemic treatment with a single BRM without requiring combination
therapy (13). Not only its antitumor activity but also its stimulating activities on
macrophages and lymphocytes suggest that BP-LPS is a low-toxic promising BRM. It
actually had prophylactic activity against microbial infection in mice (unpublished data).
Several nontoxic and immunostimulating derivatives of LPS have recently been
reported. One of them, monophosphoryl lipid A, known as a detoxified LPS, is a typical
Lipid A derivative and its antitumor activity has been extensively investigated (16). The
antitumor activity of these derivatives, however, has been evidenced only under certain
limited experimental conditions such as intratumoral administration and/or combination
treatment with other therapeutics (16). We therefore believe that BP-LPS has much
stronger antitumor activity than other nontoxic LPS- derivatives.
Endotoxins and purified LPS were extracted from several strains of *B. pertussis* and
their chemical structure was partially clarified as reviewed by Chaby and Caroff (17).
Their immunostimulatory activities were also shown to be largely similar to those of
LPS from other enterobacteria. Recently Watanabe et al. reported that there are
significant differences in the biological properties of LPS isolated from several strains
of Bordetella including Tohama strain, but they did not refer to the low toxicity of

BP-LPS is very active not only in BRM assay as described above but also in Limulus lysate assay (unpublished data) rules this out. Our preliminary studies on chemical structure indicated that BP-LPS (Tohama strain) mainly consists of rough-type LPS , but no significant structural difference from BP-LPS (strain 165) could be detected. Further studies on the chemical structure of BP-LPS will be made.

The toxicity of LPS was reported to be mediated primarily by TNF produced endogenously by LPS (15), but our experiments BP-LPS induced TNF *in vivo* as much as *E. coli*-LPS, as described. Several other intrinsic factors, such as IL-1, IFN-gamma, complement(C5) and PAF are known to potentiate or influence LPS toxicity; some of these are known to be produced by neutrophils. Here, we have shown that the inability to activate neutrophils, which appears correlated with low toxicity, is a characteristic property of BP-LPS. We can thus hypothesize that the low toxicity of BP-LPS is due to this property. This hypothesis is supported by the findings that low-toxic LPS-related derivatives, monophosphoryl Lipid A (19) or deacylated Lipid A (20) lack the activity to prime the oxidative burst of neutrophils, and the finding that some toxic reactions induced by LPS were inhibited in neutropenic mice (21).

We are not yet able to explain fully why the different responsiveness of neutrophils to LPS affect the degree of toxic reaction *in vivo*. In this context we know that neutrophil response to BP-LPS is not completely absent but is qualitatively different from those of toxic LPS (to be published elsewhere).

Finally it is our hope that low-toxic BP-LPS will be tested for use as a BRM in cancer patients in the near future, because even the toxicity of usual types of LPS are becoming to be controllable in human patients by some anti-inflammatory agents (22).

ACKNOWLEDGEMENTS

The authors wish to thank Dr. H.Minagawa, Kyowa Hakko Ind.(Tokyo) for cooperation in the preliminary stage of this work and are grateful to Dr. T.Okutomi, Mr.T.Nishizawa and Mr.M.Iguchi, the Biotechnology Research Center, Teikyo University for their kind assistance in chemical analyses of LPS.

REFERENCES

1. N. Qureshi, K.Takayama and E.Ribi, Purification and structural determination of nontoxic lipid A obtained from lipopolysaccharide of *Salmonella typhimurium*, J. Biol. Chem. 257:11808 (1982).
2. H. Minagawa, H. Kobayashi, H. Yoshida, M. Teranishi, A. Morikawa, S. Abe, H. Oshima and D. Mizuno, Intratumoral induction of tumor necrosis factor by systemic administration of Bordetella pertussis vaccine, Br. J. Cancer 62:372 (1990).
3. H. Minagawa, Y. Kakamu, H. Yoshida, F. Tomita, H. Oshima and D. Mizuno, Endogenous tumor necrosis factor induction with *Bordetella pertussis* vaccine as a triggering agent and its therapeutic effect on MM46 carcinoma-bearing mice, Jpn. J. Cancer Res. (Gann) 79:384 (1988).
4. S. Abe, K. Takahashi, M. Yamazaki and D. Mizuno, Complete regression of Lewis lung carcinoma by cyclophosphamide in combination with immunomodulators, Jpn. J. Cancer Res. (Gann) 76:626 (1985).
5. O. Westphal and K. Jann, Extraction with phenol-water and further applications of the procedure, Methods Carbohydr. Chem. 5:83 (1965).

6. M. Kohno, S. Abe, H. Nakajima, M. Yamazaki and D.-I. Mizuno, Comparative studies on immunological properties of antitumor polysaccharide TC-13, Gann 73:618 (1982).

7. M. Satoh, H. Inagawa, H. Minagawa, T. Kajikawa, H. Oshima, S. Abe, M. Yamazaki and D. Mizuno, Endogenous production of TNF in mice long after BCG sensitization, J. Biol. Res. Modif. 5:117 (1986).

8. N. Yakuwa, T. Inoue, T. Watanabe, K. Takahashi and F. Sendo, A novel neutrophil adherence test effectively reflects the activated state of neutrophils, Microbiol. Immunol. 33:843 (1989).

9. B. Behrens and G. Kärber, Wie sind Reihenversuche für biologische Auswertungen am zweckmässigsten anzuordnen?, Arch. f. exper. Path. u. Pharmakol 177:379 (1935).

10. S. Abe, O. Yoshioka, Y. Masuko, J. Tsubouchi, M. Kohno, H. Nakajima, M. Yamazaki and D. Mizuno, Combination antitumor therapy with lentinan and bacterial lipopolysaccharide against murine tumors, Gann 73:91 (1982).

11. D. Mizuno, O. Yoshioka, M. Akamatsu and T. Kataoka, Antitumor effect of intracutaneous injection of bacterial lipopolysaccharide, Cancer Res. 28:1531 (1968).

12. M. J. Berendt, R. J. North and D. P. Kirstein, The immunological basis of endotoxin-induced tumor regression; requirement for T-cell-mediated immunity, J. Exp. Med. 148:1550 (1978).

13. S. Abe, J. Tsubouchi, K. Takahashi, M. Yamazaki and D. Mizuno, Combination therapy of murine tumors with lentinan plus lipopolysaccharide plus cyclophosphamide, Gann 73:961 (1982).

14. T. Okutomi and M. Yamazaki, Augmentation of release of cytotoxin from murine bone marrow macrophages by gamma- interferon, Cancer Res. 48:1808 (1988).

15. B. Beutler, I. W. Milsark, and A. C. Cerami, Passive immunization against cachectin/tumor necrosis factor protects mice from lethal effect of endotoxin, Science 229:869 (1985).

16. J. A.Rudbach, J.L. Cantrell, J.T. Ulrich, and M. S. Mitchell, Immunotherapy with bacterial endotoxins, in: "Endotoxin," H. Friedman, T. W. Klein, M. Nakano, and A. Nowotny, ed., Plenum Press, New York (1989).

17. R. Chaby and M. Caroff, Lipopolysaccharides of *Bordetella pertussis* endotoxin in: "Pathogenesis and Immunity in Pertussis", A. C. Wardlaw and R. Parton. ed., John Wiley & Sons, Chichester(1988).

18. M. Watanabe, H.Takimoto, Y.Kumazawa and K.Amano, Biological properties of lipopolysaccharides isolated from Bordetella, in: "Endotoxin," H. Friedman, T. W.Klein, M. Nakano, and A. Nowotny, ed.,Plenum Press, New York (1989)

19. D. F.Heiman, M. E.Astiz, E. C.Rackow, D. Rhein, Y. B. Kim and M. H. Weil, Monophosphoryl lipid A inhibits neutrophil priming by lipopolysaccharide, J. Lab. Clin. Med. 116:237 (1990).

20. A. R. D. Norage and W. C.Yarbrough, Jr, A comparison of the effects of intact and deacylated lipopolysaccharide on human polymorphonuclear leukocytes, J. Immunol. 144:1404 (1990).

21. S.-W. Chang, C.O.Feddersen, P.M.Henson and N.F.Voelkel, Platelet activating factor mediats hemodynamic changes and lung injury in endotoxin-treated rats, J. Clin. Invest. 79:1498 (1987).

22. R. Engelhardt, A. Mackensen, C. Galanos and R. Andreesen, Biological response to intravenously administered endotoxin in patients with advanced cancer, J. Biol. Res. Modif. 9:480 (1990).

INTERLEUKIN-2 AND THE REGULATION OF ACTIVATED

MACROPHAGE CYTOTOXIC ACTIVITIES

Barbara J. Nelson[1], Miodrag Belosevic[2], Shawn J. Green[1],
Jim Turpin[1], and Carol A. Nacy[1]

[1]Walter Reed Army Institute of Research, Washington, DC, and
[2]University of Alberta, Edmonton, Alberta, Canada

INTRODUCTION

In 1976, Morgan and colleagues reported that conditioned media from mitogen-stimulated mononuclear cells contained a factor which maintained the exponential proliferative growth of human leukemic blood or bone marrow cells (1); the proliferative cells were identified as normal T lymphocytes (2). Isolation, characterization, and subsequent purification of this factor in the conditioned medium lead to the identification of a T cell growth factor (TCGF) now known as IL-2 (3), a glycoprotein of 15.5 kD with a slightly basic isoelectric point (4). Shortly thereafter, Taniguchi et al. (5) isolated a cDNA clone for IL-2. Although IL-2 was initially described as the ultimate mitogenic signal for both antigenically and polyclonally activated T cells, aiding in their cell cycle transition from G_1 to S phase (6), subsequent studies showed that IL-2 stimulates NK and LAK cell activity (7, 8), induces B cell differentiation and proliferation (9), and activates macrophage cytotoxicity (10). IL-2 also participates in induction of T cell synthesis of cytokines, such as IFN-γ (11) and B cell growth factor-1, or IL-4 (12).

IL-2 RECEPTOR EXPRESSION AND FUNCTION ON MACROPHAGES

Responsiveness to IL-2 depends on the expression of a receptor for IL-2 (IL-2R) on the surface of the cell. The IL-2R on T cells consists of an α (p75) chain and a β (p55) chain, or Tac antigen (13). The two chains of this heterodimer, referred to as the high affinity IL-2R, are non-covalently linked and bind IL-2 with a K_d of 10^{-11}M. The first IL-2 binding protein to be recognized was the β chain. A monoclonal antibody developed by Uchiyama et al., was instrumental in this finding (14). This anti-Tac antibody interferes with the functional effects of IL-2 on T cells and competes with IL-2 for binding to IL-2R (15). The β (Tac) subunit binds IL-2 with a low affinity, $K_d = 10^{-8}$M (13). The role of the β chain is to capture IL-2, thus functioning as a helper binding site without signaling activity. The inability of the β subunit to transduce the IL-2-dependent T cell proliferation signal by itself may be due to its structurally short cytoplasmic domain. Individual α chains bind IL-2 with an intermediate affinity of $K_d = 10^{-9}$M and are responsible for signalling cell proliferation and certain effector activities.

Microbial Infections, Edited by H. Friedman *et al.*
Plenum Press, New York, 1992

For example, NK cells are activated for cytotoxicity in response to IL-2 through signals transmitted by the α subunit, independent of Tac antigen expression (16). The p75 subunit alone can mediate rapid endocytosis of cell-bound IL-2 (17) and activate resting T cell proliferation by high concentrations of IL-2 (18). The high affinity IL-2R exhibits a combination of p55 and p75 subunit properties. The "on and off" rate for IL-2 binding to p55 is very rapid (5-10 sec), while the "on and off" rate for IL-2 binding to p75 is much slower (>40 min). The high affinity IL-2R binds IL-2 with a fast on rate similar to p55 and a slow off rate comparable to p75 (19). Thus, cooperation between the two binding sites on the α and ß chains leads to a marked enhancement, from 100-1000 fold, in the binding affinity of the heterodimer receptor.

Originally, IL-2R were thought to exist only on cells of the T lymphocyte lineage. However, IL-2R are found on NK cells (16), activated B lymphocytes (20-22), and cells of the monocytic lineage. Freshly isolated human blood monocytes are anti-Tac negative, but culture of these cells for 12-24 hr induces Tac antigen expression on one-third of the cell population (23). The number of positive cells and the number of IL-2 binding sites increases by 50% with the addition of IFN-γ, but not IL-2, and IFN-γ-activated monocytes bind IL-2. LPS also potently induces IL-2R on monocytes (24); this LPS-induced receptor has a K_d of 10^{-11}M, similar to the high-affinity receptor of T cells (13). IL-2R are also expressed on activated, but not resting human alveolar macrophages, myelomonocytic and monoblast cell lines, as well as Langerhans cells (25, 26).

To gain insight into the IL-2R expression on human monocytes, Wahl and co-workers used an IL-2R cDNA probe to examine IL-2R gene expression in resting and activated monocytes (27). Northern analysis showed both a 3.5 Kb and 1.5 Kb transcript within 4 hr after activation with LPS. These transcripts are comparable in size to those found in mitogen-stimulated T cells (28). A few hours after the detection of message, the IL-2R protein (p55) appears on the cell surface of LPS-activated monocytes. Neither species of mRNA is detected in resting monocytes. Both resting monocytes and U937 cells lack p55 β chain IL-2 receptors on their surface (29). Following treatment with IFN-γ for 12 hr, 40% of these cells react with an anti-Tac IL-2R monoclonal antibody. Up-regulation of IL-2R expression is directly related to the induction of the IL-2R gene by IFN-γ.

Induction of the p55 IL-2R subunit clearly occurs on activated human monocytes. Only recently, however, has the p75 subunit been demonstrated on these cells (30). A monoclonal antibody specific for the p75 subunit, TU27, demonstrates that the p75, unlike the p55, is constitutively expressed in high levels on resting human monocytes: neither IFN-γ nor IL-2 further upregulates expression of these p75 receptors.

A murine macrophage cell line, ANA-1, was used to analyze whether the differentiation of monocytes into macrophages had any effect on IL-2 receptor expression (31). Affinity binding and cross-linking experiments with [^{125}I] IL-2 to ANA-1 macrophages demonstrated the constitutive expression of both the p55 and p75 subunits on the surface of these cells. Thus, monocytes constitutively express the α IL-2R and can be induced to express the β IL-2R; differentiated macrophages express both.

BIOLOGICAL ACTIVITIES OF IL-2 ON MONOCYTES AND MACROPHAGES

Human monocytes respond to IL-2 alone in a concentration dependent manner for the induction of cytotoxic activity against the human T24 cells, a transitional-cell carcinoma (10). IL-2 induction of monocyte tumoricidal activity is confirmed by Espinoza-Delgado and coworkers with the tumor target, HT29, a human colon carcinoma cell line (30). Addition of TU27 monoclonal antibody to IL-2-treated monocytes blocks the development of tumoricidal activity, while anti-Tac monoclonal antibodies have no effect. This data strongly suggests that the development of cytotoxic activity against tumor cells is mediated by the p75 protein. Monocytes that are primed with low concentrations of LPS or IFN-γ are induced to transcribe IL-2R β chain mRNA and express IL-2R, but are not fully activated (27). Once primed, these cells respond to IL-2 with a marked enhancement of superoxide anion (0_2^-) production compared to cells treated with the same concentrations of LPS, IFN-γ, or IL-2 alone. Monocytes also respond to IL-2 for the induction of microbicidal activity against the extracellular protozoan target, *Giardia lamblia* (32). Activation for antimicrobial activities by human monocytes, unlike tumoricidal activity (10, 30), may require the presence of both the α and β subunit.

IL-2 induces human monocytes to secrete several cytokines. Monocytes and human alveolar macrophages secrete TNF-α after 3 hr treatment with high levels of IL-2 (2000 U/ml); TNF-α release peaks within 18 hr (33). Exposure to IL-2 also results in an increase in TNF-α mRNA accumulation by both monocytes and alveolar macrophages, as demonstrated by *in situ* hybridization studies. In addition, IL-2 induces human monocytes to express IL-1 β mRNA (34). Human monocytes do not constitutively express IL-1 ß mRNA; after a 1 hr treatment with IL-2, however, transcription occurs. The IL-1 ß mRNA expression induced by IL-2 is controlled by a protein kinase C-dependent second messenger signal transduction pathway. Incubation of monocytes with IL-2 for 24 hr results in the secretion of low levels of IL-1β (34, 35). Furthermore, IL-2 can augment the IL-1β production by IFN-γ-pretreated monocytes stimulated with LPS (35). We recently examined the regulation of TGF-β production by murine resident peritoneal macrophages (36). These cells are cultured for 24 hr in serum-free Aim-V medium alone, or with 100 U/ml of IFN-γ, IL-2, IL-4 or TNFα. All cytokines tested increase the level of TGF-β above baseline constitutive release (1-30 pg/ml/10^6 cells). IL-2, however is extremely active. The concentration of TGF-β induced by IL-2 is approximately 5000 pg/ml, as compared to an average concentration of 200 pg/ml induced by IFN-γ and IL-4. IL-2 also serves as a cofactor with LPS or IFN-γ for the induction of TNFα production by murine macrophages. After *in vitro* stimulation by LPS or IL-2, peripheral blood monocytes and peritoneal macrophages from mice treated intraperitoneally with IL-2 secrete higher concentrations of TNFα than control animals treated with saline alone (37). Narumi et al. (38) examined the induction of TNF-α and an interferon-inducible cytokine, IP-10, mRNAs by IL-2, IFN-γ, or IFN-β. None of the cytokines by themselves stimulated TNFα expression; IP-10 was modestly stimulated by either IFN-γ or IFN-β. Simultaneous addition of IL-2 and IFN-γ induced high expression of TNF-α mRNA, with enhanced expression of IP-10 mRNA. Substitution of IFN-ß for IFN-γ showed similar, but not as impressive results. The synergistic effect of IL-2 and IFN-γ is also gene selective, since no cooperation is observed for the induction of another macrophage inflammatory gene, D3.

COOPERATIVE INTERACTIONS OF IL-2 AND IFN-γ FOR CYTOTOXIC ACTIVITIES

Tumoricidal Activity

The expression of macrophage tumoricidal activity against certain targets depends on the synergistic effects of IL-2 and IFN-γ. ANA-1 macrophages (a murine cell line derived from an LPS sensitive mouse strain) treated with IFN-γ in the presence of LPS develop potent tumoricidal activity against the tumor target P815 (31). Neither IL-2 nor IFN-γ alone is able to induce the cytotoxic activity: these two cytokines cooperate, however, to induce tumoricidal activity comparable to levels of cells treated with IFN-γ and LPS. GG2EE macrophages (a murine cell line derived from an LPS-insensitive mouse strain) also do not develop tumoricidal activity after treatment with IFN-γ alone, LPS alone, or even the combination of IFN-γ and LPS. Exposure to IL-2 and IFN-γ, however, induces cytotoxic activity. IL-2 acts as a co-stimulator of macrophage-mediated tumoricidal activity regardless of the presence of extraneous LPS.

Microbicidal Activities

Macrophages activated *in vivo* during immune reactions to infectious pathogens (i.e., BCG) develop two distinct antimicrobial activities: resistance to infection and intracellular killing (39). Treatment of macrophages *in vitro* with activation signals found in culture fluids from antigen/mitogen-stimulated lymphocytes, known as lymphokines, induces these same effector activities (40). Analysis of the cytokines present in whole lymphokines which influence the development and expression of these antimicrobial activities suggests that IL-2 may play a hitherto unrecognized major role in the regulation of macrophage function.

IL-2 and Regulation of Resistance to Infection. The phenomenon of resistance to infection was originally described in 1969 by Miller and Twohy, who reported that macrophages obtained from mice inoculated with viable *Mycobacterium bovis* strain Bacillus Calmette-Guerin (BCG) ingested fewer *Leishmania donovani* amastigotes *in vitro* than macrophages from non-inoculated control mice (41). This observation was rediscovered 10 years later with several different obligate or facultative organisms (42-46). The organisms for which this effector function has been described all share the characteristic that at least part of their infectious cycle is spent sequestered within macrophages of the infected host. Macrophages treated with lymphokines *in vitro* before exposure to infectious agents are activated to display this decreased susceptibility to invasion by parasites and acquire resistance to infection. Resident peritoneal macrophages, bone-marrow derived macrophages (47), as well as macrophages obtained after *in vivo* inoculation with an inflammatory stimulus, all have the capacity to express equivalent levels of resistance to infection (48).

Analysis of factors in lymphokines that induce resistance to infection originally shows a single peak of activity at the 45-55 kDa region (40, 49). This peak of activity is physicochemically and biologically indistinguishable from IFN-γ (50). Although resident peritoneal macrophages treated with recombinant IFN-γ alone are unable to express this activity, depletion of IFN-γ from lymphokines by immunoaffinity chromatography totally eliminates lymphokine activity for inducing this effector activity (51). Since addition of recombinant IFN-γ back to IFN-γ-depleted lymphokines restores lymphokine activity, there must be other factors in lymphokines which require the cooperation of IFN-γ for the induction of resistance to infection. We treated peritoneal macrophages, bone marrow-derived macrophages, or enriched peritoneal (anti-Thy1.2$^+$ depleted) macrophage populations with a variety of recombinant or high purified cytokines: colony-stimulating factor-1 (CSF-1) granulocyte/macrophage colony-stimulating factor (GM-CSF), IL-1,-2,-3,-4, and -5, and IFNα/β, and examined their ability to induce resistance to infection by themselves or in combination with IFN-γ. Like IFN-γ, none of these other cytokines are effective alone, nor are they effective in combination with each other. However, in the presence of IFN-γ, three of the cytokines, GM-CSF, IL-2, and IL-4, are active (50, 52). Overall, it is clear from these experiments that resistance to infection involves the strict requirement of IFN-γ and the function of an additional cytokine for the induction of activity.

IL-2, then, participates with IFN-γ in the induction of a microbicidal activity against an obligate intracellular parasite, *L. major*, in activated murine resident peritoneal macrophages (50). Unlike *Giardia* studies described above (32), IFN-γ and IL-2 induction of this effector function does not involve up-regulation of receptors for these cytokines, but is related to their cooperative activities on TNF-α production (53). Peritoneal or bone-marrow derived macrophages activated with IL-2 and IFN-γ, and infected with parasites in the presence of polyclonal anti-TNF-α antibodies, are totally inhibited in expression of this effector response (50). In contrast, the effector activity of cells that were treated with a control antibody of the same isotype but irrelevant specificity, paralleled those cells activated in the absence of antibody. Thus these studies further provide evidence for the induction and secretion of bioactive TNF-α by IL-2 and IFN-γ.

To show that IL-2 directly participates in the induction of resistance to infection, we treated bone marrow macrophages with IL-2 and IFN-γ in the presence of neutralizing antibodies to IL-2. These macrophages fail to develop resistance to infection in the presence of anti-IL-2 antibodies. To determine whether IL-2 serves as a second signal for cells primed with IFN-γ, we tested pharmacologic reagents that can be substituted for priming and triggering signals in other activated macrophage effector reactions, phorbol myristate acetate (PMA) and calcium ionophore A23187 (54, 55). Only combinations of IFN-γ and IL-2 induce resistance to infection; these cytokines can not be replaced with either PMA and calcium ionophore A23187 in any combination (50). Intracellular signals, (activation of protein kinase C, redistribution of calcium) which are reported for activation of macrophage tumoricidal activity (54, 55), are not necessary for the induction of resistance to infection. Inhibitors of these metabolic pathways also have no effect on activation of macrophage resistance to infection by IFN-γ and IL-2. On the other hand, protein synthesis inhibitors block the induction of this response. In a series of signal sequence experiments, we showed that macrophages treated with either IL-2 *or* IFN-γ first, washed and exposed to the second signal, developed equivalent resistance to infection (50). Thus, IFN-γ and IL-2 serve as cofactors for the induction of this effector activity.

To further substantiate the direct interaction of IL-2 with macrophages, we treated macrophages with two different anti-IL-2 receptor monoclonal antibodies either alone or in combination with IL-2 and IFN-γ (56). One antibody, 3C7, blocks the binding of IL-2 to its 55-kDa β chain receptor on T lymphocytes (57). The other, 7D4, also binds to the β chain but at a site that does not interfere with binding of IL-2 (58). To our surprise, instead of inhibiting the induction of resistance to infection by IL-2, both antibodies could substitute for IL-2 in the reaction. Irrelevant antibodies of the same class and isotype had no effect. This data suggests that the 55-kDa β chain does indeed have a biological function other than the augmentation of binding affinity to the 75-kDa α chain (13, 16, 42). This function allows transmission of the IL-2 signal via the β chain of the IL-2 receptor on macrophages for the induction of resistance to infection.

IL-2 and the Regulation of Intracellular Killing. Macrophages treated with activation signals, such as lymphokines or IFN-γ, develop the capacity to kill and eliminate intracellular amastigotes of Leishmania (40). Unlike resistance to infection, a long-lived effector function in which the parasite is prevented from entering the cell, this capacity to kill established intracellular parasites is short-lived (59). Like resistance to infection, macrophage intracellular killing is effective against many targets: such as *Leishmania, Rickettsia, Listeria, Salmonella, Toxoplasma, and Francisella* (40, 42, 48, 60, 61). Expression of intracellular killing is limited to differentiated macrophages; although inflammatory macrophages are more responsive to signals that induce destruction of extracellular targets, such as tumor cells and skin-stage schistosomula of *Schistosoma mansoni* (62), they respond poorly to signals for induction of intracellular killing (63). Characterization of the lymphokines which activate differentiated macrophages to express intracellular killing by molecular sieve chromatography reveal three regions of activity: 115-130 kDa, 50-60 kDa, and 10-20 kDa (40, 49). Approximately 70% of the total lymphokine activity is found in the 50-60 kDa region which, coincidentally, migrates with IFN-γ. Unlike resistance to infection, IFN-γ alone is able to induce intracellular killing activities.

The major effector molecule responsible for the intracellular destruction of *L. major* amastigotes by IFN-γ-activated macrophages is nitric oxide (NO) (64). NO is formed after the oxidation of the terminal guanido nitrogen group of L-arginine. This highly reactive toxic intermediate has a half-life of milliseconds and is further oxidized to nitrites (NO_2^-) and nitrates (NO_3^-). When we add a specific inhibitor of L-arginine nitrogen oxidation (N^G-monomethyl-L-arginine) to infected macrophages treated with IFN-γ, it completely inhibits this killing response. The level of the intracellular destruction of *L. major* amastigotes correlates with NO_2^- production (64), a stable degradation product of NO, and we use the levels of NO_2^- released in culture supernatants as a quantitative indicator of macrophage intracellular killing activity for the investigation of trigger signals.

Most macrophage effector mechanisms require a two signal event, known as priming and triggering (65). While resistance to infection is signal sequence independent, intracellular killing requires strict orderly sequence of priming and triggering activation factors. However, IFN-γ alone is able to induce intracellular killing of infected macrophages without the deliberate addition of a trigger signal, such as LPS. In fact, extensive studies demonstrate that LPS is not present in the culture system or a component of the parasite membrane. These studies suggest that the amastigote itself provides the trigger signal for nitrogen oxidation by activated macrophage. The only other effective trigger signal is TNF-α, a product of macrophages (66). To determine if the amastigote-induced second signal is mediated through TNF-α, we treated infected macrophages with IFN-γ in the presence of anti-TNF-α antibodies. Anti-TNF-α antibodies, but not control antibodies, decreases both NO_2^- production and microbicidal activity by over 90%. Moreover, TNF-α added in excess restored this inhibition. Thus, TNF-α release triggered by the amastigotes acts as an autocrine factor to induce NO-dependent cytotoxicity in IFN-γ activated macrophages.

TGF-β, a 25 kDa homodimeric protein secreted as a latent protein complex by a variety of transformed and nontransformed cells, suppresses over 70% of the intracellular killing activity of *L. major* infected peritoneal macrophages treated with lymphokines or IFN-γ (67). TGF-β activity on macrophage intracellular killing occurs within the first 10 minutes: delay of TGF-β treatment for as little as 30 minutes results in a time-dependent reduction in the suppressive effects. Thus, TGF-β blocks intracellular killing in the priming stage of macrophage activation. Pretreatment of macrophages with TGF-β 1 is even more effective; treatment of cells with TGF-β for 4 hr before addition of lymphokine abolishes all macrophage activity. TGF-β, however, has no inhibitory effect on macrophage resistance to infection, and limited effects on tumoricidal activity. Since TGF-β has such profound effects on intracellular killing, we asked whether TGF-β exerts its suppressive effects on the production of NO. Addition of TGF-β to either lymphokine- or IFNγ-treated macrophages not only suppresses intracellular killing, but blocks the synthesis NO_2^- as well. The target of TGF-β suppressive effects may be either an intermediate in the biochemical pathway for production of NO from L-arginine, or production of TNF-α, the second signal for intracellular killing produced by infected macrophages.

A suppressor factor functionally similar to TGF-β, but antigenically dissimilar, is one of several cytokines released by stimulation of EL-4 thymoma cells (68) with PMA. IFN-γ (69), CSF (70), IL-4 (71), and IL-2 (72) are among the known cytokines present in these culture fluids. The EL-4(Farrar) subline used in our studies of this suppressor factor was originally selected for its capacity to secrete high concentrations of IL-2. Since IL-2 is a potent stimulant of TGF-β production by macrophages, we investigated whether the EL-4 suppressive activity was an indirect consequence of TGF-β 1 induction by the IL-2 in the EL-4 fluids (73). Infected macrophages treated with lymphokines in the presence or absence of different concentrations of IL-2 (≥10 IU/ml) are inhibited from development of intracellular killing activity; IL-2 added at the time of infection, or up to 4 hr before, suppresses induction of intracellular killing by at least 50%. Addition of anti-IL-2 antibody in the presence of lymphokines and EL-4 fluids results in an 83% reduction in suppression of intracellular killing. Thus, IL-2 is the suppressor factor in EL-4 fluids, and IL-2 mediates its action indirectly through the induction of TGF-β production by macrophages. These studies are, to our knowledge, the first demonstration that IL-2 can suppress a macrophage effector activity.

IL-2 AND REGULATION OF IMMUNE RESPONSES *IN VIVO*

Over the past twenty-five years, extensive evidence has accumulated which strongly suggests that IL-2 plays a much greater role in the regulation of immune responses than originally anticipated. Though first described as a T cell growth factor, IL-2 can also enhance T and B cell responses, NK cell responses, induce the production of other cytokines, such as TGF-β 1, TNF-α and IL-1, as well as control macrophage effector functions via IL-2 receptors on the surface of these cells. That IL-2 can regulate macrophage effector functions suggests a novel role for this cytokine in host defense against intracellular microorganisms. Lack of IL-2 production and/or Ag-specific unresponsiveness of sensitized T cells correlates with immunosuppression in a variety of microbial infections in both humans and animals. In mice, decreases in IL-2 production have been found with infections with *L. donovani, L. major, Trypanosoma cruzi, Plasmodium berghei, M. bovis,* and *histoplasma capsulatum* (74-79). Peripheral blood mononuclear cells from patients infected with *M. leprae* are unable to proliferate when stimulated *in vitro* with *M. leprae* antigen; addition of exogenous IL-2 restores this response (80, 81). *In vivo* administration of IL-2 to mice infected with an LD_{100} dose of *Toxoplasma gondi* significantly reduced the mortality of these animals (82); furthermore, *in vivo* administration of IL-2 supports the development of Ag-induced T-cell proliferation (83, 84), of cytotoxic T cells (85, 86), and of natural killer cell activity (86). Thus, defects in IL-2 production may affect both specific and nonspecific immune functions against these pathogens. The fact that IL-2 serves as a cofactor for the induction of resistance to infection, and as a suppressor factor of intracellular killing, demonstrates its potential as a macrophage regulatory agent. An obvious question is, how does this dual regulatory role influence events that occur *in vivo* during infectious disease? IL-2 may have different effects at different times during the infection process or operate in different tissue compartments in completely different ways. If the macrophage interacts with IL-2 and IFN-γ before it comes in contact with the parasite, the combination of these signals triggers the expression of resistance to infection, and the cell can no longer be infected. On the other hand, once infected, macrophages only require the presence of IFN-γ or a non-IFN-γ macrophage activation factor for the expression of intracellular killing via the synthesis of NO. However, in the waning of the immune response, the accumulation of IL-2 may serve to downregulate this macrophage effector function through the autocrine production of TGF-β. The TGF-β released completes the regulatory circuit and re-establishes the non-active yet fully responsive state of the immune system. By recognizing IL-2 as an immunomodulatory agent of activated macrophage effector functions, one can begin to examine its potential as a therapeutic agent in a number of chronic infections caused by obligate intracellular parasites.

REFERENCES

1. D. A. Morgan, F. W. Ruscett, and R. C. Gallo, Selective *in vitro* growth of T lymphocytes from normal human bone marrows, Science 193:1007 (1976).

2. F. W. Ruscett, D. A. Morgan, and R. C. Gallo, Functional and morphologic characterization of human T cells continuously grown *in vitro*, J. Immunol. 119:131 (1977).

3. K. A. Smith, Interleukin-2: Inception, Impact and Implications, Science. 240:1169 (1988).

4. R. J. Robb and K. A. Smith, heterogeneity of human T cell growth factor due to glycosylation, Mol. Immunol. 18:1087 (1981).

5. T. Taniguchi, H. Matsui., T. Fujita, C. Takaoka, N. Kashima, R. Yoshimoto, and J. Hamuor, Structure and expression of a cloned cDNA for human interleukin-2, Nature (London) 302:305.

6. B. M. Stadler, S. F. Dougherty, J. J. Farrar, and J. J. Oppenheim, Relationship of cell cycle recovery of IL-2 activity from human mononuclear cells, human and mouse T cell lines, J. Immunol. 127:1936 (1981).

7. W. Domzig, B. M. Stadler, and R. B. Herberman, Interleukin-2 dependence of human natural killer (NK) cell activity, J. Immunol. 130:1970 (1983).

8. C. S. Henney, K. Kuribayashi, D. E. Kern, and S. Gillis, Interleukin-2 augments natural killer cell activity, Nature (London). 291:335 (1981).

9. R. J. Robb, Interleukin-2: the molecule and its function, Immunol. Today 5:203. (1984).

10. M. Malkovsky, B. Loveland, M. North, G. L. Asherson, L. Gao, P. Ward, and W. Fiers, Recombinant interleukin-2 directly augments the cytotoxicity of human monocytes, Nature 325:262 (1987)

11. J. J. Farrar, W. R. Benjamin, M. L. Hilfiker, M. Howard, W. L. Farrar, and J. Fuller-Farrar, The biochemistry, biology and role of interleukin-2 in the induction of cytotoxic T cell and antibody-forming B cell responses, Immunol. Rev. 63:129 (1982).

12. M. Howard, L. Matis, T. R. Malek, E. Shevach, W. Keuk, D. Cohen, K. Nakanishi, and W. E. Paul, Interleukin 2 induces antigen-reactive T cell lines to secrete BCGF-1, J. Exp. Med. 158:2024 (1983).

13. K. Teshigawara, H-M Wang, K. Kato, and K. A. Smith, Interleukin 2 high-affinity receptor expression requires two distinct binding proteins, J. Exp. Med. 165:223 (1987).

14. T. Uchiyama, S. Border, and T. A. Waldmann, A monoclonal antibody (anti-Tac) reactive with activated and functionally mature human T cells, J. Immunol. 126:1293 (1981).

15. W. J. Leonard, J. M. Depper, T. Uchiyama, K. A. Smith, T. A. Waldman, and W. C. Green, A monoclonal antibody that appears to recognize the receptor for human T cell growth factor, Nature 300:267 (1982).

16. J. H. Kehrl, M. Dukovich, G. Whalen, P. Katz, A. S. Fauci, and W. C. Greene, Novel interleukin 2 (IL-2) receptor appears to mediate IL-2-induced activation of natural killer cells, J. Clin. Invest. 81:200 (1988).

17. R. J. Robb and W. C. Greene, Internalization of IL-2 is mediated by the beta chain of the high affinity IL-2 receptor, J. Exp. Med. 165:1201 (1987).

18. L. T. Bich-Thuy, M. Dukovich, N. J. Peffer, A. S. Fauci, J. H. Kehrl, and W. C. Greene, Direct activation of human resting T cells by IL-2: the role of an IL-2 receptor distinct from the tac protein, J. Immunol. 139:1550 (1987).

19. J. L. Lowenthal and W. C. Greene, Contrasting IL-2 binding properties of alpha (tac) and beta (p70) protein subunits of the human high affinity IL-2 receptor complex, J. Exp. Med. 166:1156 (1987).

20. T. A. Waldman, C. K. Goldman, R. J. Robb, J. M. Depper, W. J. Leonard, S. O. Sharrow, K. F. Bongiovanni, S. J. Korsmeyer, and W. C. Greene, Expression of interleukin 2 receptors on activated human B cells, J. Exp. Med. 160:1450 (1984).

21. A. W. Boyd, D. C. Fisher, D. A. Fox, S. F. Schlossman, and L. M. Nadler, Structural and functional characterization of IL 2 receptors on activated human B cells, J. Immunol. 134:2387 (1985).

22. R. Mittler, P. Rao, G. Olini, E. Westberg, W. Newman, M. Hoffman, and G. Goldstein, Activated human B cells display a functional IL 2 receptor, J. Immunol. 134:2393 (1985).

23. W. Holter, R. Grunow, H. Stockinger, and W. Knapp, Recombinant interferon-γ induces IL-2 receptors on human peripheral blood monocytes, J. Immunol. 136:2171 (1986).

24. W. Holter, C. K. Goldman, L. Casabo, D. L. Nelson, W. C. Greene, and T. A. Waldmann, Expression of functional IL 2 receptors by lipopolysaccharide and interferon-γ stimulated human monocytes, J. Immunol. 138:2917 (1987).

25. W. W. Hancock, W. A. Muller, and R. S. Cotran, Interleukin 2 receptors are expressed by alveolar macrophages during pulmonary sarcoidosis and are inducible by lymphokine treatment of normal human lung macrophages, blood monocytes, and monocyte cell lines, J. Immunol. 138:185 (1987).

26. G. Steiner, E. Tschachler, M. Tani, T. R. Malek, E. M. Shevach, W. Holter, W. Knapp, K. Wolff, and G. Stingl, Interleukin 2 receptors on cultured murine epidermal Langerhans cells, J. Immunol. 137:155 (1986).

27. S. M. Wahl, N. McCartney-Francis, D. A. Hunt, P. D. Smith, L. M. Wahl, and I. M. Katona, Monocyte interleukin 2 receptor gene expression and interleukin 2 augmentation of microbicidal activity, J. Immunol. 139:1342 (1987).

28. W. J. Leonard, M. Kronke, N. J. Peffer, J. M. Depper, and W. C. Greene, Interleukin 2 receptor gene expression in normal human T lymphocytes, Proc. Natl. Acad. Sci. USA 82:6281 (1985).

29. A. Rambaldi, D. C. Young, F. Herrmann, S. A. Cannistra, and J. D. Griffin, Interferon-γ induces expression of the interleukin 2 receptor gene in human monocytes, Eur. J. Immunol. 17:153. (1987).

30. I. Espinoza-Delgado, J. R. Ortaldo, R. Winkler-Pickett, K. Sugamura, L. Varesio, and D. L. Longo, Expression and role of p75 interleukin 2 receptor on human monocytes, J. Exp. Med. 171:1821 (1990).

31. G. W. Cox, B. J. Mathieson, S. L. Giardina, and L. Varesio, Characterization of IL-2 receptor expression and function on murine macrophages, J. Immunol. 145:1719 (1990).

32. P. D. Smith, D. B. Keister, and C. O. Elson, human host response to Giardia lamblia. I. Spontaneous killing by mononuclear leukocytes in vitro, J. Immunol. 128:1372 (1983).

33. R. M. Strieter, D. G. Remick, J. P. Lynch, R. N. Spengler, and S. L. Kunkel, Interleukin-2-induced tumor necrosis factor-alpha (TNF-α) gene expression in human alveolar macrophages and blood monocyte, Am. Rev. Resp. Dis. 139:335 (1989).

34. E. J. Kovacs, B. Brock, L. Varesio, and H. A. Young, IL-2 induction of IL-1 mRNA expression in monocytes: regulation by agents that block second messenger pathways, J. Immunol. 143:3532 (1989).

35. F. Herrmann, S. A. Cannistra, A. Lindemann, D. Blohm, A. Rambaldi, R. H. Mertelsmann, and J. D. Griffin, Functional consequences of monocyte IL-2 receptor expression: induction of IL-1β secretion by IFN-γ and IL-2, J. Immunol. 142:139 (1989).

36. C. A. Nacy, B. J. Nelson, S. J. Green, and A. I. Meierovics, Cytokine networks and regulation of macrophage antimicrobial activities, in: "Cellular and Cytokine Networks in Tissue Immunity," M. S. Meltzer and A. Montovani, eds., Wiley-Liss, Inc., New York, (1991).

37. J. S. Economou, W. H. McBride, R. Essner, K. Rhoades, S. Golub, E. C. Holmes, and D. L. Morton, Tumour necrosis factor production by IL-2-activated macrophages *in vitro* and *in vivo*, Immunology 67:514 (1989).

38. S. Narumi, J. H. Finke, and T. A. Hamilton, Interferon γ and interleukin 2 synergize to induce selective monokine expression in murine peritoneal macrophages, J. Biol. Chem. 265:7036 (1990).

39. M. G. Pappas, C. A. Nacy, Antileishmanial activities of macrophages from C3H/HeN and C3H/HeJ mice treated with Mycobacterium bovis strain BCG, Cell. Immunol. 80:217 (1983).

40. C. A. Nacy, M.S. Meltzer, E. J. Leonard, and D. J. Wyler, Intracellular replication and lymphokine-induced destruction of Leishmania tropica in C3H/HeN mouse macrophages, J. Immunol. 127:2381 (1981).

41. H. C. Miller and D. W. Twohy, Cellular immunity to Leishmania donovani in macrophages in cultures, J. Parasitol. 55:200 (1969).

42. C. A. Nacy and M. S. Meltzer, Macrophages in resistance to rickettsial infection: macrophage activation *in vitro* for killing Rickettsia tsutsugamushi, J. Immunol. 123:2544 (1979).

43. S. B. Salvin and S.-L. Cheng, Lymphoid cells in delayed-type hypersensitivity II. *in vitro* phagocytosis and cellular immunity, Infect. Immun. 3:548 (1971).

44. R. Hoff, Killing *in vitro* of Trypanosoma cruzi by macrophages from mice immunized with T. cruzi or BCG, and absence of cross immunity on challenge *in vivo*, J. Exp. Med. 142:299 (1975).

45. M. A. Horwitz and S. C. Silverstein, Activated human monocytes inhibit the intracellular multiplication of Legionnaires' disease bacteria, J. Exp. Med. 154:1618 (1981).

46. C. A. Nacy and M. S. Meltzert Macrophages in resistance to rickettsial infection: strains of mice susceptible to the lethal effects of Richettsia akari infection show defective microbicidal activity *in vitro*, Infect. Immun. 126:204 (1981).

47. M. Belosevic, D. S. Finbloom, M. S. Meltzer and C. A. Nacy, IL-2: A cofactor for induction of activated macrophage resistance to infection, J. Immunol. 145:831 (1990).

48. C. A. Nacy, C. N. Oster, S. L. James, and M. S. Meltzer, Activation of macrophages for destruction of intracellular and extracellular parasites, Contemp. Top. Immunobiol. 13:147 (1984).

49. C. A. Nacy, E. J. Leonard, and M. S. Meltzer, Macrophages in resistance to rickettsial infections: characterization of the lymphokines that induce rickettsiacidal activity in macrophages, J. Immunol. 126:204 (1981).

50. M. Belosevic, C. E. Davis, M. S. Meltzer, and C. A. Nacy, Regulation of macrophage antimicrobial activities: identification of lymphokines that cooperate with IFN-γ for induction of resistance to infection, J. Immunol. 141:890 (1988).

51. C. E. Davis, M. Belosevic, M. S. Meltzer, and C. A. Nacy, Regulation of activated macrophage antimicrobial activities: cooperation of lymphokines for induction of resistance to infection, J. Immunol. 141:627 (1988).

52. P. Ralph, C. A. Nacy, M. S. Meltzer, N. Williams, I. Nakionz, and E. J. Leonard, Colony stimulating factors and regulation of macrophage tumoricidal and microbicidal activities, Cell Immun. 76:10 (1983).

53. C. A. Nacy, A. I. Meierovics, M. Belosevic, and S. J. Green, TNF-α: central regulatory cytokine in the induction of macrophage antimicrobial activities, Pathobiol. 59:182 (1991).

54. T. A. Hamilton, D. A. Becton, S. D. Somers, and D. O. Adams, Interferon γ modulates protein kinase C activity in murine peritoneal macrophages, J. Biol. Chem. 260:1378 (1985).

55. D. A. Becton, D. O. Adams, and T. A. Hamilton, Characterization of protein kinase C activity in interferon γ-treated murine peritoneal macrophages, J. Cell. Physiol. 125:485 (1986).

56. M. Belosevic and C. A. Nacy, Interleukin-2, Anti-interleukin-2 receptor antibody, and activation of macrophages, Cell. Immunol. 128:635 (1990).

57. G. Ortega, R. J. Robb, E. M. Shevach, and T. R. Malek, The murine IL-2 receptor I. Monoclonal antibodies that define distinct functional epitopes on activated T cells and react with activated B cells, J. Immunol. 133:1970 (1984).

58. T. R. Malek, R. J. Robb, and E. M. Shevach, Identification and initial characterization of a rat monoclonal antibody reactive with the murine interleukin-2 receptor ligand complex, Proc. Nat. Acad. Sci. USA 80:5694 (1983).

59. C. N. Oster and C. A. Nacy, Macrophage activation to kill Leishmania tropica: kinetics of macrophage response to lymphokines that induce microbicidal activities against Leishmania tropica amastigotes, J. Immunol. 132:1492 (1984).

60. R. M. Crawford, A. H. Fortier, M. Belosevic, and C. A. Nacy, Macrophage Killing Mechanisms, in: "Immunopharmacology," T. J. Rogers and S. C. Gelman, eds., Telford Press, NJ (1990).

61. A. H. Fortier, S. J. Green, T. Polsinelli, and C. A. Nacy, Murine peritoneal macrophage interactions with Francisella tularensis live vaccine strain: characterization of growth and interferon-γ induced inhibition of growth in vitro (in preparation).

62. L. P. Ruco and M. S. Meltzer, Macrophage activation for tumor cytotoxicity: increased lymphokine responsiveness of peritoneal macrophages during acute inflammation, J. Immunol. 120:1054 (1978).

63. D. L. Hoover and C. A. Nacy, Macrophage activation to kill Leishmania tropica: defective intracellular killing of amastigotes by macrophages elicited with sterile inflammatory agents, J. Immunol. 132:1487 (1984).

64. S. J. Green, M. S. Meltzer, J. B. Hibbs, Jr., and C. A. Nacy, Activated macrophages destroy intracellular Leishmania major amastigotes by an L-arginine-dependent killing mechanism, J. Immunol. 144:278 (1990).

65. L. P. Ruco and M. S. Meltzer, Macrophage activation for tumor cytotoxicity: development of macrophage cytotoxic activity requires completion of a sequence of short-lived intermediary reaction, J. Immunol. 121:2035 (1978).

66. S. J. Green, R. M. Crawford, J. T. Hockmeyer, M. S. Meltzer, and C. A. Nacy, Leishmania major amastigotes initiate the L-arginine-dependent killing mechanism in IFN-γ-stimulated macrophages by induction of tumor necrosis factor-α, J. Immunol. 145:4290 (1990).

67. B. J. Nelson, P. Ralph, S. J. Green, and C. A. Nacy, Differential susceptibility of activated macrophage cytotoxic effector reactions to the suppressive effects of transforming growth factor-β1, J. Immunol. 146:1849 (1991).

68. C. A. Nacy, Macrophage activation to kill Leishmania tropica: characterization of a T cell derived factor that suppresses lymphokine-induced intracellular destruction of amastigotes, J. Immunol. 133:448 (1984).

69. M. S. Meltzer, W. R. Benjamin, and J. J. Farrar, Macrophage activation for tumor cytotoxicity: induction of macrophage tumoricidal activity by lymphokines from EL-4 fluids, a continuous T cell line, J. Immunol. 129:2802 (1982).

70. M. L. Hilfiker, R. N. Moore, and J. J. Farrar, Biological properties of chromatographically separated murine thymoma derived interleukin 2 and colony-stimulating factor. J. Immunol. 127:1983 (1981).

71.	M. Howard, J. Farrar, M. Hilfiker, B. Johnson, and W. Paul, Identification of a T-cell derived B-cell growth factor distinct from interleukin 2, J. Exp. Med. 155:914 (1982).

72.	J. J. Farrar, J. Fuller-Farrar, P. L. Simon, M. L. Hilfiker, B. M. Stadler, and W. L. Farrar, Thymoma production of T cell growth factor (interleukin 2), J. Immunol. 125:2555 (1982).

73.	B. J. Nelson, J. Rossio, and C. A. Nacy, The EL-4 suppressor factor that blocks development of macrophage intracellular killing activities works through the induction of TGF-β-1, (in preparation).

74.	N. E. Reiner and J. H. Finke, Interleukin 2 deficiency in murine leishmaniasis donovani and its relationship to depressed spleen cell responses to phyto-hemagglutinin, J. Immunol. 131:1487 (1983).

75.	E. Cillari, F. Y. Liew, and R. Lelchuk, Suppression of interleukin 2 production by macrophages in genetically susceptible mice infected with Leishmania major, Infect. Immun. 54:386 (1983).

76.	A. Harell-Bellani, A. Joskowitz, D. Fradelizi, and H. Elsen, Modification of T cell proliferation and interleukin 2 production in mice infected with Trypanosoma cruzi, Proc. Natl. Acad. Sci. USA 80:3466 (1983).

77.	R. Lelchuk, R. Rose, and J. H. L. Playfair, Changes in the capacity of macro-phages and T cells to produce interleukins during murine malaria infection, Cell. Immunol. 84:253 (1983).

78.	V. Coolizi, In vivo and in vitro administration of interleukin 2-containing preparation reverses T cell responsiveness in Mycobacterium bovis BCG-infected mice. Infect. Immun. 45:25 (1984).

79.	S. R. Watson, S. K. Schmitt, D. E. Hendricks, and W. E. Bullock, Immunoregula-tion in disseminated murine histoplasmosis: disturbances in the production of interleukins 1 and 2, J. Immunol. 135:3487 (1985).

80.	G. Kaplan, D. E. Weinstein, R. M. Steinman, W. R. Lewis, U. Elvers, M. E. Patattoyo, and Z. A. Cohn, An analysis of in vitro T cell responsiveness in lepromatous leprosy, J. Exp. Med. 162:917 (1985).

81.	A. Harengowoin, T. Godal, A. S. Mustafa, A. Belehn, and T. Yemaneberhan, T cell conditioned media reverse T cell unresponsiveness in lepromatous leprosy, Nature 303:342 (1985).

82.	S. D. Sharma, J. M. Hofflin, and J. S. Remington, In vivo recombinant interleukin 2 administration enhances survival against a lethal challenge with Toxoplasma gondii, J. Immunol. 135:4160 (1985).

83.	M. A. Cheever, P. D. Greenberg, A. Fefer, and S. Gillis, Augmentation of the anti-tumor therapeutic efficacy of long-term cultured T lymphocytes by in vivo administration of purified interleukin 2, J. Exp. Med. 155:968 (1983).

84.	M. Malkovsky, P. D. Medawar, F. R. S. Hunt, L. Palmer, and C. Dore, A diet enriched in vitamin A acetate or in vivo administration of interleukin 2 can counteract a tolerigenic stimulus, Proc. R. Soc. Lond. (Biol.) 220:439 (1984).

85.	H. Wagner, C. Hardt, K. Heeg, M. Rolinghoff, and K. Pfizenmaier, T cell-derived helper factor allows in vivo induction of cytotoxic T cells in nu/nu mice, Nature 284:278 (1980).

86.	S. N. Hafeneider, P. J. Conlon, C. S. Henney, and S. Gillis, In vivo interleukin 2 administration augments the generation of alloreactive T lymphocytes and resident natural killer cells, J. Immunol. 130:222 (1983).

PROTECTIVE EFFECTS OF CYTOKINES IN

MURINE SALMONELL

Masayasu Nakano, Kazuyasu Onozuka, Hiromi Yamasu,
Wang Fu Zhong and Yasunobu Nakano

Department of Microbiology, Jichi Medical School
Tochigi, Japan

INTRODUCTION

Salmonella typhimurium as well as *Salmonella enteritidis* are thought to be
facultative pathogens and are natural causative agents of systemic infection in mice (1).
When a virulent strain of Salmonella infects mice, the organisms multiply in the
phagocytic cells, especially in the spleen and liver, and can kill the mice within a week.
Phagocytic cells play an important role in the innate resistance of mice to infection by
Salmonella. The outcome of an infection is influenced by a number of factors, ranging
from how the microorganisms are handled by phagocytic cells during the earliest phase
of infection, to the appearance of specific cellular and humoral immune responses in
the latter phases of infection. The initial control of *S. typhimurium* proliferation in the
host presumably depends on an inherent property of macrophages (2-4) and on
macrophage activation (5).

The role of macrophages activated by cytokines such as interferon (IFN)-γ is well
known in the host defense mechanisms against neoplasia and infection (6,7). The most
efficient way to render macrophages cytotoxic is to activate them with IFN-γ in
combination with a small amount of bacterial lipopolysaccharide (LPS) (6). We earlier
reported on the ability of recombinant murine (rMu) IFN-γ to activate anti-Salmonella-
activity in normal mice and in beige mutant (bg/bg) mice with Chediak-Higashi
syndrome, and on its cooperative effect with LPS (8). We have also reported the
protective effect of rMu tumor necrosis factor (TNF)-α/cachectin in murine salmonellosis and its cooperative effect with IFN-γ (9). These findings suggest involvement of
cytokines in the host-resistance mechanism against Salmonella infection. In the present
study, we investigated whether or not other cytokines are capable of enhancing
resistance in mice against Salmonella.

MATERIALS AND METHODS

Mice

C3H/HeN and C3H/HeJ mice of both sexes were used between 8 to 12 wks of
age.

Microbial Infections, Edited by H. Friedman *et al.*
Plenum Press, New York, 1992

Cytokines

rMu-IFN-γ (5 x 10^6 U/mg), recombinant human (rHu) TNF-α (3 x 10^6 JRU/mg), rHu-TNF-β (6.8 x 10^8 U/mg), rHu- interleukin (IL)-1β(2 x 10^7 U/mg) and rHu-IL-6 (4 x 10^6 U/mg) were given to us by Shionogi Pharmaceutical Co., Ltd, Dainippon Pharmaceutical Co., Ltd., Eisai, Tsukuba Institute, Ohtsuka Pharmaceutical Co., Ltd., and Ajinomoto Co., Ltd., respectively. The lyophilized cytokines were dissolved in endotoxin-free physiological saline with 5% fetal calf serum (FCS), and the diluted solution was injected intraperitoneally (i.p.) into the mice (0.2 ml/mouse). The solution, except TNF-β, contained no endotoxin, as determined by a Limulus amebocyte lysate assay kit (Endospecy test; Seikagaku Kogyo, Tokyo). The TNF-β contained very small amounts of LPS (930 pg/mg).

Bacterial Strain and Infection

S. typhimurium LT2 strain was cultivated overnight at 37°C on nutrient agar and then suspended in physiological saline. Mice were challenged i.p. with 1 x 10^6 colony forming units (CFU) of the bacteria (0.2 ml/mouse). Mortality of the mice (10 per group) was scored during the 3 week period following infection.

Quantitative Cultivation of Inoculated Organisms

To determine viable numbers of infecting organisms in the peritoneal cavities, bacteria were counted at 24 hr after infection. The detailed procedure was described previously (8,9). Three mice per group were used for each determination of viable numbers of bacteria, and the arithmetic mean and standard error were calculated.

White Cell Counts

White cell numbers in the peritoneal cavity were counted by a hemocytometer. The smears were stained with a Giemsa solution and the white cell percentage was calculated by microscopic observation.

Killing Rate

The peritoneal exudate cells were taken from the peritoneal cavities of the mice by washing them with RPMI-1640 medium supplemented with 2.5% FCS, and the cells were then cultured in a 24 well plate (10^6 cells/ml/well) in an atmosphere of 5% CO_2-air for 40 min. The non-adherent cells were washed out, and 10^6 CFU of *S. typhimurium* was added to the adherent cell cultures and phagocytized for 30 min. The cultures were repeatedly washed to remove unphagocytized organisms. In order to disrupt the phagocytic cells, the culture was divided in half and 0.05% Triton-X-100 solution was added to the first aliquot. Fresh culture medium was added to the other aliquot which was then incubated for an additional 3 hr. At the end of the incubation period, Triton-X-100 solution was also added to the second aliquot. CFU in the disrupted cell cultures were determined by quantitative cultivation of the bacteria in the solution. The killing rate was calculated by the following formula: [1 - (No. of CFU at 3 hr/No. of CFU at 30 min)] x 100.

RESULTS

Enhancement of Bactericidal Activity by TNF-α or TNF-β.

C3H/HeJ mice were injected with rHu-TNF-α, rHu-TNF-β or rMu- TNF-α. Another group of C3H/HeJ mice was injected with rHu-TNF-β together with rMu-IFN--γ. After 6 hr, all mice were challenged with *S. typhimurium* LT2 organisms, and 24 hr after infection the bacterial number in the peritoneal cavities of the mice was assessed. As shown in Fig. 1, the bacterial number was significantly reduced in the mice injected

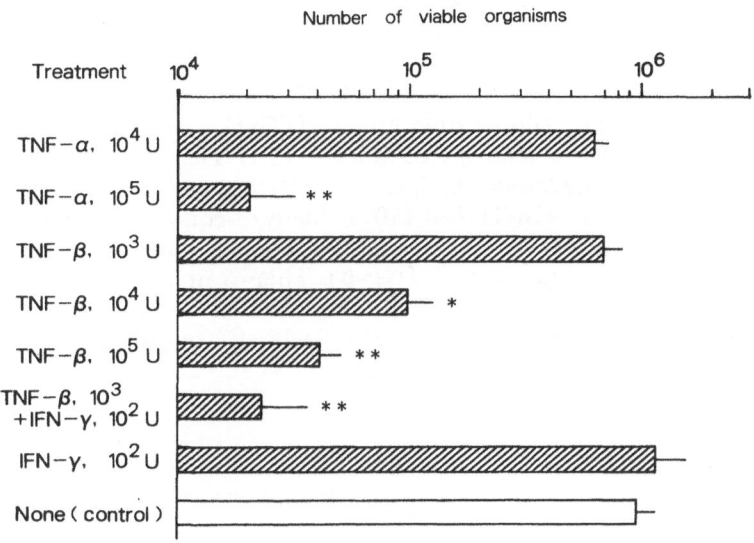

Fig. 1. Effects of TNF-α, TNF-β or TNF-β plus IFN-α on bactericidal enhancement in the peritoneal cavities of C3H/HeJ mice. All mice except the control group were injected i.p. with either rHu-TNF-α, rHu-TNF-β or rHu-TNF-β plus rMu-IFN-γ. After 6 hr, the mice were challenged i.p. with 10^6 organisms of *S. typhimurium* LT2. CFU in the peritoneal cavities of the mice were assessed 24 hr after infection. *, $\underline{P}$ < 0.05 and **, $\underline{P}$ < 0.01, significantly different from the control group.

with 10^5 U of rHu- TNF-α, or with 10^4 U more of rHu-TNF-β. Furthermore, a cooperative effect of ineffective doses of TNF-β and IFN-γ was observed in the mice (Fig. 1), and was reflected in the survival rate (Fig. 2). The results indicate that TNF-β is similar to TNF-α in its ability to enhance the host defense system against Salmonella. Neutrophils and macrophages were counted in the peritoneal cavities 6 hr after the injection of 10^5 U rHu-TNF-α, and their average number of neutrophils and macrophages was double that in the untreated control mice. Furthermore, in the *in vitro* experiment, adherent phagocytic cells (macrophages) taken from the peritoneal cavities of the rHu-TNF-α-treated mice engulfed approximately 4 times more of the Salmonella

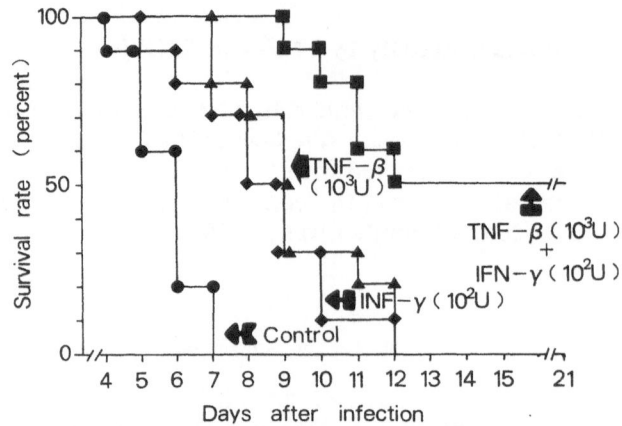

Fig. 2. The cooperative effect of TNF-β and IFN-γ on the protective resistance in C3H/HeJ mice against *S. typhimurium* infection. Mice were injected i.p. with rHu-TNF-β (10^3 U/daily/3 consecutive days) and/or rMu-IFN-γ (10^2 U/once at the time of the last injection of TNF-β). Then 6 hr after the last injection of TNF-β, they and untreated control mice were challenged i.p. with 10^6 organisms of *S. typhimurium* LT2 strain.

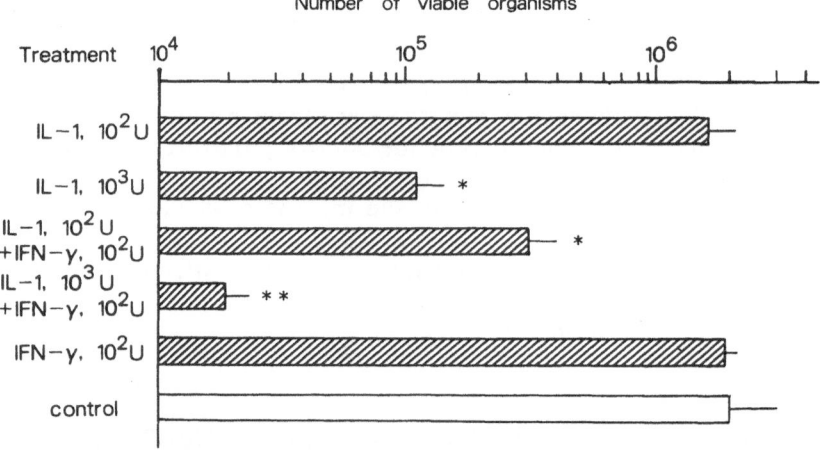

Fig. 3. Effects of IL-1 or IL-1 plus IFN-γ on bactericidal enhancement in the peritoneal cavities of C3H/He mice. The mice were injected i.p. with rHu-IL-1, rMu-IFN-γ or rHu-IL-1 plus rMu- IFN-γ 6 hr in advance and then challenged i.p. with 10^6 organisms of *S. typhimurium* LT2. CFU in the peritoneal cavity were counted 24 hr after the challenge. *, $P < 0.05$ and **, $P < 0.01$: significantly different from the control group.

organisms than those of the untreated control mice (8.1×10^3 vs. 2.2×10^3 organisms/-well), and killed the organisms more efficiently (killing-rate: 68% vs. 50%).

Enhancement of Bactericidal Activity By IL-1

IL-1 is thought to be a group of inflammatory cytokines, and therefore we examined the effect of rHu-IL-1β. As shown in Fig. 3, the bactericidal capacity in the peritoneal cavities of mice that had been previously injected with 10^3 U IL-1 was obviously augmented, whereas 10^2 U IL-1 was not effective. However, when 10^2 U IL-1 was injected together with an ineffective dose of IFN-γ, we achieved the desired effect. These findings suggest that IL-1β is capable of augmenting bactericidal capacity in the host and that it has a cooperative effect with IFN-γ.

Enhancement of Bactericidal Activity By IL-6

IL-6 is a multipotential cytokine produced by macrophages. The enhancement of bactericidal capacity by rHu-IL-6 and its cooperative effect with rHu-TNF-α were seen in C3H/HeJ mice (Fig. 4).

DISCUSSION

In this study, we demonstrated the effectiveness of some rHu-cytokines against the infection of *S. typhimurium* in an animal model. Since the effect of IFN-γ is species specific, we used rMu-IFN-γ in combination with rHu-cytokines. The cytokines used in the experiments contained non-detectable or negligible amounts of LPS, but as a further precaution we used LPS-non-responsive C3H/HeJ mice to insure against the effect of LPS.

TNF-α is a product of activated macrophages that was originally found to mediate antitumor effects. More recently it has been found to display various other biologic activities beyond tumor cytotoxicity (10). As previously reported (8,9), the reduction of Salmonella in the peritoneal cavities of mice previously treated with rMu-IFN-γ or rMu-TNF-α was due to bacterial killing *in situ*, but not due to the rapid diffusion of the infecting organism from the peritoneal cavity into other areas. Mice injected with both rHu-TNF-α and rMu-IFN-γ underwent a reduction in CFU of the infecting *S. typhimurium* (Fig. 1). The enhancement of bactericidal activity seemed to be due to the rapid recruitment of phagocytic cells, and to the augmentation of the phagocytic and killing abilities of the macrophages.

TNF-β is a cytokine produced by T lymphocytes. It is molecularly different from TNF-α, but both TNF-β and -α bind to the same receptors on the cell surface (11), and both activate macrophages (12) as well as neutrophils (13,14). In fact, rHu-TNF-β appeared just as effective in enhancing the bactericidal activity as did rHu-TNF-α (Fig. 1). In our previous paper (9), we demonstrated the cooperative effect of TNF-α and IFN-γ in protecting against Salmonella. The bactericidal activity in the peritoneal cavities was obviously elevated when a small amount of TNF-β and a very low dose of IFN-γ were given together (Fig. 1). The survival rate of the infected mice was also increased (Fig. 2).

It has been reported that the injection of rHu-IL-1 protects mice from a lethal injection of *P. aeruginosa* (15,16), *Klebsiella pneumoniae* (15,17) or *Listeria monocytogenes* (18), and that rHu-IL-6 has a macrophage-activating function (19). In our experiments, an administration of IL-1β or IL-6 enhanced bactericidal activity (Figs. 3 and 4). Cooperative effects of low doses of IL-1 plus IFN-γ and of IL-6 plus TNF-α were also observed.

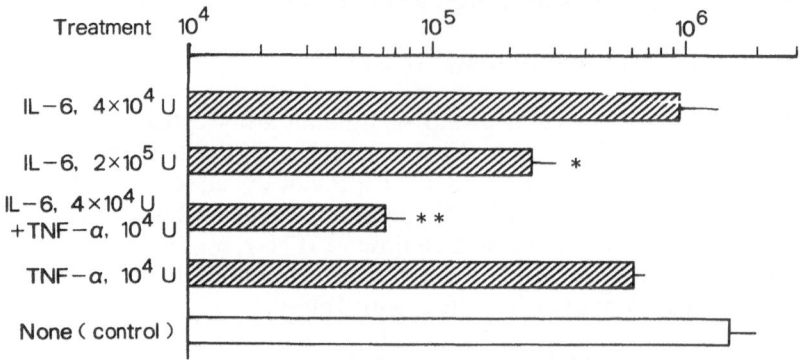

Fig. 4.   Effects of IL-6 or IL-6 plus IFN-γ on bactericidal enhancement in the peritoneal cavities of C3H/HeJ mice.   The
animals   were injected i.p. with rHu-IL-6 or rHu-IL-6 plus
rHu- TNF-α, and 6 hr after the last injection they were
challenged i.p. with 10^6 organisms of *S. typhimurium* LT2.
CFU in the peritoneal cavity   were counted 24 hr after the
challenge.   *, P < 0.05   and **, P < 0.01: significantly
different from the control group.

Bactericidal and protective activity to IFN-γ against Salmonella is increased in the
presence of LPS (8) or TNF-α (9).   Anti-TNF-α antibody can abolish the cooperative
effect of IFN-γ and LPS on bactericidal activity, suggesting the participation of
LPS-induced TNF-α (9).   In the present study, we demonstrated the cooperative effects
of IL-1 plus IFN-γ and IL-6 plus TNF-α. These findings suggest that cooperation among
cytokines is an ordinary phenomenon in the host-defense system. For   prophylaxis and
therapy against infections, adequate combinations of very low doses of   cytokines may
produce a better effect and/or may mitigate the side-effects in hosts.

ACKNOWLEDGEMENT

We thank the companies that provided us with recombinant cytokines.

REFERENCES

1.   F. M. Collins, Vaccines and cell mediated immunity, Bacteriol. Rev. 38:371
      (1974).
2.   C. E. Hormaeche, J. Brock, and P. Pettifor, Natural resistance to mouse typhoid:
      possible role for the macrophage,in: "Genetic Control of Natural Resistance
      to Infection and Malignancy," E. Skamene, P. A. L. Kongshavn and M. Landy,
      eds., Academic Press, New York, p.121 (1980).
3.   A. D. O'Brien and E. S. Metcalf, Control of early *Salmonella typhimurium* growth
      in innately Salmonella-resistant mice does not require functional T lympho-
      cytes, J. Immunol. 129:1349 (1982).

4. J. T. van Dissel, P. C. J. Leijh, and R. van Furth, Differences in initial rate of intracellular killing of *Salmonella typhimurium* by resident peritoneal macrophages from various mouse strains, J. Immunol. 134:3404 (1985).

5. M. Nakano, K. Onozuka, T. Saito-Taki, and N. Minato, Recovery of immune response in immunodeficient mice after administration of lipopolysaccharide, in: "Bacterial Endotoxin," J. Y. Homma, S. Kanegasaki, O. Lüderitz, T. Shiba, and O. Westphal, eds., Verlag Chemie, Weinheim, p.281 (1984).

6. D. O. Adams and T. A. Hamilton, The cell biology of macrophage activation, Ann. Rev. Immunol. 2:283 (1984).

7. E. M. DeMaeyer and J. DeMaeyer-Guignard, in: "Interferons and Other Regulatory Cytokines, "John Wiley & Sons, Inc., New York (1988).

8. H. Matsumura, K. Onozuka, Y. Terada, Y. Nakano, and M. Nakano, Effect of murine recombinant interferon-γ in the protection of mice against Salmonella, Int. J. Immunopharm. 12:49 (1990).

9. Y. Nakano, K. Onozuka, Y. Terada, H. Shinomiya, and M. Nakano, Protective effect of recombinant tumor necrosis factor-α in murine salmonellosis, J. Immunol. 144:1935 (1990).

10. L. J. Old, Tumor necrosis factor (TNF), Science 230:630 (1985).

11. B. B. Aggarwal, T. E. Eessalu, and P. E. Hass, Characterization of receptors for human tumor necrosis factor and their regulation by γ-interferon, Nature 318:665 (1985).

12. S. Heidenreich, M. Weyers, J. H. Gong, H. Sprenger, M. Nain and D. Gemsa, Potentiation of lymphokine-induced macrophage activation by tumor necrosis factor-α. J. Immunol. 140:1511 (1988).

13. J. M. Klebanoff, M. A. Vadas, J. M. Harlan, L. H. Sparks, J. R. Gamble and J. M. Agosti, Stimulation of neutrophils by tumor necrosis factor, J. Immunol. 136:4220 (1986).

14. J. W. Larrick, D. Graham, K. Toy, L. S. Lin, G. Senyk, and B. M. Fendly, Recombinant tumor necrosis factor causes activation of human granulocytes, Blood 69:640 (1987).

15. Y. Ozaki, A. Ohashi, A. Minami, and S. Nakamura, Enhanced resistance of mice to bacterial infection induced by recombinant human interleukin-1α, Infect. Immun. 55:1436 (1987).

16. J. W. M. Van der Meer, M. Barza, S. M. Wolff, and C. A. Dinarello, A low dose of recombinant interleukin 1 protects granulocytopenic mice from lethal gram-negative infection, Proc. Natl. Acad. Sci. USA 85:1620 (1988).

17. J. W. M. Van der Meer, The effects of recombinant interleukin-1 and recombinant tumor necrosis factor on nonspecific resistance to infection, Biotherapy 1:19 (1988).

18. C. J. Czuprynski and J. F. Brown, Recombinant murine interleukin-1α enhancement of nonspecific antibacterial resistance, Infect. Immun. 55:2061 (1987).

19. I. E. A. Flesch and S. H. E. Kaufmann, Stimulation of antibacterial macrophage activities by B-cell stimulatory factor 2 (interleukin-6), Infect. Immun. 58:269 (1990).

LEGIONELLA PNEUMOPHILA INFECTION AND CYTOKINE PRODUCTION

Thomas W. Klein, Yoshimasa Yamamoto, Scott Wilson,
Cathy Newton, and Herman Friedman

Department of Medical Microbiology and Immunology
University of South Florida College of Medicine
Tampa, Florida 33612

INTRODUCTION

Legionella pneumophila has a relatively low attack rate for humans and generally causes disease either after exposure to high doses of bacteria or infection of an immunosuppressed host (1). This low attack rate suggests, among other things, that Legionella is relatively immunogenic or reactogenic in humans and other animals. Reports appearing soon after the epidemic of 1976 suggested that humans exposed to Legionella readily produce high titers of circulating antibody (2). Later, it was reported that sensitized lymphocytes appeared in the blood during acute legionellosis, suggesting the development of cell-mediated immunity, and that normal donors displayed a rather vigorous lymphocyte proliferation response to Legionella antigens (3). We reported that peripheral blood cells from normal donors readily produced cytokines such as interferon γ and tumor necrosis factor when cultured with either Legionella antigens or living Legionella organisms (4), suggesting that human leukocytes vigorously react to Legionella antigens in a non-immune fashion resulting in the upregulation of the cytokine network. Similar results were obtained employing mouse leukocyte cultures (5). Not only are Legionella bacteria reactogenic, perturbing host cells to respond and become activated, but also it appears that living, virulent bacteria are more reactogenic than either avirulent organisms (6) or killed bacteria (7).

Legionella pneumophila is a facultative intracellular bacterium which grows in cultures of guinea pig and human mononuclear phagocytes. Macrophages obtained from most mouse strains fail to support the growth of Legionella (8, 9). However, cells from A/J mice do permit the replication of Legionella, and these animals are moderately more susceptible to infection with Legionella bacteria (10). Although macrophages from these animals are permissive and, therefore, tend to promote susceptibility to infection, lymphocytes from these animals are fully responsive to stimulation by Legionella antigens (11) and their leukocytes respond with cytokine production (10). It would appear from these studies that A/J mice are similar to humans in that both have lymphocytes which respond to Legionella antigens but, on the other hand, both possess macrophages which support the growth of the pathogen. Lymphocytes, therefore, could have an important role in determining the outcome of infection, and it would be expected that, if the functioning of these cells was suppressed

Microbial Infections, Edited by H. Friedman *et al.*
Plenum Press, New York, 1992

(i.e., immunosuppression), disease would ensue. In the present report, we show that virulent Legionella, which readily grow in macrophages from A/J mice, are more immunostimulatory than avirulent cells which do not grow in macrophage cultures. Infection of cultures with virulent Legionella leads to an augmentation of interleukin 1 (IL-1) production, measured by both a bioassay and Northern blot analysis. Furthermore, enhanced production of IL-1 by virulent organisms was not due simply to the increased accumulation of bacteria in the macrophage cultures, suggesting that factors other than antigenic concentration contribute to increased cytokine production.

MATERIALS AND METHODS

Animals and Bacteria

A/J female mice were obtained from Jackson Laboratories (Bar Harbor, ME) and were used in these studies at 8-12 wk of age. The mice were housed and cared for in our animal facility, which is fully accredited by the American Association for Accreditation of Laboratory Animal Care. A virulent strain of *Legionella pneumophila*, serogroup 1, was obtained from a case of legionellosis and cultured on buffered charcoal yeast extract medium (BCYE; Becton Dickinson), as previously described (6, 8). An avirulent strain of Legionella was obtained by passage of the virulent parent strain on supplemented Mueller-Hinton agar, using batch passing technique (12). For infection of macrophages, the bacteria were grown on BCYE agar for 48 hr at 37°C, harvested into pyrogen-free saline, and diluted to a working dilution with tissue culture medium (see below).

Peritoneal Macrophages

Peritoneal exudate cells (PEC) were removed from A/J mice 4 days after intraperitoneal injection with 3 ml thioglycolate broth (3%). PEC were suspended in RPMI 1640 tissue culture medium (Gibco Laboratories, Madison, WI) containing 10% fetal calf serum (FCS) to a concentration of 2×10^6 cells/ml and allowed to adhere to 24 well tissue culture plates (Costar, Cambridge, MA) for two hr at 37°C. The resulting macrophage monolayers were washed with Hank's balanced salt solution (Gibco) to remove nonadherent cells and infected with various concentrations of either virulent or avirulent Legionella for 30 min at 37°C. Following infection, the cultures were washed to remove nonadhered or noningested bacteria and then incubated for various periods of time. These cultures were subsequently analyzed for either the number of colony-forming units (CFU) or IL-1 activity. The number of viable bacteria (CFU) was determined by lysing the macrophages with 0.1% saponin, diluting the lysates, and plating on BCYE agar (8). In some experiments, erythromycin (Sigma Chemical, St. Louis, MO) was added to the cultures at a concentration of 0.5 μg/ml.

Interleukin 1 Assay

Culture supernatants were removed, membrane filter sterilized, and assayed for interleukin 1 (IL-1) activity as previously described (13). The assay used was the C_3H/HeJ mouse thymocyte co-mitogenic assay. Filter sterilized supernatants were diluted 1/2 and added to thymocyte cultures along with a submitogenic concentration of concanavalin A (Con A; Sigma). The cells were cultured for 48 hr, pulsed with ^3H-thymidine for 18 hr, and the extent of thymidine incorporation determined by scintillation counting. The IL-1 activity is reported as counts per minute (cpm) of incorporated ^3H-thymidine.

Northern Blot Analysis

PEC were cultured in tissue culture grade petri dishes (100 mm) at a concentration of 2×10^7 cells/plate. The cultures were incubated for 2 hr, rinsed with Hank's balanced salt solution to remove nonadherent cells, infected with Legionella, and then prepared for RNA isolation. RNA was isolated by the guanidine thiocyanate/cesium chloride method (14). The cultures were first treated with 1.5 ml 4M guanidine thiocyanate and the cell lysate loaded onto a 2 ml cushion of cesium chloride (5.7 M CsCl, 100 mM sodium acetate, 5 mM EDTA, pH 5). The RNA was purified by ultracentrifugation for 18 hr. The RNA pellet was solubilized in TE buffer containing 0.1% SDS and extracted with an equal volume of chloroform, followed by precipitation with three volumes of ethanol and 0.1 volume of 3 M sodium acetate at -80°C. The pelleted RNA was dried and dissolved in 50 μl of TE. The RNA was subjected to formaldehyde/1.2% agarose electrophoresis (14). Separated RNA bands were transferred to Hybond-N membranes by capillary blotting. Probing for IL-1 β mRNA was done with a ^{32}P, end-labelled, 42 base pair oligonucleotide probe provided by M. Nakano, Jichi Medical School, Japan. The IL-1 β probe was hybridized for 16 hr at 55°C, followed by multiple washes with various strength SSC/0.1% SDS at both room temperature and 55°C. The membranes were also probed with a ^{32}P labelled gamma actin probe to estimate the accuracy of RNA transfer to the membrane. This 1.3 kb probe was supplied by Dr. Patrick Lai, Tampa Bay Research Institute, St. Petersburg, FL, and was cloned as previously described (15). The gamma actin probe was hybridized for 20 hr at 60°C, followed by multiple washes and autoradiography.

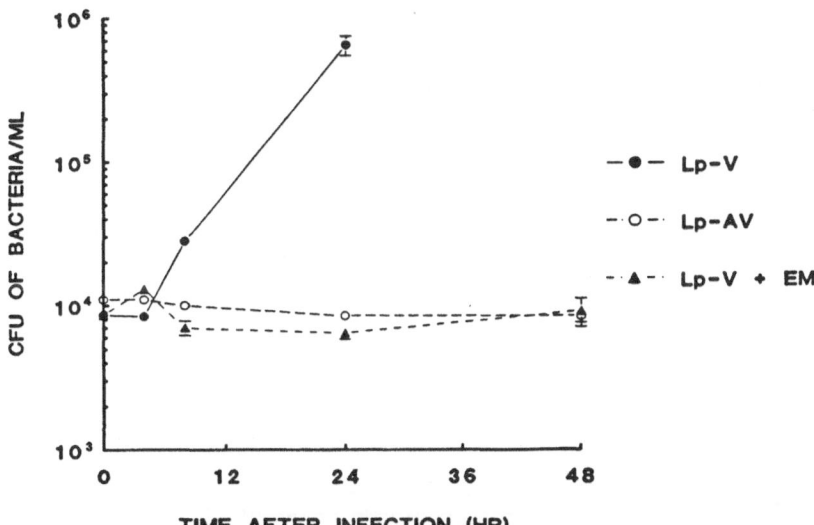

Fig. 1. Colony-forming unit (CFU) production by virulent *Legionella*. Macrophage cultures from A/J PEC were established in 24-well culture plates, infected for 30 min. with either virulent *Legionella* (Lp-V) or avirulent *Legionella* (Lp-AV), washed to remove free bacteria, and subsequently tested for the number of cell-associated CFU's. Time 0 is immediately after the 30 min. infection period and washing. The data are representative and are expressed as the mean CFU ± S.D. of triplicate cultures.

RESULTS AND DISCUSSION

CFU Production by Virulent and Avirulent Legionella

The virulence of Legionella can be defined either by morphological characteristics on plating medium and lethality for guinea pigs (12) or by intracellular growth and replication in cultured macrophages (8). In order to study characteristics associated with Legionella virulence, we isolated an avirulent strain by 5 passages on BCYE, followed by 5 passages on supplemented Mueller Hinton agar (12). Colonies which grew on supplemented Mueller Hinton and were presumed to be avirulent were isolated and used to infect PEC cultures obtained from A/J mice. Figure 1 shows that parental bacteria (LP-V) rapidly proliferated in the first 24 hr and increased several logs over the number of bacteria taken up by the macrophages. However, although the strain derived by passage and believed to be avirulent (LP-AV) was taken up by the macrophages to the same extent at the start of culture, these bacteria did not proliferate in the macrophage cultures. Interestingly, they were not killed by the macrophages, with their numbers remaining constant throughout the 2-day incubation. The avirulent strain was also observed to be nonlethal when injected into guinea pigs at high doses (unpublished observation). These results suggested that the culture adapted strain had lost the ability to multiply inside the macrophages.

IL-1 Induction by Virulent and Avirulent Legionella

Several lines of evidence suggested that either living or virulent Legionella were more immunostimulatory than killed or avirulent bacteria (6, 7, 16). Virulent organisms, freshly isolated from a fatal case of legionellosis, were much more effective

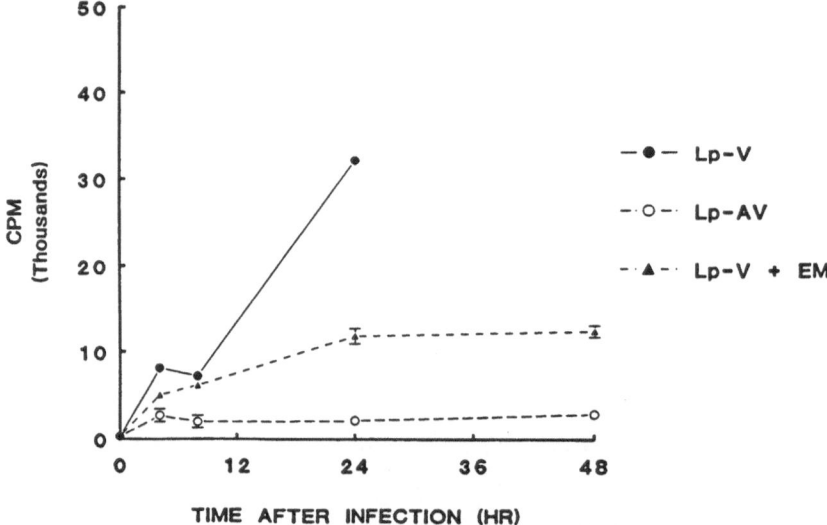

Fig. 2. Interleukin 1 activity in the culture supernatants from macrophages infected with virulent (Lp-V) and avirulent (Lp-AV) *Legionella*. Macrophages were infected as described in Fig. 1. Culture supernatants were harvested at the times indicated and assayed for IL 1 activity by the thymocyte co-mitogenic assay. Data are representative and are expressed as mean counts per minute ± S.D. of triplicate macrophage cultures.

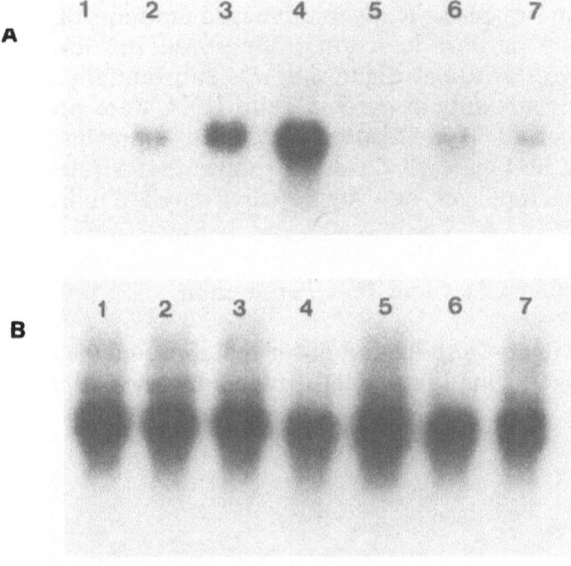

Fig. 3. Infection with virulent *Legionella* increases cellular IL 1ß mRNA. Macrophage cultures were established in petri dishes, infected with either virulent or avirulent *Legionella*, RNA extracted at various times following infection, and the RNA analyzed by Northern blotting and probing with ^{32}P-labelled probes specific for IL 1ß and γ actin. Panel A probed for IL 1ß: lane 1--control; lane 2--virulent, 3 hr.; lane 3--virulent, 6 hr.; lane 4--virulent, 20 hr.; lane 5--avirulent, 3 hr.; lane 6--avirulent, 6 hr.; and lane 7--avirulent, 20 hr. Panel B probed for γ actin: lanes as in Panel A.

in sensitizing animals for a DTH skin test response and a lymphocyte proliferation response than were low virulence organisms passed multiple times on plating medium (6). Furthermore, viable bacteria induced higher levels of IL-1 in macrophage cultures than killed bacteria (16), and viable Legionella were more potent inducers of immune phenomena in a rat infection model (7). This suggested that virulent organisms might be more reactogenic than avirulent organisms, and that the former might therefore be more potent inducers of IL-1 in infected macrophage cultures. Figure 2 shows that indeed this is the case when supernatant IL-1 is measured by the thymocyte co-mitogen assay. The data show that infecting the PEC cultures with equal numbers of either virulent or avirulent bacteria leads to divergent results in the quantity of IL-1 activity associated with the supernatants. Cultures infected with the avirulent organisms produced barely detectable amounts of IL-1 activity while those infected with virulent organisms produced substantial amounts within 24 hr following infection. Because these thymocyte assay results might be influenced by cytokines other than IL-1, we isolated RNA from the macrophage cultures and probed with a specific IL-1 β probe to verify an upregulation in the quantity of IL-1 messenger RNA. Figure 3, Panel A shows that infection with virulent organisms leads to increased amounts of specific IL-1 β message at 3 hr, 6 hr, and 20 hr post infection. Moreover, the levels of specific message following infection with avirulent organisms was substantially less. Figure 3, Panel B shows that equivalent amounts of gamma actin RNA were present, verifying that the level of RNA transferred to each membrane was approximately the same. Taken together, the results in Figures 1, 2, and 3 show that virulent organisms grow and replicate inside of macrophages, and suggest that they are more potent inducers of the IL-1 β.

Erythromycin Effects on CFUs and IL-1 Production

The difference in IL-1 production observed above could be due to an increase in the intracellular antigenic load in the cells infected with virulent organisms because of

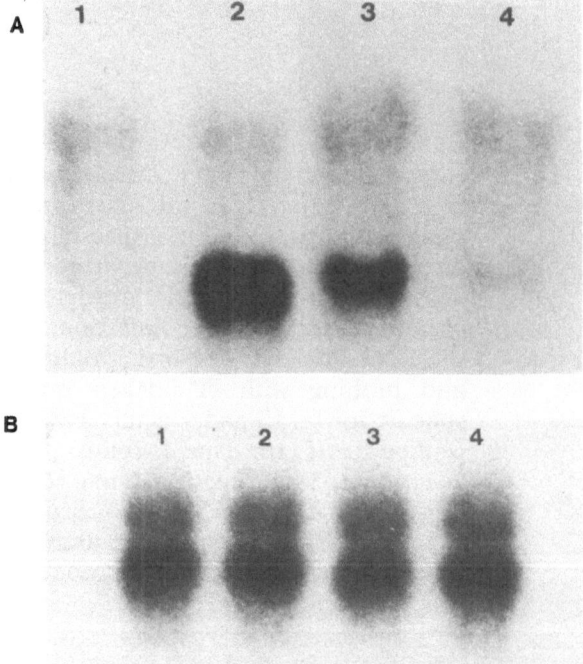

Fig. 4. Erythromycin treatment suppresses the accumulation of cellular IL 1ß mRNA. Macrophage cultures were established in petri dishes, infected with either virulent or avirulent *Legionella*, incubated for 24 hr. in medium ± erythromycin, and the RNA extracted and analyzed by Northern blotting and probing with ³²P-labelled probes specific for IL 1ß and γ actin. Panel A probed for IL 1ß: lane 1--control; lane 2--virulent, 24 hr.; lane 3--virulent plus erythromycin, 24 hr.; lane 4--avirulent, 24 hr. Panel B probed for γ actin: lanes as in Panel A.

the enhanced intracellular proliferation of this strain. To test this possibility, we treated the infected cultures with the antibiotic erythromycin, which has been reported to reversibly suppress the intracellular multiplication of Legionella (17, 18). Figure 1 shows that erythromycin treatment of PEC cultures following infection completely suppressed the proliferation of virulent organisms to the level observed in cultures infected with avirulent organisms. There was no growth of the virulent organisms throughout the 2 day culture. However, antibiotic treatment did not completely suppress supernatant IL-1 activity (Figure 2) or the accumulation of IL-1 β mRNA (Figure 4) to the level observed in avirulent infected cultures. These results show that inhibiting the multiplication of Legionella suppressed the induction of IL-1 but did not totally eliminate the upregulation of this cytokine. It is possible, therefore, that the increased number of intracellular bacteria in macrophages infected with virulent organisms is only partially responsible for the enhancement of cytokine production. Other factors associated with the virulent organisms, such as unique cell surface ligands or the intracellular release of unique metabolic products, possibly contribute to a unique array of intracellular cytokine-upregulating signals. We speculate that the extent of IL-1 gene expression occurring during Legionella infection is regulated by signals generated at the cell surface and by additional signals resulting from the intracellular metabolism, growth, and replication of the invading parasite. Additional studies are required to unravel the complexity of this host-parasite interaction.

ACKNOWLEDGEMENTS

We would like to thank Ms. Judy Flynn for assistance in the preparation of this manuscript. This work was supported by NIAID grant AI 16618.

REFERENCES

1. P. H. Edelstein, Environmental aspects of *Legionella*, ASM News 51:460 (1985).
2. C. E. Farshy, G. C. Klein, and J. C. Feeley, Detection of antibodies to Legionnaires disease organism by microagglutination and micro-enzyme-linked immunosorbent assay tests, J. Clin. Microbiol. 7:327 (1978).
3. J. F. Plouffe and I. M. Baird, Lymphocyte blastogenic responses to *L. pneumophila* in acute legionellosis, J. Clin. Lab. Immunol. 7:43 (1982).
4. D. K. Blanchard, H. Friedman, T. W. Klein, and J. Y. Djeu, Induction of interferon-gamma and tumor necrosis factor by *Legionella pneumophila*: Augmentation of human neutrophil bactericidal activity, J. Leukocyte Biol. 45:538 (1989).
5. D. K. Blanchard, J. Y. Djeu, T. W. Klein, H. Friedman, and W. E. Stewart II, Protective effects of tumor necrosis factor in experimental *Legionella pneumophila* infections of mice via activation of PMN function, J. Leukocyte Biol. 43:429 (1988).
6. T. W. Klein, H. Friedman, and R. Widen, Relative potency of virulent versus avirulent *Legionella pneumophila* for induction of cell-mediated immunity, Infect. Immun. 44:753 (1984).
7. S. J. Skerrett and T. R. Martin, Alveolar macrophage activation in experimental legionellosis, J. Immunol. 147:337 (1991).
8. Y. Yamamoto, T. W. Klein, C. A. Newton, R. Widen, and H. Friedman, Differential growth of *Legionella pneumophila* in guinea pig versus mouse macrophage cultures, Infect. Immun. 55:1369 (1987).

9. S. I. Yoshida and Y. Mizuguchi, Multiplication of *Legionella pneumophila* Philadelphia-1 in cultured peritoneal macrophages and its correlation to susceptibility of animals, Can. J. Microbiol. 32:438 (1986).

10. Y. Yamamoto, T. W. Klein, C. A. Newton, R. Widen, and H. Friedman, Growth of *Legionella pneumophila* in thioglycolate-elicited peritoneal macrophages from A/J mice, Infect. Immun. 56:370 (1988).

11. T. W. Klein, D. K. Blanchard, Y. Yamamoto, C. Newton, R. Widen, and H. Friedman, Role of cytokines in resistance to infection with *Legionella pneumophila*, Adv. in Biosci., 68:259 (1988).

12. C. E. Catrenich and W. Johnson, Virulence Conversion of *Legionella pneumophila*: a one-way phenomenon, Infect. Immun. 56:3121 (1988).

13. T. W. Klein, C. A. Newton, D. K. Blanchard, R. Widen, and H. Friedman, Induction of interleukin 1 by *Legionella pneumophila* antigens in mouse macrophage and human mononuclear leukocyte cultures, Zbl. Bakt. Hyg., A 265:462 (1987).

14. J. Sambrook, E. F. Fritsch, and T. Maniatis, "Molecular Cloning: A Laboratory Manual," Cold Spring Harbor Laboratory, Cold Spring Harbor, NY (1989).

15. E. A. Fyrberg, B. J. Bond, N. D. Hershey, K. S. Mixter, and N. Davidson, The actin genes of drosophila: protein coding regions are highly conserved but intron positions are not, Cell 24:107 (1981).

16. R. H. Widen, T. W. Klein, C. A. Newton, and H. Friedman, Induction of interleukin 1 by *Legionella pneumophila* in murine peritoneal macrophage cultures, Proc. Soc. Exp. Biol. Med. 191:304 (1989).

17. M. A. Horwitz and S. C. Silverstein, Intracellular multiplication of Legionnaires' disease bacteria (*Legionella pneumophila*) in human monocytes is reversibly inhibited by erythromycin and rifampin, J. Clin. Invest. 71:15 (1983).

18. S. I. Yoshida and Y. Mizuguchi, Antibiotic susceptibility of *Legionella pneumophila* Philadelphia-1 in cultured guinea-pig peritoneal macrophages, J. Gen. Microbiol. 130:901 (1984).

CYTOKINE ACTIVATION OF KILLER CELLS IN MYCOBACTERIAL

IMMUNITY

D. Kay Blanchard

University of South Florida College of Medicine
Department of Medical Microbiology and Immunology
Tampa, FL

ABSTRACT

Mycobacterium avium-intracellulare (MAI) is an ubiquitous soil contaminant that rarely causes disseminated disease in adults, regardless of immunological status. In AIDS patients, however, this microorganism invades virtually every tissue and organ, and most conventional chemotherapeutic agents are usually ineffective against MAI. We report here that monocytes, in which MAI has established an intracellular parasitic stage, appear to be under the control of natural killer (NK) cells. Autologous large granular lymphocytes (LGL), purified from human peripheral blood mononuclear cells (PBMC), were capable of efficiently lysing MAI-infected monocytes ln a 5 hr ^{51}Cr-release assay. More importantly, interleukin 2 (IL-2) was able to activate the LGL to a high degree of lysis of infected monocytes. Additionally, 3 to 4 days of incubation of LGL with MAI resulted in the induction of killer cells capable of killing bacterially-infected monocytes, as well as tumor cells. Northern blot analysis of RNA from MAI-stimulated LGL revealed specific messages for both IL-2 receptor proteins (p55 and p70). Thus, MAI can directly activate killer cells, which may therefore play a role in containment of MAI infection by lysis of parasitized monocytes before the bacteria can multiply and spread to other sites.

INTRODUCTION

The NK cell system is a natural resistance mechanism believed to provide an early line of defense against tumor cells and viral infections (1). The ability of NK cells to be activated by IL-2 to lyse otherwise NK-resistant target cells has proved to be a key property that has been applied in the clinic for immunotherapy of cancer patients (2). The usefulness of NK cells apparently can be extended to the control of intracellular bacteria. In this regard, we have demonstrated that human NK cells could effectively lyse autologous monocytes that are infected intracellularly with *Legionella pneumophila*, the causative agent of pulmonary Legionnaires' disease (3). This cytolytic activity was markedly enhanced by activation of the NK cells with recombinant IL-2, which resulted in LAK cells. In a similar study, we demonstrated that MAI-infected monocytes were also susceptible to lysis by NK and LAK cells (4), a phenomenon which has been

Microbial Infections, Edited by H. Friedman *et al.*
Plenum Press, New York, 1992

confirmed by others (5,6). Thus, bacterially-infected cells appear to be successfully lysed by NK or LAK cells, demonstrating a potential role for natural cytotoxic activity in the resistance of intracellular bacterial infections.

In this report, we compare the ability of cytokines to activate NK cells, or LGL, to kill tumor cells and MAI-infected monocytes. A second aim of this study was to examine the direct effects of MAI on NK or LAK activity, since many bacterial products are known to activate their cytotoxic ability. In this regard, MAI was found to increase the expression of IL-2 receptors on stimulated cells. Classically, IL-2 has been considered to be a lymphocyte activating and growth promoting factor, and has been widely studied on T cells and NK cells (7). Three classes of IL-2 receptors exist, binding IL-2 with low, intermediate, or high affinity (8). Low affinity receptors contain the 55 kDa IL-2 receptor alpha chain (p55) without the 70-75 kDA IL-2R-beta (p75); intermediate affinity receptors contain IL-2R-beta without IL-2R-alpha; high affinity receptors contain both IL-2R-alpha and IL-2R-beta (9,10). Whereas intermediate affinity receptors have been found on unstimulated T cells and LGL, high and low affinity receptors are expressed preferentially on activated cells (11), and in this study, MAI-activated LGL.

MATERIALS AND METHODS

Preparation of Human Leukocytes

Leukocyte buffy coats, obtained from normal volunteers at the Southwest Florida Blood Bank, were diluted 1:2 in PBS and layered on 10 ml of Ficoll-Hypaque solution (Pharmacia, Piscataway, NJ) (4). After centrifugation at 400 g for 20 min at room temperature, the band of PBMC at the interphase was collected and washed twice with PBS. The washed PBMC were suspended in RPMI 1640 medium (GIBC0 Laboratories, Grand Island, NY) containing 5% heat-inactivated human AB serum (Flow Laboratories, McLean, VA, 2mM L-glutamine, 10 W ml penicillin, 5 mM HEPES buffer (GIBC0), and 5 x 10-5 M 2-mercaptoethanol (Sigma Chemical Co., St. Louis, M0), and will subsequently be referred to as complete medium.

Preparation of Monocytes

PBMC were allowed to adhere to gelatin-coated tissue culture flasks for 1 hr at 37°C. Nonadherent cells (NAC) were recovered by vigorous washing of the flasks with warm medium, and the adherent cells were cultured for 5 days with fresh medium. Adherent cells were removed by gentle scraping with a cell scraper (COSTAR, Cambridge, MA) after the addition of cold PBS.

Discontinuous Percoll Density Gradient Centrifugation. The separation of LGL from T cells was accomplished by the use of a discontinuous Percoll density gradient. NAC were further depleted of adherent cells and B cells by incubation on nylon wool columns for 30 min at 37°C. The cells passing through the columns were then placed on a six-step discontinuous density gradient that varied in 2.5% concentrations from 40 to 52.5% Percoll as previously described (12). After centrifugation at 550 g for 30 min at room temperature, the bands of lymphocytes were collected and examined for LGL morphology on Giemsa-stained cytocentrifuged slides. In this series of experiments, fractions 2 and 3 contained 60-80% LGL and represented 10-15% of PBL.

Activation of Effector Cells

LGL were incubated at a concentration of 2×10^6 cells/ml with 100 U/ml of recombinant human IL 2, which was kindly provided by Hoffman-LaRoche, Inc. (Nutley, NJ), as previously described (4). Other cytokines used were tumor necrosis factor (TNF), from Genentech, at 1000 U/ml, and IL-6, from Genetics Institute, also at 1000 U/ml. For activation with bacteria, MAI was added at a concentration of 10 bacteria per LGL in medium containing antibiotics to prevent microbial growth. LGL were optimally activated for 3-4 days in 25 cm2 tissue culture flasks, and were then washed twice in medium and readjusted to the original cell concentration.

Infection of Monocytes with MAI

MAI used in this study was obtained from the Tampa Branch of Florida State Laboratories, Tampa, FL, and was determined to be serotype 8. For infection, MAI way cultured for 14 days in Middlebrook 7H11 agar medium (GIBC0 Laboratories, Madison, WI) from a stock culture and recovered by washing with sterile PBS. The suspension was vigorously vortexed. Monocytes were inoculated with MAI by incubating the monolayer in a 75 cm2 tissue culture flask with 2 ml of bacterial suspension estimated to contain 10 bacteria per monocyte for 2 hr, washing, and incubating in medium alone for 5 days. The medium used for infected monocytes did not contain penicillin or streptomycin.

Measurement of Cytotoxicity

A 5 hr ^{51}Cr-release assay was used to measure the cytotoxicity of IL 2-activated LGL against autologous MAI-infected-monocytes and FMEX, which is an NK-insensitive melanoma cell line (13). Monocytes were labelled with 200 uCi of sodium ^{51}Cr chromate (Amersham Corp., Arlington Heights, IL) for 1 hr in 0.5 ml of medium. Target cells were then washed twice and then added to effector cells at 5×10^3 cells/well, regulating in effector:target ratios ranging from 20:1 to 2.5:1 in a final volume of 0.2 ml in each well. After 5 hr incubation at 37°C, the culture supernatants were harvested by removing 0.1 ml of the fluid to be counted in a gamma counter. The percentage of specific lysis was calculated by the formula: [(experimental cpm - spontaneous cpm) maximal cpm incorporated] x 100. All determinations were done in triplicate. The SEM of all assays was calculated and was typically 5% of the mean or less. Lytic units (LU) were calculated and were defined as the number of effector cells required to lyse 20% of 5×10^3 target cells.

Northern Blot Analysis

Total cellular RNA from LGL was isolated according to the method of Chomczynski and Sacchi (14). RNA samples were denatured in a glyoxal-dimethyl sulfoxide mixture and then fractionated on a 0.8% agarose gel. Fractionated RNA was then transferred to Nytran filters (Schleicher & Schuell, Keene, NH) and stained with methylene blue-acetate to determine presence and integrity of transferred RNA (15). Prehybridization was performed at 45°C in a 50% deionized formamide solution (16) and hybridization was at the same temperature for 18 hr with the cDNA fragment of human IL-2R-beta (17) or human IL-2R-alpha (18), which were generous gifts from Dr. Warren J. Leonard. Each probe was labelled with dCTP-^{32}P using random primed labelling. After hybridization, blots were washed at 60°C with 1X SSC with 0.2% SDS and 1 mM EDTA. The membrane was then exposed to film with an intensifying screen at -70°C for 5 days.

RESULTS AND DISCUSSION

We have previously reported that MAI-infected monocytes are susceptible to the lytic effects of autologous killer cells (4). While treatment of LGL with IL-2 was shown to augment their activity, the direct effect of MAI on those cells was not determined. This seemed to be of some importance since once bacterially-laden monocytes were lysed by NK/LAK cells, MAI antigens could then stimulate other NK cells or LGL to continue escalating the lytic processes. Thus, the effect of MAI stimulation on LGL was determined (Table 1). As expected, incubation of LGL for 4 days with IL-2 resulted in the induction of both antitumor and anti-monocyte activities. The use of FMEX as an indicator target cell for LAK activity (since these tumor cells are resistant to NK lysis), confirmed that LAK cells were generated during the incubation period. Interestingly, the co-culture of LGL with MAI also generated anti-tumoricidal activity, but proportionately more anti-monocyte activity was induced. These results would seem to suggest that subpopulations of killer cells exist, with differential responses to either cytokine or antigenic stimuli, although further investigation is required to substantiate this hypothesis.

Since MAI is known to stimulate the production of a variety of cytokines, it is possible that the effect of this bacterium on LGL is mediated by these potent hormone-like molecules. We have recently reported that MAI can induce LGL to produce both TNF and IL-6 (19,20), so the effect of these cytokines on the induction of tumoricidal and anti-monocyte activities were determined. As shown in Table 2, neither TNF or IL-6, or the combination of these two cytokines, affected LGL activities. However, there was a synergistic induction of both antitumor and anti-monocyte lysing with the combination of IL-2 and TNF. A recent report has demonstrated that TNF up-regulates IL-2 receptor expression (21), which is a possible explanation for the increased activities noted here. At any rate, none of the cytokine combinations tested were as successful in the induction of anti-monocyte activity as was intact MAI.

Table 1.   Induction of Cytolytic Activity by MAI

| | % Specific lysis of target cells at E/T ratio: | | | | | | | |
| | FMEX | | | | Monocytes + MAI | | | |
Effector treatment	20/1	10/1	5/1	(LU)	20/1	10/1	5/1	(LU)
LGL	3	3	2	(4)	10	7	4	(29)
LGL + IL-2	52	43	32	(503)	20	16	10	(195)
LGL + MAI	27	18	10	(114)	21	16	13	(199)

LU, lytic units

Table 2. Induction of Cytolytic Activity of LGL by Cytokines

Treatment of LGL	Target cells (LU)	
	FMEX	Monocytes + MAI
Medium control	2	14
IL-2	333	159
TNF	2	10
IL-2 + TNF	439	329
IL-6	3	11
IL-2 + IL-6	312	213
TNF + IL-6	3	14
MAI	241	571

LU, lytic units

Since activation of cytolytic capabilities of both NK and T cells is associated with expression of the IL-2R-alpha (p55 or TAC antigen), we then explored the possibility that MAI could up-regulate the expression of this receptor (Fig 1). Flow cytometric analysis of unstimulated LGL demonstrated that no Tac receptors are detected, as expected. Also expected, IL-2 was found to induce the expression of this receptor. More importantly, co-culture of LGL with MAI was also found to result in the expression of Tan antigen, indicating that this bacterium is capable of providing a potent signal to the cell, resulting in its activation. The mechanism whereby this signal is transduced is still unknown. To determine whether the induction of Tac antigen was seen at a message level, total RNA from MAI-stimulated and unstimulated LGL was extracted and probed for the presence of both p55 and p75 subunits of the IL-2 receptor. As seen in Fig 2, unstimulated LGL were found to contain constitutive levels of p75 message, as previously reported. However, very little mRNA for the Tac antigen was detected. Upon stimulation with MAI, mRNA for both subunits of the IL-2 receptor were significantly increased, although p55 seemed to be more dramatically affected.

The importance of NK/LAK cells during intracellular bacterial infection was first suggested in our studies of *L. pneumophila* (3), and we confirmed that activation of LAK cells is also possible against a more clinically important microorganism, MAI (4). MAI presents as an obligate intracellular parasite and establishes itself in the host monocyte which has phagocytized the bacterium as part of the early immune response. Beside the ability to inactivate and kill the host monocyte it infects, MAI has posed a medical problem because of its resistance to a number of antibiotics (2). Our finding that IL-2 activated LGL to kill autologous infected monocytes indicates that the host has an alternative defense mechanism to deal with intracellular bacteria, which may be a useful parameter to augment for treatment of MAI infections. While the direct

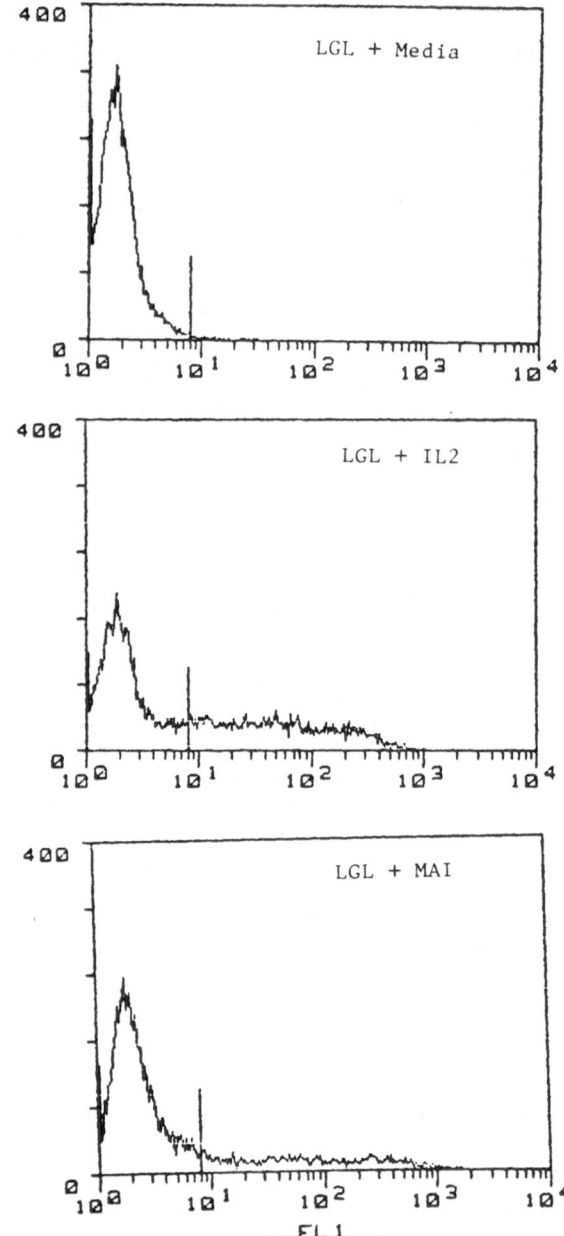

Fig. 1. Expression of IL-2 receptors (P55) on LGL. LGL were stained with anti-IL-2R antibodies (Becton-Dickinson) after 4 days of incubation with the indicated stimulator and analyzed by Flow cytometry on FACScan.

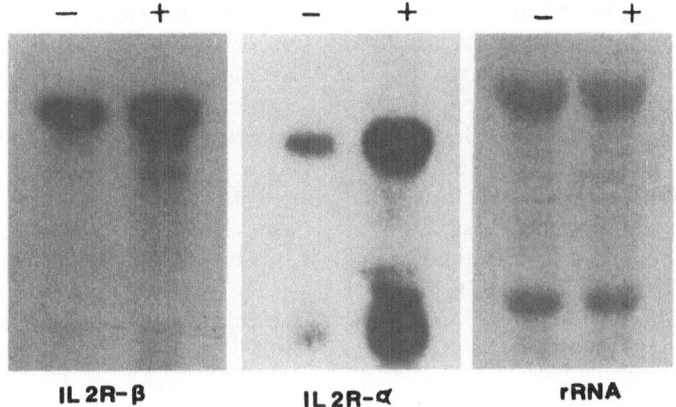

IL 2R-β	IL 2R-α	rRNA

Fig. 2.   Northern blot analysis of LGL. LGL were incubated in medium control (-) or in the presence of MAI (+) for 4 days. Total RNA was isolated and 20 μg samples were fractionated and probed for the presence of the indicated message. Ribosomal RNA (rRNA) is depicted by methylene blue staining to demonstrate that equivalent amounts of RNA were analyzed.

interaction of LGL with MAI indicate a complex relationship exists, further study is required to fully understand the effects of this opportunistic pathogen on cells of the immune response.

REFERENCES

1. R. M. Welsh, CRC Crit. Rev. Immunol. 5:55 (1984).
2. G. A.Woods and J. A. Washington II, Rev. Infect. Dis. 9:275 (1987).
3. D. K. Blanchard, W. E. Stewart II, T. W. Klein, H. Friedman, and J. Y. Djeu, J. Immunol. 139:551 (1987).
4. D. K. Blanchard, M. B. Michellini-Norris, H. Friedman, and J. Y. Djeu, Cell. Immunol. 119:402 (1989).
5. P. Katz, H. Yeager, G. Whalen, M. Evans, R. P. Swartz, and J. Roecklein, J. Clin. Immunol. 10:71 (1990).
6. L. E. M. Bermudez, andL. S. Young, J. Immunol. 146:265 (1991).
7. K. A. Smith, Ann. Rev. Immunol. 2:319 (1984).
8. T. A. Waldmann, Ann. Rev. Biochem. 58:875 (1989).
9. M. Sharon, R. D. Klausner, B. R. Cullen, R. Chizzonite, and W. J. Leonard, Science (Wash. DC) 234:859 (1986).
10. M. Tsudo, R. W. Kozak, C. K. Goldman, and T. A. Waldmann, Proc. Natl. Acad. Sci. USA 83:9694 (1986).
11. M. Sharon, J. P. Siegal, G. Tosato, J. Yodoi, T. L. Gerrard, and W. J. Leonard, J. Exp. Med. 167:1265 (1988).
12. T. Timonen, and E. Saksela, J. Immunol. Meth. 36:285 (1980).
13. T. Lindmo, C. Davies, O. Fodstad, and A. C. Morgan, Int. J. Cancer 34:504 (1985).
14. P. Chomczynskl, and N. Sacchi, Anal. Biochem. 162:156 (1987).
15. T. Maniatis, E. F. Fritsch, and J. Sambrook, in: "Molecular Cloning", A Laboratory Manual. Cold Spring Harbor Laboratory, New York, (1982).

16. Y. Nagamine, M. Sudol, and E. Reich, Cell 32:1181 (1983).

17. J. R. Gnarra, H. Otani, M. G. Wang, O. W. McBride, M. Sharon, and W. J. Leonard, Proc. Natl. Acad. Sci. USA 87:3440 (1990).

18. W. J. Leonard, J. M. Depper, G. R. Crabtree, S. Rudikoff, J. Pumphrey, R. J. Robb, M. Kronke, P. B. Svetlik, N. J. Peffer, T. A. Waldmann, and W. C. Greene, Nature 311:626 (1984).

19. B. M. Michellini-Norris, D. K. Blanchard, H. Friedman, and J. Y. Djeu, J. Leukocyte Blol. (in press).

20. D. K. Blanchard, M. B. Michelini-Norris, C. A. Pearson, C. S.Freitag, and J. Y. Djeu, Blood 77:2218 (1991).

21. S. Choualb, J. Bertoglio, J.-Y. Blays, C. Marchiol-Forunigault, and D. Fradelizl, Proc. Natl. Acad. Sci. USA 85:6875 (1988).

T CELL-MONOCYTE INTERACTIONS INDUCED BY *LISTERIA*

MONOCYTOGENES

M. Mielke, H. Hahn

Institut für Med. Mikrobiologie und Infektionsimmunologie
FU Berlin (FRG)

Much of our current understanding of cell-mediated immunity to facultative intracellular pathogens stems from studies of the immune response of mice to *Listeria monocytogenes* - a model first introduced by Mackaness in 1962 (1, for reviews see 2-4). The specific cellular immune response to this and other facultative intracellular bacteria results in inflammatory and protective phenomena, the most prominent ones being

a) DTH, granuloma formation, and splenomegaly which in turn are
   correlated with
b) the eradication of bacteria from infected organs.

In each case of infection with facultative intracellular bacteria, it has been found that macrophages of infected animals show greatly enhanced microbicidal properties (1,5,6). This enhanced bactericidal capacity is nonspecific, as evidenced by resistance to superinfection with unrelated bacteria (7,8), and is lost soon after the infection is terminated. However, it can be regenerated in an accelerated fashion by challenging the host with the homologous organism (9). The terms 'cellular resistance' (nonspecific) and 'cellular immunity' (specific) are used to define these two juxtaposed phenomena (10). Mackaness (11), using the murine listeriosis model, showed that macrophage activation is induced by specific lymphocytes. His conclusions were based on the results of adoptive transfer experiments in which viable spleen cells from immune animals could transfer specific cell-mediated immunity to non-immune recipients. The role of lymphocytes was substantiated further by the use of antilymphocyte globulin which suppressed immunity to Listeria (12). After the discovery that the lymphocyte surface antigen Thy-1 is a marker for thymus-dependent T cells, and antisera to this antigen had become available (13), it was shown that lymphocytes from Listeria-immune donor mice which transfer not only protection to Listeria (14-16), but also confer DTH (17), are T cells. Experiments in T cell deficient nude mice further substantiated the assumed distinction between the mechanisms of non-specific resistance and the nature of specific cellular immunity. Thus, these animals are able to restrict replication of viable bacteria during the early phase of infection, but are incapable of forming granulomas and developing DTH and finally succumb to chronic infection (18,19). The most characteristic function of sensitized T cells then seems to be the attraction and accumulation of monocytes at foci of infection. The need for macrophages to cooperate with T cells in the expression of immunity is also emphasized by adoptive immunization studies which revealed that adoptive immunity cannot be expressed in recipients that have been treated with agents known to suppress the recipient's production and function of monocytes and macrophages (20-22).

The response to allogeneic MHC antigens has long provided a useful model for the study of T cell functions and finally led to the insight that all immune functions mediated by T cells are associated with cell surface structures codified in the mouse by the H2-gene complex. Using a set of mice differing at certain H2-loci, Zinkernagel and others (23) showed that interactions of Listeria-specific effector T cells with macrophages are MHC-class II-restricted, i.e., histocompatibility of cell donors and recipients at the H2-I locus was necessary and sufficient for antibacterial protection to be transferable by T cells. Furthermore, data obtained using in vitro systems (24-26) demonstrated that secondary stimulation by heat killed Listeriae (HKL) of specific T cell functions, including antigen-specific proliferation and the production of lymphokines, is also dependent on identity in class II MHC products on antigen presenting macrophages and responding T cells. In 1983, Cheers and Sandrin (27) re-evaluated the H2 restriction of immunity to Listeria monocytogenes postulated by Zinkernagel et al. employing a different set of mice. These authors convincingly demonstrated that transfer of protection was only possible between strains of mice compatible at the K end of the H2-complex (class I), independent of any compatibility in the I-A region (class II). This result fell in line with earlier observations by Jungi (28) who had found a class I restriction of adoptive protection to exist in the rat model of listeriosis.

The conflicting data led to the hypothesis that differently regulated T cell subpopulations might be involved in the expression of Listeria-specific immunity.

T Cell Subpopulations in Immunity to *Listeria monocytogenes*

Experiments by Cantor and Boyse (29,30) provided the first evidence that T cells indeed form a heterogeneous population. The subsequent discovery (31) of an association between MHC-restriction, phenotype, and functional capacity of T cell subsets prompted us, amongst others, to start a new series of analyses asking in what way, and to what extent, different T cell subsets contribute to protective immunity and DTH. This analysis was facilitated by the development of a monoclonal antibody (mAb), GK1.5, with specificity to L3T4 (32) because of the expression of L3T4 (CD4) and Lyt-2 (CD8) cell surface molecules separates murine T cells into two non-overlapping subsets. Four major experimental strategies were followed by us:

- Immunohistology
- Selective depletion of T cell subsets in adoptive transfer systems
- Analysis of T cell lines and clones and
- *In vivo* depletion of T cell subsets.

Immunohistology

The involvement of both, the L3T4+ and Lyt2+, T cell subsets in Listeria-induced immunity has been demonstrated by immunohistologic analysis of their distribution in granulomatous lesions in livers of infected mice (33). However, whereas the fraction of cells in granulomas bearing the L3T4 marker increased from 60 to 96 hrs after adoptive transfer of immune T lymphocytes, the relative amount of Lyt-2+ cells remained constant, an observation that points to different mechanisms of activation.

Adoptive Transfer Experiments

Using subset-specific mAbs, Kaufmann et al. (34) showed that the capacity of peritoneal exudate T lymphocyte enriched cells to adoptively mediate protection against *Listeria monocytogenes* and DTH to listerial antigens was markedly reduced by pretreatment of cells with either anti-L3T4 or anti-Lyt-2 antibodies plus complement, but that it could be restored by the admixture of the two selected T cell subsets. Therefore, a cooperation between specific L3T4+ and Lyt-2+ cells in mediating protection and DTH was assumed. Further evidence for a crucial role of Lyt-2+ cells in anti-listerial protection was obtained by transferring spleen cells from immune donors treated only by in vitro-incubation with the appropriate monoclonal antibody. In these experiments (35), we were able to show that the strong protective capacity of splenic equivalents can be abrogated by anti-Thy-1 or anti-Lyt-2 mAb treatment, whereas there was only a slight inhibition after anti-L3T4 treatment. Similar results (36) were recently reported from other laboratories. In the rat model of listeriosis Chen Woan et al. (37,38) selected thoracic duct lymphocytes on the basis of their W3/25 (CD4) or OX8 (CD8)-phenotype. In these experiments, the mediators of DTH were identified as W3/25+ OX8 T cells, whereas the protective T cells had the W3/25 OX8+ phenotypes.

Propagation of Specific T Cell Lines and Clones

Kaufmann (39,40) succeeded in the production of Listeria-specific murine T cell clones of both the L3T4+ and the Lyt-2+ phenotypes. These T cell clones as well as the rat W3/25+ T cell clones established by Stolpmann (41) have been extensively characterized *in vitro* (42-44).

Although T cell clones were highly active *in vitro*, intravenous transfer of the cells resulted in relatively low protection against a systemic infection with L. monocytogenes. Marked local protection, however, could be obtained by injecting the cells together with *L. monocytogenes* subcutaneously into the footpad of animals. The reduced level of systemic protection was explained by the fact that intravenously transferred cells have an altered migration pattern as revealed by Cr-labelling studies. Lyt-2+ clones (40) have so far been shown to have a specific MHC-restricted and non-restricted cytolytic activity against Listeria-infected target cells and to produce IFN-a as well. Based on this observation, Kaufmann proposed a hypothesis according to which cytolytic cells would establish direct contact with infected target cells, lyse them, and in this way release intracellular bacteria. The latter would subsequently be taken up by invading monocytes better equipped to kill the bacteria than the cells originally infected. Most recently (45), it has been reported that L3T4+ clones could also specifically lyse Listeria-infected Ia-positive bone-marrow macrophages *in vitro*. Thus, with the exception of the target cell spectrum, at the clonal level, there seems at present to be no clear-cut distinction between phenotype and function of the T cell subsets involved in immunity against Listeria infection.

In Vivo Serotherapy with Monoclonal Antibodies Specific for T Cell and Monocyte Surface Structures

As described here, the current view of immunoregulation is based mainly on studies of lymphocyte subsets either in vitro or by adoptive transfer to naive recipients. Recently, it has been shown (46) that unmodified mAbs of the IgG2b type can be extremely effective in depleting cells in vivo and therefore can be used for the selective manipulation of the immune response. We, therefore, treated mice with three different mAbs specific for Thy-l, L3T4, and Lyt-2, respectively, in order to assess the relative contribution of T cell subsets to DTH, protection, and granuloma formation in murine listeriosis (35,47). Irrespective of whether mAbs were administered during the induction or the effector phase of DTH, the latter was affected by the application of anti-Thy-1 or anti-L3T4 mAb only. The interrelatedness of DTH and protection against facultative intracellular bacteria has long been a subject of considerable controversy. The fact that animals can be desensitized without loss of resistance or, conversely, can be rendered hypersensitive without causing a corresponding increase in resistance argued against the widely accepted view that both phenomena depend on the same mechanism. The possibility of ablating DTH reaction by the administration of anti-L3T4, but not by anti-Lyt-2, mAb *in vivo* offered the opportunity to investigate whether DTH and protection 10 are indeed mediated by different T cell subsets, as was assumed by the studies of Chen-Woan et al. (37,38) in the rat model. Therefore, the effect of mAbs, administered during primary and secondary infection, on DTH, protection and granuloma formation was determined. Our data (35,47) demonstrate that in primary infection, bacterial

Table 2. Relation between T cell-subsets and characteristic immune phenomena

DTH	Splenomegaly	Granulomatous Inflammation	Protection
CD4+	CD4+	CD4+	CD4+
			CD8+

clearance is inhibited by anti-Thy-1.2 mAb treatment and, to a lesser extent, by anti-Lyt-2 mAb treatment. Surprisingly, anti-L3T4 mAb treatment, although preventing expression of DTH, had no significant effect on the resolution of infection. Treatment with mAbs shortly before secondary infection of animals which had been strongly immunized 4 wks previously, yielded a comparable pattern. However, the effect of the anti-Lyt-2 mAb treatment was more pronounced. Anti-Thy-1.2 mAb treatment abolished, and anti-Lyt-2 mAb treatment markedly reduced, established protective immunity, whereas anti-L3T4 mAb treatment had a negligible effect. On the basis of observations in nude (48) and SCID (49) mice that, despite macrophage activation, adequate protection cannot be achieved in the absence of granuloma formation, we investigated (47) the effects of mAb treatment on this essential feature of the host response. The examination of liver sections from immune animals 48 hr after reinfection showed complete absence of mononuclear foci in anti-Thy-1.2-treated animals. The typical Listeria induced lesions in such treated animals were polymorphonuclear infiltrations similar to those that can be observed in the very early phase of primary infection. Most surprising was the fact that livers of anti-L3T4 treated (CD8-sufficient) animals were free of any granulomatous lesions in the presence of protection. On the contrary, anti-Lyt-2-treated (CD4-sufficient) animals were able to form granulomas (Table 2).

The rapidly expressed CD8+ T cell-mediated protection that can be observed in a secondary infection in the absence of granuloma formation, therefore, was argued to be the result of a rapid interaction with resident macrophages, shown to be activated by their elevated expression of MHC class II products in the absence of CD4+ T cells (Mielke and Niedobitek, unpublished results). In order to prove this assumption, we employed the monoclonal antibody 5C6 (50-52), which has been shown to inhibit the recruitment of myelomonocytic cells into inflammatory foci and to exacerbate a primary Listeria infection by blocking complement receptor type 3 (CR3, Mac-1, CDllb), a member of the leukocyte integrin family. If Listeria-primed CD8+ T cells act independently of invading monocytes by rapidly activating resident macrophages for efficient listericidal activity, this antibody would have minimal effects on the outcome of a secondary infection in actively immunized mice. Our recent study (53), however, demonstrates that the inhibition of CR3-dependent migration of phagocytes dramatically increased the susceptibility to *L. monocytogenes*, not only in primary but also in secondary infection, where potent, granuloma independent antibacterial mechanisms are expressed by CD8+ T cells. However, two elements of T cell mediated immunity could be demonstrated even in the face of 5C6-treatment and therefore seem to be CR3/CDllb-independent, i.e. a) acquired immunity against low doses of viable Listeriae and b) granuloma formation elicited in immune mice by heat killed bacteria (HKL).

In conclusion, the results of studies performed using the Listeria model and presented in this review may be summarized as follows:

i) Listeria-primed CD8+ T cells mediate protection by mechanisms dependent on a rapid interaction with phagocytes invading infected tissues within the first 24 h after reinfection in a CR3/CDllb-dependent, but CD4+ T cell-independent manner. This CD8+ T cell monocyte-interaction does not result in granuloma formation, DTH or splenomegaly.

ii) Listeria-specific CD4+ T cells act via a) less protective early and b) a delayed but longlasting interaction with monocytes which results in DTH and granuloma formation and which is CR3/CDllb-independent.

What Is the Murine Listeriosis a Model for?

When the medically important facultative intracellular bacteria are grouped according to the duration of the natural course of infection in experimental animals, three groups can be formed with Listeria at the one and mycobacteria at the other end (Table 1). This, when taken together with the distinct composition of cell walls and different evasion mechanisms displayed, points to different mechanisms underlying the eradication of these bacteria. In fact, this is most impressively substantiated by differences in the susceptibility of infected animals to CR3-blocking antibodies. Whereas mice infected with *L. monocytogenes* are highly susceptible to mAb 5C6 (52,53), there is no effect on brucellosis (Mielke, Rosen, unpublished data) or BCG infection (Rosen, unpublished data). Consequently, in face of the minor medical importance of listeriosis in man, the question arises whether the Listeria model is a relevant model at all or, at least, what it is a model for.

Listeria Monocytogenes is A Replicating Antigen to which Man and Mouse Respond in the Same Manner.

Listeria monocytogenes is known as a pathogenic microorganism that causes septicaemia which results in the infection of liver, lung, spleen and central nervous system. All strains virulent in man have been shown to be virulent in mice, too, and in both species corticosteroid therapy or pregnancy increase the susceptibility to infection. Autoptic studies of human Listeriosis (54) demonstrated that, like in the mouse (55), Listeria-induced lesions range from

- "foci of necrosis without any inflammatory cell response to
- micro abscesses with mixed, but primarily polymorphonuclear, inflammatory cell infiltrates to
- granuloma-like lesions composed of histiocytes, monocytes and lymphocytes."

Table 1. Facultative intracellular bacteria, grouped according to the duration of the natural course of infection in experimental animals

---

Listeria

- Yersinia
- Legionella
- Rickettsia
- Salmonella typhi/paratyphi
- Bartonella
- Chlamydia
- Francisella tularensis

- Brucella
- Borrelia
- Treponema

- Nocardia
- Mycobacterium tuberculosis, leprae
- Mycobacterium avium-intracellulare

Since both, the causative agent and the induced responses are similar in mouse and man, it is tempting to speculate that the mechanisms in the "black box" in between are similar, too (Fig. 1). We conclude that the experimental Listeria infection of mice is an excellent model for the study of microbial-induced T cell-mediated inflammatory host responses. However, the eradication of the bacteria from infected organs seems to be not simply a result of macrophage activation. As long as the ultimate protective effector mechanisms are unknown, conclusions concerning protective mechanisms, therefore, should not be extrapolated from the Listeria-model to other infections by facultative intracellular bacteria (Fig. 2).

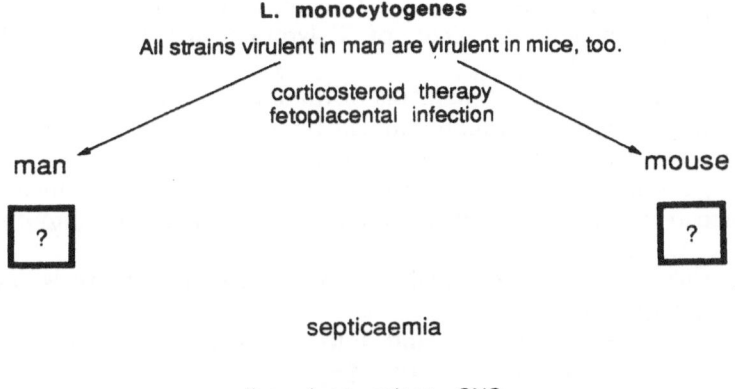

L. monocytogenes

All strains virulent in man are virulent in mice, too.

corticosteroid therapy
fetoplacental infection

man

mouse

?

?

septicaemia

liver   lung   spleen   CNS

Fig. 1.   Is the murine Listeriosis model a relevant model at all?

- mechanisms of bacterial virulence (invasion, intracellular survival, intercellular
      spread)

- induction of an acute-phase-response by gram-pos. bacteria
- recruitment of polymorphonuclear phagocytes

- mechanisms of monocyte-dependent innate resistance mediated by
      - resident macrophages
      - recruitment of monocytes
      - NK cell activity

- induction of specific CD4+, CD8+, and CD4-CD8- T cells
- effector mechanisms of specific T cells
         - protection (macrophage/ monocyte activation? lysis of infected cells?)
         - inflammation (granuloma formation, DTH, splenomegaly)

- longterm T cell memory

Fig. 2.   What is the murine Listeriosis model a model for?

REFERENCES

1. G. B. Mackaness, Cellular resistance to infection. J. Exp. Med. 116:381 (1962).
2. H. Hahn, S. H. E. Kaufmann, The role of cell mediated immunity in bacterial infections. Rev. Inf. Dis. 3:1221 (1981).
3. R. J.North, Immunity of *Listeria monocytogenes*. in: "Immunology of Human Infection", Part 1, p. 201. Nahmias, A.J., O'Reilly, R.J. eds., Plenum, New York (1981).
4. S. H.E.Kaufmann, Immunity against intracellular bacteria: Biological effector functions and antigen specificity of T lymphocytes, Cur. Top. Microbiol. Immunol. 138:141 (1988).
5. M. B. Lurie, Studies on the mechanisms of immunity in tuberculosis: The fate of tubercle bacilli ingested by mononuclear phagocytes derived from normal and immunized animals. J. Exp. Med. 130:247 (1942)
6. J. Ruskin, , McIntosh, H. and J.S. Remington, Studies on the mechanisms of resistance to phylogenetically diverse intracellular organisms. Immunol. 103:252 (1969).
7. G. B. Mackaness, The immunological basis of acquired cellular resistance. J. Exp. Med. 120:105 (1964).
8. R. V. Blanden, G.B. Mackaness, and F.M. Collins, Mechanisms of acquired resistance in mouse typhoid. J. Exp. Med. 124:585 (1966).
9. R. J. North and J.F.Deissler, Nature of memory in T cell-mediated immunity: Cellular parameters that distinguish between the active immune response and a state of memory. Infect. Immun. 12:761 (1975).
10. A. S. Armstrong and C.P. Sword, Cellular resistance in listeriosis. J. Infect. Dis. 114:258 (1964).
11. G. B. Mackaness, The influence of immunologically committed lymphoid cells on macrophage activity *in vivo*. J. Exp. Med. 129:973 (1969).
12. G. B. Mackaness and W.C. Hill, The effect of anti-lymphocyte globulin on cell-mediated resistance to infection. J. Exp. Med. 129:993 (1969).
13. M. C. Raff, Theta isoantigens as a marker of thymus-derived lymphocytes in mice. Nature 224:378 (1969).
14. F. C. Lane and E.R. Unanue, J. Exp. Med. 135:1104 (1972).
15. R. V. Blanden and R.E. Langman, Cell-mediated immunity to bacterial infection in the mouse. Thymus derived cells as effectors of acquired resistance to *Listeria monocytogenes*. Scand. J. Immunol. 1, 379 (1972).
16. R. J. North, Importance of thymus-derived lymphocytes in cell-mediated immunity to infection. Cell. Immunol. 7:166 (1973).
17. S. Youdim, O. Stutman and R.A. Good, Thymus dependency of cells involved in transfer of delayed hypersensitivity to *Listeria monocytogenes* in mice. Cell. Immunol. 8:395 (1973).
18. C. Cheers and R. Waller, Activated macrophages in congenitally athymic nude mice and lethally irradiated mice. J. Immunol. 115, 844 (1975).
19. P. Emmerling, H. Finger and J. Bockemuhl, *Listeria monocytogenes* infection in nude mice. Infect. Immun. 12:437 (1975)
20. D. D. McGregor and F.T. Koster, The mediator of cellular immunity. IV. Cooperation between lymphocytes and mononuclear phagocytes. Cell. Immunol. 2:317 (1971).
21. H. Hahn, Requirements for a bone marrow-derived component in the expression of cell-mediated antibacterial immunity. Infect. Immun. 11:949. (1975).
22. H. Hahn, Effects of dextran sulfate 500 on cell mediated resistance to infection with Listeria monocytogenes in mice. Infect. Immun. 10:1105 (1974).

23. R. M. Zinkernagel, A. Althage, B. Adler, R.V. Blanden, W.F. Davidson, U. Kees, M.B.C. Dunlop and D.C. Shreffler, H-2 restriction of cell-mediated immunity to an intracellular bacterium. Effector T cells are specific for Listeria antigen in association with H-2I region-coded-self-markers. J. Exp. Med. 145, 1353. (1977).

24. A. G. Farr, M.E. Dorf and E.R. Unanue, Secretion of mediators following T lymphocyte/macrophage interaction is regulated by the major histocompatibility complex. Proc. Natl. Acad. Sci. I.S.A. 74:3542. (1977).

25. A. G. Farr, J.-M. Kiely and E.R. Unanue, Macrophage-T cell interactions involving *Listeria monocytogenes* - role of H-2 gene complex. J. Immunol. 122:2395 (1979).

26. A. G. Farr, W.J. Wechter, J.-M. Kiely and E.R. Unanue, Induction of cytocidal macrophages after *in vitro* interactions between Listeria-immune T cells and Macrophages - role of H-2. J. Immunol. 122:2405 (1979).

27. C. Cheers and M.S. Sandrin, Restriction in adoptive transfer of resistance to *Listeria monocytogenes*. II. Use of oncogenic and mutant mice shows transfer to be H-2K restricted. Cell. Immunol. 78:199 (1983).

28. T. W. Jungi, H.W. Kunz, T.J. Gill III and R. Jungi, Genetic control of cell-mediated immunity in the rat. II. Sharing of either the RT6.A or RT1.B locus is sufficient for transfer of antimicrobial resistance. J. Immunogenet. 9:433 (1982).

29. H. Cantor and E.A. Boyse, Functional subclasses of T lymphocytes bearing different Ly antigens. I. The generation of functionally distinct T cell subclasses is a differentiative process independent of antigen. J. Exp. Med. 141:1376 (1975a).

30. H. Cantor and E.A. Boyse, Functional subclasses of T lymphocytes bearing different Ly antigens. II. Cooperation between subclasses of Lyt cells in the generation of killer activity. J. Exp. Med. 141:1390. (1975b).

31. S. Swain, T cells subsets and the recognition of MHC class. Immunol. Rev. 74:129 (1983).

32. D. P. Dialynas, D.B. Wilde, P. Marrack, A. Pierres, K.A. Wall, W. Havran, G. Otten, M.R. Loken, M. Pierres, H. Kappler, and F.W. Fitch, Characterization of the murine antigenic determinant, designated L3T4a, recognized by monoclonal antibody GK1.5: Expression of L3T4a by functional T cell clones appears to correlate primarily with class II MHC antigen-reactivity. Immunol. Rev. 74:29 (1983).

33. H. Naher, U. Sperling, L. Takacs, H. Hahn, Dynamics of T cells of L3T4 and Lyt2 phenotype within granulomas in murine listeriosis. Clin. Exp. Immunol. 60:559 (1985).

34. S. H. E. Kaufmann, E. Hug, U. Vath, and U. Muller, Effective protection against Listeria monocytogenes and delayed-type hypersensitivity lo listerial antigens depend on cooperation between specific L3T4+ and Lyt2+ T cells. Infect. Immun. 48:263 (1985).

35. M. E. A. Mielke, S. Ehlers, H. Hahn, T cell subsets in delayed-type hypersensitivity, protection, and granuloma formation in primary and secondary Listeria infection in mice; superior role of Lyt2+ cells in acquired immunity. Infect. Immun. 56:1920 (1988).

36. For review of literature see: The patterns of cytokines produced by the spleens of naive and CD4+, CD8+, or totally T cell-depleted immunized Listeria-infected mice. This issue.

37. M. Chen-Woan, D.H. Sajewskit and D.D. McGregor, T-cell cooperation in the mediation of acquired resistance to *Listeria monocytogenes*. Immunology 56:33 (1985).

38. M. Chen-Woan, D.D. McGregor and S.K.Noonan, Isolation and characterization of protective T cells induced by *Listeria monocytogenes*. Infect. Immun. 52:401. (1986).

39. S. H. E. Kaufmann and H. Hahn, Biological functions of T cell lines with specificity for the intracellular bacterium *Listeria monocytogenes in vitro* and *in vivo*. J. Exp. Med. 155:1754 (1982).

40. S. H. E. Kaufmann, E. Hug and G. de Libero, *Listeria monocytogenes*-reactive T lymphocyte clones with cytolytic activity against infected target cells. J. Exp. Med. 164:363 (1986).

41. R. M. Stolpmann, H. Naher, H. Osawa, T. Herrmann, H. Hahn and T. Diamanstein, Production of Listeria-specific rat T-cell clones and role of Interleukin-2 receptors in regulation of Listera-dependent T-cell clone growth *in vitro*. Infect. Immun. 47:822 (1985).

42. S. H.E. Kaufmann, H. Hahn, R. Berger, and H. Kirchner, Interferon-production by *Listeria monocytogenes* specific T cells active in cellular antibacterial immunity. Eur. J. Immunol. 13:265 (1983).

43. U. Sperling, S.H.E. Kaufmann and H. Hahn, Production of macrophage-activating and migration inhibition factors *in vitro* by serologically selected and cloned *Listeria monocytogenes*-specific T cells of the Lytl + 2 phenotype. Infect. Immun. 46:111 (1984).

44. R. M. Stolpmann, U. Sperling, and H. Hahn, Characterization of three different rat T-cell clones with specificity to *Listeria monocytogenes*: Phenotype, specific proliferation, lymphokine production, and protective capacity in *vivo*. Cell. Immunol. 101:548 (1986).

45. S. H.E. Kaufmann, E. Hug, U. Vath and G. de Libero, Specific lysis of *Listeria monocytogenes*-infected macrophages by class II-restricted L3T4 + T cells. Eur. J. Immunol. 17:237 (1987).

46. S. P. Cobbold, A. Jayasuriya, A. Nash, T.D. Prospero and H. Waldmann, Therapy with monoclonal antibodies by elimination of T-cells subsets *in vivo*. Nature 312:548 (1984).

47. M. E.A. Mielke, G. Niedobitek, H. Stein, H. Hahn, Acquired resistance to *Listeria monocytogenes* is mediated by Lyt2 + T cells independently of the influx of monocytes into granulomatous lesions. J. Exp. Med. 170:589 (1989).

48. M. F. Newberg, R.J. North, On the mechanisms of T cell-independent anti-Listeria resistance in nude mice. J. Immunol. 124:571 (1980).

49. G. J. Bancroft, K.C.F. Sheehan, R.D. Schreiber, E.R. Unanue, Tumor necrosis factor is involved in the T cell-independent pathway of macrophage activation in SCID mice. J. Immunol. 143:127 (1989).

50. H. Rosen, S. Gordon, Monoclonal antibody to the murine type 3 complement receptor inhibits adhesion of myelomonocytic cells and inflammatory cell recruitment *in vivo*. J. Exp. Med. 166:1685 (1987).

51. G. Rosen, G. Milon, S. Gordon, Antibody to the murine type 3 complement receptor inhibits Z lymphoyte dependent recruitment of myelomonocytic cells *in vivo*. J. Exp. Med. 169:535 (1989).

52. H. Rosen, S. Gordon, R.J. North, Exacerbation of murine listeriosis by a monoclonal antibody specific for the type 3 complement receptor of myelo-monocytic cells. J. Exp. Med. 170:27 (1989).

53. M. E.A. Mielke, H. Rosen, S. Borche, C. Peters, H. Hahn, Listeria-induced T cell-monocyte interactions in protective immunity and granuloma formation depend on two distinct but TNF- and a-IFN-mediated mechanisms. Dissociation on the basis of CDllb-dependency. Submitted for publication.

54. E. C. Klatt, Z. Pavlova, A.J. Teberg, M.L. Jonekura, Epidemic perinatal Listeriosis at autopsy. Hum. Pathol. 17:1278 (1986).

55. M. E.A. Mielke, T cell subsets in granulomatous inflammation and immunity to *L. monocytogenes* and *B. abortus*. <u>Behring Inst. Mitt.</u> 88:99 (1991).

ANTIGEN SPECIFIC SUPPRESSOR T CELLS RESPOND

TO CYTOKINES RELEASED BY T CELLS

Christopher E. Taylor

Laboratory of Immunogenetics
National Institute of Allergy and Infectious Diseases
National Institutes of Health
Twinbrook II Research Facility
12441 Parklawn Drive
Rockville, Maryland 20852

INTRODUCTION

Previous studies have demonstrated that suppressor T (Ts) cells are capable of down-regulating the antibody response to pneumococcal polysaccharide Type III (SSS-III) (1-11), by limiting the expansion of antigen stimulated B lymphocytes. Such studies have also shown that during the course of a normal antibody response both Ts cells and amplifier T cells are activated not by antigen but by cell associated antibody on the surface of immune B cells (10, 12, 13,). However, the critical events that occur during activation as well as the events following activation, such as the clonal expansion of precursor T cells leading to fully functional Ts cells, have not yet been examined in sufficient detail. It is not clear which growth factors are required either for the activation of resting Ts cells or for the expansion of these cells. Therefore, studies were conducted to examine the cytokine requirement for the functional activity of Ts cells.

MATERIALS AND METHODS

Mice

Female BALB/cByJ mice (8 to 10 weeks old), were obtained from the Jackson Laboratory, Bar Harbor, Maine, and used in the experiments to be described. Mice were maintained in accordance with the guidelines established by the National Institutes of Health, Animal Care Committee.

Antigens and Immunization Procedure

The immunological properties of the preparation of Type III pneumococcal polysaccharide (SSS-III) and dextran B-1355 used and the method by which they were prepared have been described (2, 10, 14). For immunization, mice were given a single intraperitoneal (i.p.) injection of an optimally immunogenic dose (0.5 μg) of SSS-III in 0.5ml of saline. The magnitude of the antibody response produced was determined 5 days after immunization.

Immunological Methods

Numbers of plaque-forming cells (PFCs) making antibody specific for SSS-III were detected by means of a well-established slide version of the technique of localized hemolysis-in-gel, 5 days after immunization (i.p.) with SSS-III, i.e., at the peak of the antibody response (14-17). Here PFC's making antibody of the immunoglobulin M (IgM) class (≥ 90% of all PFC found) were detected using indicator sheep erythrocytes (SRBC) coated with SSS-III by the $CrCl_3$ method (14-17). Polyethylene glycol (average molecular weight 6,000 to 7,500; J.T. Baker Chemical Co., Phillipsburg, N.J.) was added to the reaction mixture (melted agarose) at a final concentration of 0.25% (wt/vol) to improve the quality of the plaques obtained. Corrections were made (by subtraction) for the small number of background SRBC-specific PFCs present, so that only values for PFC making antibody specific for SSS-III (SSS-III-specific PFC) are considered in this work. The values obtained which are log normally distributed (18) are expressed as the geometric mean (anti-log) of the log_{10} number of PFCs per spleen for groups of similarly treated mice. This provides a reasonably good measure of the magnitude of the total antibody response produced, since SSS-III-specific PFC are detected only in the spleens of immunized mice (1, 16, 19) and the magnitude of the PFC response produced is directly related to the serum antibody titer (1, 16, 19).

Student's t test was used to assess the significance of the differences observed. Differences were considered to be significant when probability (P) values of <0.05 were obtained.

Preparation and Treatment of Spleen Cells

Spleen cells were obtained from donor mice 16-24 hr after prior exposure (priming) of the animals (see below) to a subimmunogenic dose (0.005 μg) of SSS-III; they were resuspended at a density of 10^8 cells per ml in Medium 199. Then, 2.5ml aliquots were incubated in the presence of each lymphokine for 30 min at 4°C. In all cases spleen cells were washed three times with Medium 199 after lymphokine treatment and then resuspended at appropriate densities for injection (i.v.) into recipient mice. Recipient mice were immunized (i.p.) simultaneously with 0.5 μg SSS-III. Previous studies have consistently shown, that spleen cells harvested 16-24 hr after priming with SSS-III are rich in Ts activity [3-5, 20-22]. Suppression can be transferred with at least $20x10^6$ whole spleen cells or with 10^5 purified T cells recovered from plates treated with monophosphoryl lipid A or MPL (4). We have observed in the present studies as well as in past (4) that $5x10^6$ or $2x10^6$ primed spleen cells not treated with rIL2 do not transfer significant suppression.

Lymphokines and Antibodies

Recombinant IL2, recombinant IL4, and recombinant IL5 (rIL2, rIL4, rIL5, respectively) and recombinant interferon gamma (IFN gamma) were obtained from Genzyme Corporation, Boston, MA. Human recombinant IL6 (rIL6) was a generous gift of Dr. Jin Kim. (Genentech Inc., San Francisco, CA). Anti-IL-2 receptor antibody (aIL2R) was obtained from Boehringer Mannheim Biochemicals (Indianapolis, IN).

Table 1. Effect of rIL2 on the Antibody Response to SSS-III

Amount of [a] rIL2	SSS-III-Specific[b] PFC/Spleen	p Value[c]
None	4.202 ± 0.092 (15,928) n = 8	
2x10³ units	4.247 ± 0.094 (17,667) n = 9	p > 0.05
2x10⁴ units	3.990 ± 0.109 (9,767) n = 9	p < 0.05
2x10⁵ units	3.800 ± 0.130 (6,302) n = 9	p < 0.05

[a] Recombinant IL2 (rIL2) was given, i.v., at the time of immunization with 0.5μg SSS-III (i.p.).

[b] $\log_{10}$ SSS-III-specific PFC/Spleen ± SEM for groups of [n] mice, 5 days after immunization with SSS-III; geometric means (antilogs) are in parentheses.

[c] Probability (P) values relative to controls immunized with SSS-III but not given rIL2.

Table 2. Effect of Interferon (IFN) Gamma on the Antibody Response to SSS-III

Amount (U) of[a] rIFN gamma	$\log_{10}$ SSS-III-Specific[b] PFC/Spleen ± SEM	p Value[c]
None	4.298 ± 0.098 (19,854) n = 8	
1 x 10⁴	3.851 ± 0.108 (7,097) n = 10	p < 0.01
1 x 10³	3.783 ± 0.104 (6,061) n = 10	p < 0.01
200	4.266 ± 0.054 (18,455) n = 10	p > 0.05

[a] Recombinant IFN (rIFN) gamma was given (i.v.) at the time of immunization with 0.5 ug SSS-III.

[b] Measured 5 days after immunization with SSS-III; geometric means (antilogs) are in parentheses.

[c] Probability (P) values relative to controls immunized with SSS-III but not given rIFN gamma.

Table 3. Effect of Interleukins 5 or 6 on the Antibody Response to SSS-III

Amount (U) of[a] rIL5, rIL6	Log$_{10}$ SSS-III-Specific[b] PFC/Spleen ± SEM	p Value[c]
None	3.970 ± 0.070 (9,333) n = 8	
1 x 10^3 (rIL5)	3.481 ± 0.068 (3,029) n = 9	p < 0.001
1.5 x 10^4 (rIL5)	3.524 ± 0.115 (3,343) n = 9	p < 0.01
1 x 10^3 (rIL6)	4.257 ± 0.074 (18,075) n = 9	p < 0.05
1 x 10^4 (rIL6)	4.140 ± 0.082 (13,798) n = 9	p < 0.05

[a] Mice were given rIL5 or rIL6, i.v., at the time of immunization with 0.5 ug SSS-III.

[b] Measured 5 days after immunization with SSS-III; geometric means (antilogs) are in parentheses.

[c] Probability (P) values relative to controls immunized with SSS-III but not given rIFN gamma.

RESULTS

Effect of Selected Recombinant Cytokines on the Antibody Response to SSS-III

We first examined the effects of selected cytokines on the antibody response to SSS-III, *in vivo*. Mice were given different amounts of each cytokine at the time of immunization with SSS-III. Five days following immunization the antibody response produced was determined and compared to that of immunized mice not given cytokine. The data show that the administration of 2 x 10^4 or 2 x 10^5 units of rIL2 (Table 1), 1 x 10^3 or 1 x 10^4 units of IFN gamma (Table 2), 1 x 10^3 or 1 x 10^4 units of rIL5 (Table 3) resulted in significant (p < 0.05) suppression of the antibody response relative to controls not receiving cytokines. The injection of other doses of the aforementioned cytokines did not cause any significant (p > 0.05) alteration of the antibody response to SSS-III. On the other hand, the injection of 1 x 10^3 or 1 x 10^4 units of rIL6 caused a significant (p < 0.05) enhancement of the antibody response to SSS-III (Table 3).

Previous studies have shown that the administration of a subimmunogenic dose of SSS-III leads to a state of immunological unresponsiveness known to be mediated by antigen specific Ts cells (2, 3, 7). In view of the results obtained with rIL2 in the preceding experiment, we conducted an experiment to examine whether one can inhibit the generation of Ts cells *in vivo* by treatment with monoclonal antibody specific for the IL2 receptor (anti-IL2R antibody). Here, two groups of animals were primed with 0.005 μg SSS-III and a control group received saline. One of the primed groups received anti-IL2R antibody at 24, 48 and 72 hr after priming. Then, all animals including the control group, were immunized with an optimally immunogenic dose of SSS-III (i.e. 0.5 μg SSS-III), at 72 hr after priming with 0.005 μg SSS-III. Five days after immunization the antibody response was assessed. The results (Table 4) show that, mice primed with 0.005 μg SSS-III alone gave a significantly (p < 0.001) lower antibody response compared with control animals, as expected. On the other hand, the antibody response of mice primed with SSS-III and then treated with anti-IL2R antibody did not differ significantly (p > 0.05) from the controls.

Effect of rIL2 on the Ability of Spleen Cells, from Primed Animals, to Transfer Suppression

Although the administration of rIL2 significantly decreased the antibody response to SSS-III *in vivo* (Table 1), one could argue that such suppression might be due to a direct effect of rIL2 on the function of antigen stimulated B cells or that rIL2 may act to increase the activity of suppressor T (Ts) cells, which are known to be activated early during the course of a normal antibody response to SSS-III (1-3, 8, 9, 20), and that the Ts cells in turn will cause suppression of the clonal expansion of antibody producing cells. Consequently, experiments were conducted to examine whether rIL2 can alter the capacity of spleen cells from primed animals (with enriched Ts cell activity) to transfer suppression. Several studies have shown that the transfer of substantial suppression of the antibody response to SSS-III requires approximately $10-20 \times 10^6$ spleen cells from primed animals and that the transfer of fewer cells ($< 10 \times 10^6$ spleen cells) is usually without effect (3, 4, 9). Such results were interpreted as indicating that the frequency of SSS-III specific Ts cells in the spleen is low and that it requires at least 10×10^6 spleen cells to transfer suppression. Therefore, the ability to transfer suppression with 10^5 or less spleen cells from primed animals, as demonstrated in other studies, for instance, using MPL adherent primed spleen cells is indicative of an enrichment in Ts cell activity (4). Thus, we wanted to determine whether one could enrich for Ts cell activity by rIL2 treatment *in vitro*.

Table 4. Effect of Anti-IL2 Receptor Antibody on the Expression of Low Dose Paralysis to SSS-III

Pretreatment of Donors SSS-III[a]	Anti-IL2R Ab.[b]	SSS-III-Specific[c] PFC/Spleen	p Value[d]
−	−	3.960 ± 0.056 (9128) n = 10	
+	−	3.298 ± 0.106 (1,988) n = 9	p < 0.001
+	+	3.826 ± 0.155 (6,698) n = 7	p > 0.05

[a] Mice were primed (i.p.) with 0.005 µg of SSS-III, 72 hrs before immunization (i.p.) with 0.5 µg SSS-III.

[b] One group of animals received (i.p.) monoclonal anti-IL2 receptor antibody (anti-IL2R Ab.) (4µg/mouse) at 24, 48 and 72 hrs after priming with 0.005 µg SSS-III.

[c] Log_{10} SSS-III-specific PFC/Spleen ± SEM for groups of (n) mice 5 days after immunization with 0.5 µg SSS-III; geometric means (antilogs) are in parentheses.

[d] Probability (P) values relative to control mice immunized with SSS-III, but not pretreated with SSS-III or anti-IL2R Ab.

In this experiment spleen cells from donors primed 16-24 hr earlier with 0.005 µg SSS-III were treated *in vitro* with rIL2 before cell transfer. Again, all mice were immunized simultaneously with 0.5 µg SSS-III at the time of cell transfer and assayed for numbers of PFCs, 5 days after immunization. The results (Table 5) show that, as expected, the transfer of 20×10^6 primed spleen cells not treated with rIL2 caused a significant (p < 0.001) suppression of the antibody response to SSS-III relative to controls not receiving transferred cells. Although significant (p < 0.001) suppression of antibody response was noted in mice that received 2×10^6 or 2×10^5 spleen cells treated with rIL2 (Table 5), the transfer of 2×10^6 primed cells was without effect in the absence of rIL2 treatment. Significant suppression (p < 0.05) of the antibody response was also noted using 2×10^6 cells when the cells were treated with IL2 concentrations of 250, 500 or 1000 units.

Spleen Cells from Naive Animals or from Dextran Primed Mice Do Not Cause Suppression after rIL2 Treatment

It was possible that portions of the rIL2 used for the treatment of spleen cells *in vitro* may be non-specifically adsorbed to the donor cells. Thus, the following experiments were done to obtain information relevant to this issue. In the first case spleen cells were obtained from either SSS-III primed donors or from naive donors (given an equivalent volume of saline) and treated separately *in vitro* with rIL2. Then,

Table 5. Effect of rIL2 on the Ability of Primed Spleen Cells to Transfer Suppression

Expt. #	Treatment of[a] Spleen Cells with rIL2	SSS-III-Specific[b] PFC/Spleen	p Value[c]
1	NCT[d]	4.333 ± 0.017 (21,542) n = 8	
	20x10⁶ cells (No treatment)	3.966 ± 0.057 (9,254) n = 8	< 0.001
	2x10⁶ cells (10⁴ IL2 units)	3.910 ± 0.013 (8,122) n = 7	< 0.01
	2x10⁵ cells (10⁴ IL2 units)	4.036 ± 0.073 (10,855) n = 8	< 0.01
2	NCT[d]	4.275 ± 0.082 (18,821) n = 10	
	10x10⁶ cells (No treatment)	4.005 ± 0.099 (10,115) n = 10	p < 0.05
	2x10⁶ cells (No treatment)	4.299 ± 0.038 (19,907) n = 10	p > 0.05

[a] Spleen cells were obtained from donor mice primed earlier with 0.005µg SSS-III for 16-24 hrs; in some cases the cells were treated with rIL2 for 30 min at 4°C. Then, spleen cells were injected (i.v.) into recipient mice, and immunized simultaneously (i.p.) with 0.5 µg SSS-III.

[b] Log₁₀ PFC/Spleen ± SEM for groups of [n] mice, 5 days after immunization with SSS-III; geometric means (antilogs) are in parentheses.

[c] Probability (P) values based on comparisons with control mice immunized with SSS-III, but not given donor cells.

[d] NCT - no cells transferred.

2 x 10^6 cells in each case were transferred to recipient mice at the time of immunization with 0.5 µg SSS-III. The results show (Table 6) that whereas rIL2 treated spleen cells from primed donors caused a significant (p < 0.001) suppression of the antibody response, rIL2 treated cells from naive mice did not cause suppression of the antibody response to SSS-III when compared to controls not receiving transferred cells. These experiments suggest that the transfer of suppression of antibody response using smaller numbers of rIL2 treated spleen cells is not due to residual IL2 on the transferred cells. In a separate experiment donor mice were primed with either 0.005 µg SSS-III or

Table 6. Specificity of Transfer of Suppression

Expt. No.	Treatment of[a] Transferred Cells	Treatment of Donor Mice	SSS-III-Specific[b] PFC/Spleen	p Value[c]
#1	NCT[d]		4.319 ± 0.037 (20,838) n = 10	
	500u IL2	SSS-III	3.973 ± 0.040 (9,390) n = 9	p < 0.001
	250u IL2	SSS-III	3.909 ± 0.073 (8,107) n = 9	p < 0.001
	250u IL2	Saline	4.257 ± 0.031 (18,057) n = 9	p > 0.05
#2	NCT[(c)]		3.963 ± 0.099 (9,183) n = 10	
	500u IL2	SSS-III	3.702 ± 0.053 (5,034) n = 9	p < 0.05
	500u IL2	Dextran	3.970 ± 0.121 (9,331) n = 10	p > 0.05

[a] Spleen cells were obtained 16-24 hrs after priming donor mice with 0.005 µg of SSS-III, 0.005 µg dextran or saline. Spleen cells were treated with rIL2 for 30 min at 4°C. Then, 2x10^6 cells were injected, (i.v.), into recipient mice and immunized simultaneously with 0.5µg SSS-III.

[b] Log_{10} SSS-III-specific PFC/Spleen ± SEM for groups of [n] mice, 5 days after immunization with SSS-III; geometric means (antilogs) are in parentheses.

[c] Probability (P) values based on comparisons with mice immunized with SSS-III, but not given primed spleen cells.

[d] NCT - no cells transferred.

0.005 µg of dextran (a dose that has been shown to generate dextran specific Ts cells). Then, as above, 2 x 10^6 cells were transferred to recipient mice at the time of immunization with 0.5 µg SSS-III. The results show (Table 6) that while the antibody response of mice receiving rIL2-treated spleen cells from SSS-III primed donors was lower relative to controls, the response of mice receiving dextran primed spleen cells was not altered. Thus, these experiments indicate that the effect of rIL2 on the transfer of SSS-III-primed spleen cells is not due to nonspecific adsorption of rIL2 or its elution after cell transfer (i.e. carry-over effects).

DISCUSSION

The exact mechanisms by which suppressor T (Ts) cells, involved in the regulation of the antibody response to pneumococcal polysaccharide type III (SSS-III), are activated and expanded remain to be elucidated. In the present studies we examined the effects of a few selected cytokines on the antibody response *in vivo* and the effects of rIL2 on the ability of Ts cells to transfer suppression of the antibody response to recipient mice immunized with SSS-III.

We found that the *in vivo* administration of rIL2, rIL5, or rIFN gamma at the time of immunization resulted in a substantial suppression of the antibody response to SSS-III (Table 1). It is conceivable that the suppression of the antibody response observed could be the result of either the direct interaction of each cytokine with antigen stimulated B cells or by an indirect effect mediated by Ts cells whose activity has been enhanced following cytokine treatment. Indeed, prior studies show that Ts cells are activated early (5-24 hr after immunization) during the course of a normal antibody response to SSS-III (3, 5, 9); thus, it is possible that the administration of these cytokines at the time of immunization, as was done in this work, (Table 1, 2 & 3) may have resulted in the expansion of a population of activated Ts cells. It is possible that IL6 on the other hand may act directly on antigen stimulated B cells providing the necessary signals and ultimately leading to increased antibody production.

Earlier studies done in this laboratory show that one can induce an antigen-specific form of unresponsiveness, known to be mediated by Ts cells, by low dose priming (i.e. prior exposure to a subimmunogenic dose of 0.005 µg SSS-III). Indeed, the administration of anti-IL2 receptor antibody at the time of low dose priming with SSS-III, resulted in a partial reversal of the reduced antibody response that is normally expected (Table 4). In effect, this experiment (Table 4) is consistent with previous observation that mitotic inhibitors abrogate the generation of Ts cells *in vivo* after low dose priming (8). The partial reversal of the low dose priming effect by anti-IL2 receptor antibody may suggest that other lymphokines are required for the induction of full expression of Ts cell activity *in vivo*. In fact the suppression of the antibody response to SSS-III by rIL5 or rIFN gamma (Tables 2 & 3) lend support to this concept.

To examine whether rIL2 can act directly on Ts cells, we tested the effects of rIL2 on the capacity of spleen cells from SSS-III primed mice to transfer suppression of the antibody response to SSS-III. We found that the transfer of suppression with rIL2-treated spleen cells could be carried out using 100-fold fewer primed spleen cells compared to cell transfer experiments using primed spleen cells not treated with rIL2 (Table 5); this indicates a substantial increase in Ts cell activity. In earlier studies we observed that because of the low frequency of putative Ts cells one needs approximately $10\text{-}20 \times 10^6$ spleen cells from SSS-III primed mice to transfer suppression. In fact, only under conditions of enrichment, such as using monophosphoryl lipid A - adherent T cells was it possible to transfer suppression using such small numbers of cells (4, 5). These studies therefore suggest that precursor Ts cells once activated express IL2 receptors, that are capable of binding available IL2; this results in clonal expansion of precursor Ts cells as reflected by an increase in Ts cell activity. Because it was also possible that the added IL2 might be non-specifically adsorbed to normal or activated spleen cells during the *in vitro* treatment, we also treated spleen cells from naive animals or mice primed with a non-cross-reactive antigen (dextran) with rIL2. The results show that the transfer of similar numbers of such rIL2 treated spleen cells did not suppress the antibody response to SSS-III in recipient animals (Table 6). In other studies (21) we found that treatment with rIL4, rIL5, or rIFN gamma also resulted in a significant increase in Ts cell activity; however, treatment with rIL6 was without effect. Although some of these lymphokines appear to influence the development of Ts cells from inactive precursors their role in the induction of Ts cells remain to be determined.

Experiments are in progress to determine which lymphokines are produced by cloned Ta or Ts cells and how these lymphokines affect the overall antibody response to SSS-III.

In other experimental systems it has been shown that IL2 can modulate Ts cell activity (22-26). For instance, it has been demonstrated that T cells capable of inhibiting CTL activity can adsorb IL2 from preparations containing IL2 activity (25). Also, it has been demonstrated that high concentrations of IL2 can induce non-specific Ts cells in mixed lymphocyte reactions (23). Furthermore, it was found that there was a deficient Ts cell activity in old animals and that the addition of exogenous IL2 restored functional suppressor T cell activity (24). In view of the above studies and the experiments reported here, it is clear that the ability of Ts cells to respond to IL2 has important clinical implications in IL2 based immunotherapeutic procedures. In some of these procedures IL2 is used *in vitro* to expand lymphokine activated killer (LAK) cells or tumor infiltrating lymphocytes (TIL); in other instances, IL2 is given *in vivo* with other lymphokines to boost the immune response (20). Recently, it was observed that a considerable improvement in the immune response was possible when cyclophosphamide was used to eliminate Ts cell activity (27). Studies conducted in our laboratory show that Ts activity can be removed from populations of spleen cells by adherence to MPL-coated plates. Thus, it seems likely that experimental approaches to expand the pool of LAK or TIL cells might be improved if suspensions containing such cells are first treated to remove Ts cell activity which can also expand in the presence of IL2.

SUMMARY

These studies were conducted to examine the cytokine requirement for clonal expansion of regulatory T cells. It was observed that the *in vivo* administration of recombinant IL2 (rIL2), rIL5 or interferon (IFN) gamma at the time of immunization with pneumococcal polysaccharide Type III (SSS-III) resulted in substantial suppression of the antibody response in each case. Using our well established cell transfer system we found that such suppression of the antibody response could be transferred using 10-100-fold fewer primed spleen cells providing these cells were treated *in vitro* with rIL2 before cell transfer; spleen cells from unimmunized mice or from mice primed with an unrelated antigen and then treated with rIL2 did not cause suppression of the antibody response to SSS-III, thereby eliminating the possibility of non-specific carry-over effects induced by rIL2. We also found that the *in vivo* administration of anti-IL2 receptor antibody inhibited the generation of Ts cells *in vivo*.

Spleen cells from SSS-III primed animals treated with rIL4, rIL5 and IFN gamma - but not rIL6 - likewise are able to transfer suppression of the antibody response with fewer cells than that required using primed cells not treated with cytokines. Thus, these studies indicate that Ts cell activity is greatly influenced by cytokines. The studies also suggest that these cytokines may be required during the activation and/or clonal expansion of Ts cells.

REFERENCES

1. P. J. Baker, Regulation of the magnitude of the antibody response to bacterial polysaccharide antigens by thymus-derived lymphocytes, Infect. Immun. (in press) (1990).
2. P. J. Baker, D. F. Amsbaugh, P. W. Stashak, G. Caldes, and B. Prescott, Regulation of the antibody response to Type III pneumococcal polysaccharide, Rev. Infect. Dis. 3:332 (1981).

3. P. J. Baker, D. F. Amsbaugh, P. W. Stashak, G. Caldes, and B. Prescott, Direct evidence for the involvement of thymus-derived (T) suppressor cells in the expression of low-dose paralysis to Type III pneumococcal polysaccharide, J. Immunol. 128:1059 (1982).

4. P. J. Baker, K. R. Hasløv, M. B. Fauntleroy, P. W. Stashak, K. Myers, and J.T. Ulrich, Enrichment of suppressor T cell activity by means of binding to monophosphoryl lipid A, Infect. Immun. 58:726 (1990).

5. P. J. Baker, J. R. Hiernaux, M. B. Fauntleroy, B. Prescott, J. L. Cantrell, and J.A. Rudbach, Inactivation of suppressor T cell activity by nontoxic monophosphoryl lipid A. Infect. Immun. 56:1076 (1988).

6. P. J. Baker, B. Prescott, P. W. Stashak, and D. F. Amsbaugh, Characterization of the antibody response to Type III pneumococcal polysaccharide at the cellular level. III. Studies on the average avidity of the antibody produced by specific plaque-forming cells, J. Immunol. 107:719 (1971).

7. P. J. Baker, N. D. Reed, P. W. Stashak, D. F. Amsbaugh, and B. Prescott, Regulation of the antibody response to Type III pneumococcal polysaccharide. I. Nature of regulatory T cells, J. Exp. Med. 137:1431 (1973).

8. P. W. Stashak, C. E. Taylor, G. Caldes, B. Prescott, and P. J. Baker, Cyclic expression of low-dose paralysis, Cell. Immunol. 77:143 (1983).

9. C. E. Taylor, D. F. Amsbaugh, P. W. Stashak, G. Caldes, B. Prescott, and P. J. Baker, Cell surface antigens and other characteristics of T cells regulating the antibody response to type III pneumococcal polysaccharide, J. Immunol. 130:19 (1983).

10. C. E. Taylor, P. W. Stashak, G. Caldes, B. Prescott, T. E. Chused, A. Brooks, and P. J. Baker, Activation of antigen-specific suppressor T cells by B cells from mice immunized with Type III pneumococcal polysaccharide, J. Exp. Med. 158:703 (1983).

11. C. E. Taylor, P. W. Stashall, G. Caldes, B. Prescott B. J. Fowlkes, and P. J. Baker, Lectin induced modulation of the antibody response to type III pneumococcal polysaccharide, Cell. Immunol. 83:26 (1984).

12. P. J. Baker, C. E. Taylor, M. B. Fauntleroy, P. W. Stashak, and B. Prescott, The role of antigen in the activation of regulatory T cells by immune B cells, Cell. Immunol. 96:376 (1985).

13. K. L. Elkins, P. W. Stashak, and P. J. Baker, Metabolic activity is necessary for activation of T suppressor cells by B cells, J. Immunol. 144:2859 (1990).

14. P. J. Baker, P. W. Stashak, and B. Prescott, The use of erythrocytes sensitized with purified pneumococcal polysaccharides for the assay of antibody and antibody-producing cells, Appl. Microbiol. 17:422 (1969).

15. P. J. Baker and P. W. Stashak, Quantitative and qualitative studies on the primary antibody response to pneumococcal polysaccharide at the cellular level, J. Immunol. 103:1342 (1969).

16. P. J. Baker, P. W. Stashak, D. F. Amsbaugh, and B. Prescott, Characterization of the antibody response to Type III pneumococcal polysaccharide at the cellular level. I. Dose-response studies and the effect of prior immunization on the magnitude of the antibody response, Immunology 20:469 (1971).

17. P. J. Baker, P. W. Stashak, D. F. Amsbaugh, and B. Prescott, Characterization of the antibody response to Type III pneumococcal polysaccharide at the cellular level. II. Studies on the relative rate of antibody synthesis and release by antibody-producing cells, Immunology 20:481 (1971).

18. C. F. Gottlieb, Applications of transformation to normalize the distribution of plaque-forming cells, J. Immunol. 113:51 (1974).

19. J. M. Jones, D. F. Amsbaugh, P.W. Stashak, B. Prescott, P. J. Baker, and D. W. Alling, Kinetics of the antibody response to Type III pneumococcal polysaccharide. III. Evidence that suppressor cells function by inhibition of the recruitment and proliferation of antibody-producing cells, J. Immunol. 116:647 (1976).

20. M. S. Pulley, V. Nagendran, J. M. Edwards, and D. C. Dumonde, Intravenous, intralesional and endolymphatic administration of lymphokines in human cancer, Lymphokine Res. 5:157 (1986).

21. C. E. Taylor, M. B. Fauntleroy, P. W. Stashak, and P. J. Baker, Antigen specific suppressor T cells respond to recombinant interleukin 2 and other lymphokines, Infect. Immunol. 59:575-579 (1991).

22. J. H. Holda, T. Varies and H. N. Claman, Natural suppressor activity in graft-vs-host spleen and normal bone marrow is augmented by IL2 and interferon-gamma, J. Immunol. 137:3538 (1986).

23. T. Oh-Ishi, C. K. Goldman, J. Mishti and T. A. Waldmann, The interaction of interleukin 2 with its receptor in the generation of suppressor T cells in antigen-specific and antigen-nonspecific systems in vitro, Clin. Immunol. Immunopath. 52:447 (1989).

24. M. L. Thoman and W. O. Weigle, Deficiency in suppressor T cell activity in aged animals. Reconstitution of this activity by interleukin 2, J. Exp. Med. 157:2184 (1983).

25. C. C. Ting, S. S. Yang and M. E. Hargrove, Induction of suppressor T cells by interleukin 2, J. Immunol. 133:261 1984.

26. H. D. Volk and T. Diamantstein, IL-2 normalizes defective suppressor T cell function of patients with systemic lupus erythematosus in vitro, Clin. Exp. Immunol. 66:525 (1986).

27. M. S. Mitchell, M. S., R. A. Kempt, W. Harel, H. Shau, W. D. Boswell, S. Lind, and E. C. Bradley, Effectiveness and tolerability of low dose cyclophosphamide and low dose intravenous interleukin-2 disseminated melanoma, J. Clin. Oncol. 6:409 (1988).

K. L. Elkins, P. W. Stashak, and P. J. Baker, Transferred B cells from auto-immune NZB/N mice fail to activate T suppressor cells, Cell. Immunol. 110:14 (1987).

"INVERSE" EFFECTS OF CORTISONE IN EXPERIMENTAL

INFECTION OF MICE

Gunther Gillissen

Department of Medical Microbiology
Faculty of Medicine
University of Aachen
Federal Republic of Germany

INTRODUCTION

Hydrocortisone (HC) depresses many immune response parameters (1). However, adrenalectomy is followed by a failure of immune response against Meth A, a sarcoma in mice which is abrogated by administration of corticosterone (2). We examined, therefore, whether HC has a beneficial effect under certain conditions in experimental infections in mice.

MATERIALS AND METHODS

Inbred male BALB/cABOM mice (Bomholdgárd, Ry, Denmark) were used for all assays. Experimental infection (n = 30 per group) was accomplished in case of *E. coli* ATCC 8739 by intraperitoneal injection of 3×10^8 carefully washed bacteria per 0.5 ml saline (approximately ID75), taken from the logarithmical growth phase. In case of *Staphylococcus aureus* DSM 349, either 1×10^9 bacteria per 0.25 ml saline were given intravenously or 8×10^8 per 0.5 ml saline intraperitoneally. LPS 055:B5 (Difco) was injected in doses of 20 µg per 0.5 ml saline intraperitoneally or intravenously. Hydrocortisone "Hoechst"[R] crystalline suspension of 1 mg per 0.5 ml, was injected in equal parts into both hind legs from 1 day before infection or sensitization, respectively. Footpad swelling reaction (FPSR)) as a model for cellular immunity (3) with sheep red blood cells as antigen was performed under optimal conditions (4). The mean (n = 5) of the difference in skin thickness when evaluated with ODI-test OOT (Kroeplin, Schlüchtern, FRG) before and 24 hr after challenge was taken for statistical comparison of groups with Student's t-Test.

RESULTS

HC and Infection with *E. coli*

HC treatment 24 hr before intraperitoneal infection with *E. coli* reduced mortality rate from 80 to 40 per cent or in an identical assay from 60 to 40 percent (Figure 1).

Microbial Infections, Edited by H. Friedman *et al.*
Plenum Press, New York, 1992

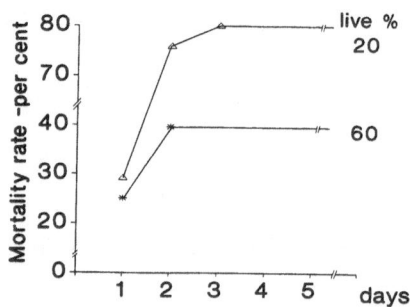

Fig. 1. Experimental infection with *E. coli* ATCC 8739 (3×10^8) injected intraperitoneally.
Δ = control; ∗ = HC-1 mg 24 hr before infection; n = 30 per group.

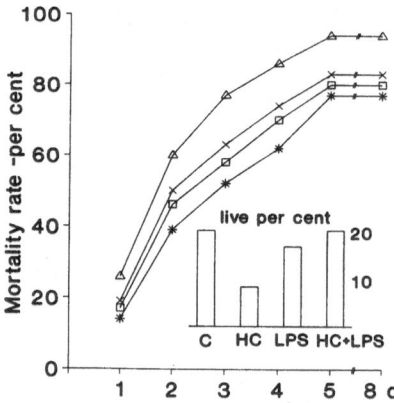

Fig. 2. Experimental infection with *Staph. aureus* DSM 349 (1×10^9) injected intravenously.
□ = control; Δ = HC-1 mg 24 hr before infection; x = LPS 055:B_5-20 µg injected intraperitoneally after infection; ∗ = treatment with HC + LPS; n = 30 per group.

HC, Endotoxin and Infection with Staphylococci

In order to elucidate the significance of LPS in this test system similar experiments were performed with Staphylococci and endotoxin under HC treatment (Figure 2). If bacteria were injected intravenously, HC distinctly increased mortality rate which was, however, abolished by a concomitant intraperitoneal application of endotoxin at the time of infection. Endotoxin alone increased slightly the mortality rate but was reduced by an additional HC-treatment. The combination of HC and endotoxin led, thus, to better results than the application of either substance. Comparable effects could also be observed in cases where all animals died. Here the effect of different treatments was characterized by a prolongation to reach the end point of 100 per cent dead, namely after 5, 6 and 8 days with HC or endotoxin alone and with HC + LPS respectively. In further experiments, however, using the same experimental model but changing the mode of application of infecting bacteria and of endotoxin (Table 1), HC or endotoxin always increased the mortality rate where the combined treatment was most effective in this respect.

Effect of HC and Endotoxin on the FPSR

Dependent on the day of injection relative to sensitization with SRBC, 20 µg LPS given intraperitoneally induced a biphasic effect on the FPSR, i.e., a depression or an

Table 1. Influence of HC and LPS on Experimental Infections with *Staphylococcus aureus* dependent on the mode of application (n = 30 per group).

	Mortality (per cent)
Staphylococcus aureus and LPS iv	
Control	76.7
HC	86.7
LPS	90.0
HC + LPS	100.0
Staphylococcus aureus and LPS ip	
Control	63.3
HC	66.7
LPS	76.7
HC + LPS	86.7
Staphylococcus aureus ip; LPS iv	
Control	63.3
HC	66.7
LPS	80.0
HC + LPS	86.7

enhancement when injected on day 0 and +2 respectively (Figures 3A and 3B). If LPS was injected intravenously the reaction was increased with a maximum by an application on day 0. A comparable LPS effect was observed under HC treatment (LPS ip/iv + HC vs. HC alone). For evaluating the HC effect under LPS, treatment groups injected with HC + LPS were compared with those treated with LPS alone in the same manner (HC alone reduced the FPSR by 75 per cent). In case of intraperitoneal LPS application on day 0, the groups pretreated with HC showed a depressed reaction but an enhancement if LPS was injected on day +2. In case of intravenous LPS application, HC showed no effect, whether LPS was given on day -3 or later up to day +2.

DISCUSSION

Large doses of glucocorticosteroids induce immune depression and impair resistance to microbial infections (1), whereas small doses did not (5). In our own experiments it could be shown, however, that HC may also have a beneficial effect on the course of infection with intraperitoneally injected *E. coli*. Similar "inverse" effects of HC were described by Y. Hiramoto et al. (2): glucocorticoid deficiency (adrenalectomy) compromised antitumor immune response being restored by HC treatment which was suggested to be due to its role in maturation of immunocompetent cells. More detailed information was obtained by *in vitro* experiments.

A proliferation promoting effect of HC on peripheral blood mononuclear cells (PBMC) was shown in the presence of suboptimal PHA-concentrations but only after the addition of IL-2. IL-2 but also T cells seem to be important (6).

Also in antibody production of PBMC *in vitro*, HC may induce an enhancement which was, however, dependent on the kind of mitogen used for activation (6). HC depressed IgG as well as IgM production of not stimulated cells but was stimulatory for both isotypes using PWM activated cells. With SAC as mitogen, HC increased only IgG production whereas IgM production was decreased.

In order to explain the beneficial effect of HC on the experimental infection with intraperitoneally given *E. coli*, the additional influence of endotoxin has also to be taken into account because LPS, as well as therapeutically used antibiotics, may modify immune response, where the dose and application time relative to infection are important parameters (7-11). In experimental infections with *Staphylococcus aureus*,

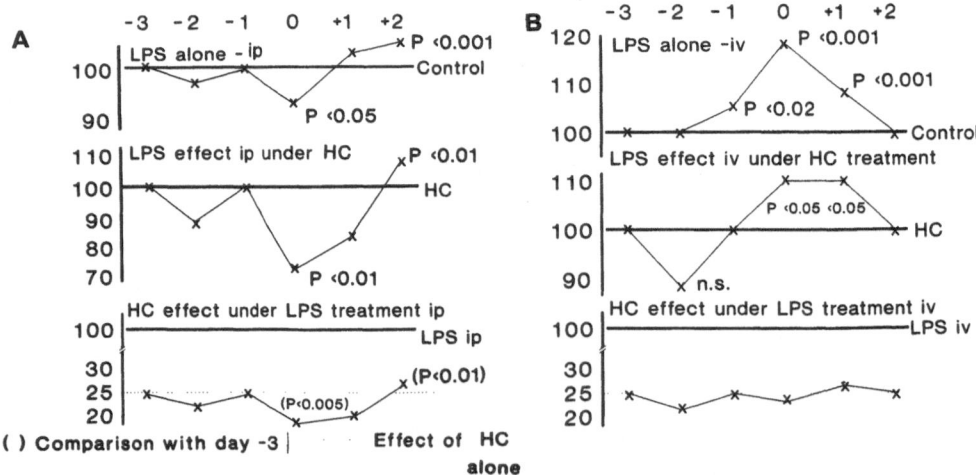

Fig. 3. Influence of LPS and HC on the FPSR (Effect in percent of controls).

however, it also could be shown that the way of application plays a role because a certain beneficial effect of HC was only seen in case of intravenous infection with LPS given intraperitoneally.

The significance of the way and the time of LPS application could also be demonstrated in the FPSR. In this model, intraperitoneal and intravenous injections of LPS induced inverse effects which are principally the same under HC treatment. A significant HC effect, however, under LPS treatment could only be observed if LPS was injected intraperitoneally changing from an inhibition to an enhancement when given on day 0 and +2 respectively. This result may be compared in a way with that of infection with Staphylococci.

It is a reasonable assumption to attribute beneficial or harmful effects of LPS alone or under HC treatment to cytokine production in sequence (5). Endotoxin (12) or antibiotic pretreated bacteria (10) induce the release of tumor necrosis factor (TNF) and IL-1, both cytokines shown to enhance phagocytosis (13, 14), intracellular killing (15, 16) resistance to infection (5, 17) and mediate in small doses beneficial effects of endotoxin (18). On the other hand TNF production after LPS application is only of short duration (12) and is blocked as well as IL-1 release by HC (19, 20) without abrogating the protective LPS effect (5). A possible effect of endogenous cytokine was, therefore, discussed.

LPS challenge is also followed by an increased serum IL-6 level mediating the manifestation of TNF and IL-1 actions *in vivo* (21). In contrast to TNF and IL-1, the release of IL-6 following upon LPS challenge was not modified by HC whereas in cell cultures IL-6 production was blocked which supports the suggestion that the target cell for *in vivo* LPS induced IL-6 production was different (5). HC may, therefore, not only decrease but also enhance mechanisms of immune responses under certain experimental conditions, but for the moment there exists no definite explanation yet at least for the *in vivo* phenomena.

SUMMARY

Hydrocortisone (HC) ameliorated the course of experimental infection with *E. coli* given intraperitoneally. In experiments with *Staphylococcus aureus* and treatment with endotoxin on the day of infection, HC had also a beneficial effect, however only when LPS was injected intraperitoneally and bacteria intravenously.

With the footpad swelling reaction as a model for cellular immunity it could be shown that HC alone reduced the reactivity by 75 per cent. Under endotoxin treatment, however, HC induced further inhibition when LPS was given intraperitoneally the day of sensitization but changing to a stimulatory effect when LPS was given two days later. In case of intravenous application of LPS, HC had no effect.

REFERENCES

1. A. S. Fauci, Glucocorticoid action, in: "Monographs on Endocrinology," Vol. 12, J. D. Baxter and G. G. Rousseau, eds., Springer Verlag Berlin, Heidelberg, New York (1979).
2. Y. Hiramoto and K. Sugimachi, Effect of glucocorticoid deficiency after adrenectomy on antitumor immunity, Cancer Immunol. Immunother. 25:157 (1987).
3. C. Papadimitriou, H. Hahn, H. Näher, S. H. E. Kaufmann, Cellular immune response to sheep erythrocytes: Interrelationship between proliferation of popliteal lymph node cells and footpad swelling, Immunobiology 164:361 (1983).

4. T. E. Miller, G. B. Mackeness, and Ph. Lagrange, Immunopotentiation with BCG. II. Modulation of the response to sheep red blood cells, J. Natl. Cancer Inst. 51:1669 (1973).

5. M. Parant, Possible mechanisms in endotoxin induced modulation of nonspecific immunity, in: "Cellular and Molecular Aspects of Endotoxin Reactions, A. Novotny, J. J. Spitzer, E. J. Ziegler, eds., Excerpta Medica Amsterdam, New York, Oxford (1990).

6. E. Stüttgen, H. Cleusters, and S. Markos-Pusztai, Differential effects of hydrocortisone (HC) on mitogen induced proliferation and Ig synthesis of lymphocytes depend on the interaction of monocytes and activated T-cells, XXI Annual Congress of the Society of Immunology, Sept. 14 - 16, Aachen, FRG (1990).

7. D. Rowley, Rapidly induced changes in the level of non-specific immunity in laboratory animal, Brit. J. Exp. Pathol. 37:223 (1956).

8. L. E. Cluff, Effects of lipopolysaccharides (endotoxins) on susceptibility to infections, in: "Microbial Toxins," Vol. 5, S. Kadis, G. Weinbaum, S. D. Ajl, eds., Academic Press, New York (1971).

9. L. Chedid and M. Parant, Role of hypersensitivity and tolerance in reactions to endotoxins, in: "Microbial Toxins," Vol. 5, S. Kadis, G. Weinbaum, S. D. Ajl, eds., Academic Press, New York (1971).

10. H. Friedman, A. Szentivanyi, T. W. Klein, and W. Warren, Enhancement by ampicillin or cyclacillin pretreatment of E. coli of phagocytosis, blastogenesis and production of lymphokines, interferons and tumor necrosis factor, in: "The Influence of Antibiotics on the Host-Parasite Relationship III," G. Gillissen, W. Opferkuch, G. Peters, and G. Pulverer, eds., Springer Verlag Berlin, Heidelberg, New York, Tokyo (1989).

11. G. Gillissen, Side effects of antibiotics on immune response parameters and their possible implications in antimicrobial chemotherapy, Zbl. Bakt. Hyg. A 270:171 (1988).

12. D. C. Morrison and J. L. Ryan, Bacterial endotoxins and host immune responses, Adv. Immunol. 28:293 (1979).

13. S. J. Klebanoff, M. A. Vadas, J. M. Harlan, L. H. Sparks, J. R. Gamble, J. M. Agosti, and A. M. Walterdorph, Stimulation of neutrophils by tumor necrosis factor, J. Immunol. 136:4220 (1986).

14. D. K. Blanchard, J. Y. Djeu, T. W. Klein, H. Friedman, and W. E. Stewart, Protective effects of tumor necrosis factor in experimental Legionella pneumophila infections of mice via activation of PMN function, J. Leukocyte Biol. 43:429 (1988).

15. C. Jupin, M. Parant, and L. Chedid, Involvement of reactive oxygen metabolites in the candidacidal activity of human neutrophils stimulated by muramyl dipeptide or tumor necrosis factor, Immunobiol. 180:68 (1989).

16. C. Jupin, M. Parant, and L. Chedid, Effect of muramyl peptides and tumor necrosis factor on oxidative responses of human blood phagocytes, Immunol. Lett. 22:187 (1989).

17. M. Parant, F. Parant, M. -A. Vinit, and L. Chedid, Action protectrice du "tumor necrosis factor" (TNF) obtenu par récombinaison génétique contre l'infection expérimentale bactérienne ou fongique, C. R. Acad. Sci. Paris Ser. III 304:1 (1987).

18. D. L. Fraker, M. C. Stovroff, M. J. Merino, and J. A. Norton, Tolerance to tumor necrosis factor in rats and the relationship to endotoxin tolerance and toxicity, J. Exp. Med. 168:95 (1988).

19. A. Waage, Production and clearance of tumor necrosis factor in rats exposed to endotoxin and dexamethasone, Clin. Immunol. Immunopathol. 45:348 (1987).

20. M. J. Staruch and D. D. Wood, Reduction of serum interleukin 1-like activity after treatment with dexamethasone, J. Leukocyte Biol. 37:193 (1985).
21. M. R. Shalaby, A. Waage, W. L. Aarden, and T. Espevik, Endotoxin, tumor necrosis factor-alpha and interleukin 1 induce interleukin 6 production *in vivo*, Clin. Immunol. Immunopathol. 53:488 (1989).

BACTERIAL CELL SURFACE BIOLOGICAL RESPONSE MODIFIERS

AND THEIR SYNTHETIC COUNTERPARTS

Shozo Kotani

Osaka College of Medical Technology
Osaka 530, Japan

INTRODUCTION

Many biological response modifiers (BRM) are located in bacterial cell surface layers where the parasitic bacteria, indigenous as well as pathogenic, come into contact with the host (1-5). Table 1 lists the cell surface BRMs whose bioactive centers have been chemically well-defined: cell wall peptidoglycans in essentially all bacterial species, endotoxic lipopolysaccharides and so-called "amino lipids" in gram-negative bacteria, lipoteichoic acids in some gram-positives, and cord factor in mycobacteria. Synthetic counterparts and their analogs or derivatives are already or soon available for these immunomodulators.

The present paper will be presented in four parts: (a) the development of a new-type of influenza virosome vaccine using a lipophilic $\underline{N}$-acetylmuramyl-L-alanyl-D-isoglutamine [MDP], namely 6-$\underline{O}$-(2-tetradecylhexdecanoyl)-MDP [B30-MDP], (b) the adjuvant effects of B30-MDP and a synthetic, low toxicity lipid A analog on immune responses of rabbits and mice to *Plasmodium falciparum* major merozoite surface protein antigen, (c) the adjuvant effects of B30-MDP in induction of tumor-specific immunity in guinea pigs, and (d) the immunopharmacological activity of streptococcal lipoteichoic acids and their glycolipid moiety to elicit "low toxicity" TNF and other cytokines and to induce tumor regression.

DEVELOPMENT OF F-HANA/B30-MDP INFLUENZA VIROSOME VACCINE (6-9)

A mass immunization study of school children in Japan revealed that the currently used influenza split virus vaccine has the advantage of safety, but it is less immunogenic than intact virus particle vaccine (10). One of the ways to raise the low immunogenicity is through the use of larger amounts of protective antigens such as hemagglutinin and neuraminidase (HANA). However, this approach tends to suffer from the increase of side effects. Another alternative is with the aid of an adjuvant which is safely applicable to humans.

B30-MDP examined in this study is one of our candidates for such a vaccine adjuvant (11-13). Fig. 1 shows that B30-MDP has hydrophobic and hydrophilic regions in the molecule. This compound, unlike its parent molecule, MDP, has the advantage

Table 1. Bacterial Cell Surface Layer BRMs and Structure of their Bioactive Center

Group	BRM	Bioactive center
ESSENTIALLY ALL	Cell wall peptidoglycan	N-Acetylmuramyl-L-alanyl-D-isoglutamine (MDP) (1)
GRAM-NEGATIVE	LPS(lipid A)	Polyacylated β(1-6)-glucosamine disaccharide bis-phosphate (2)
	Amino lipid	N$^{\alpha}$-(3-Acyloxyacyl)-L-ornithine (3)
GRAM-POSITIVE	Lipoteichoic acid	Diacylkojibiosyl-sn-glycerol, diacyl(phosphatidylkojibiosyl)glycerol(4)
ACID-FAST	Cord factor	Trehalose dimycolate (5)

that it exerts full adjuvant activity without requirement of a specified vehicle such as irritating water-in-mineral oil emulsion (11). The outline of the preparation method of a new type of influenza subunit vaccine, f-HANA/B30-MDP virosome is presented in Table 2. Purified, free hemagglutinin-neuraminidase (f-HANA) preparation, B30-MDP and cholesterol in appropriate ratios were mixed in the presence of n-octylglucoside. Removal of the detergent by dialysis resulted in stable liposomes (virosomes). Electron microscopy reveals that resulting virosomes were not only similar to influenza virus particles in size, but also had f-HANA projections attached to their surfaces.

Panel a of Fig. 2 demonstrates that in mice, a reference subunit vaccine produced only low levels of HI antibody to a homologous A/Bangkok virus, a mixture of f-HANA/B30-MDP potentiated by 2 to 3 fold the antibody response as compared with the reference f-HANA vaccine. Still greater Hi antibody production was detected with B30-MDP virosomes. Serum hemagglutination inhibition (HI) titer of 64 was detectable as early as one week after immunization. The antibody levels continued to increase, reaching 512 by 6 weeks post-immunization. As shown in panel b of Fig. 2, low but greatly heightened antibody production by f-HANA/B30-MDP virosomes was also noted to a variant virus, A/Philippines/2/82.

Table 2. Preparation of Influenza B30-MDP Virosome Vaccine - an Example - (6)

1. To 2 ml of 0.01 M phosphate-buffered saline (PBS, Ph 7.2) containing purified f-HANA (95 µg), a 6%(w/v) solution of n-octylglucoside was added together with B30-MDP (2 mg) and cholesterol (2 mg).
2. The mixture was sonicated using a Micro-Ultrasonic Cell Disrupter.
3. The sonicated mixture was thoroughly dialyzed against PBS to remove n-octylglucoside. A change in the mixture from clear to opaque indicated that liposomes (virosomes) were formed.

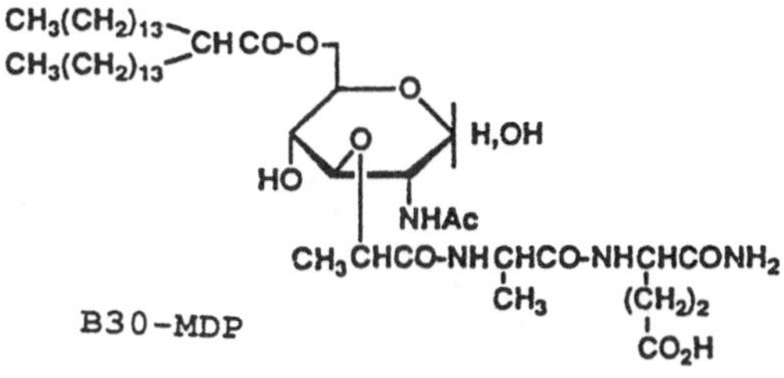

Fig. 1. Structure of MDP and B30-MDP.

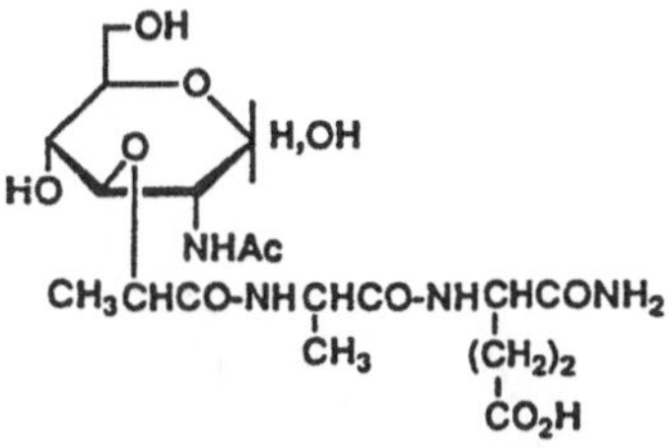

The curve in Fig. 3 shows the time-course of immune responses of mice immunized with each vaccine during a 6 month period. Identical HI titer was detected in all the mouse groups one week after the vaccination. Thereafter, the virosome vaccine produced much higher antibody levels than the reference f-HANA vaccine or a mere mixture of f-HANA and B30-MDP; these high HI antibody levels persisted at least for 6 months.

Cell-mediated immunity is known to play a significant role in combatting a host with an influenza viral infection (14). Thus B30-MDP virosome vaccine was examined for its ability to induce cellular immunity. Guinea pigs vaccinated with f-HANA/B30-MDP virosomes, like those immunized with f-HANA emulsified in Freund's complete adjuvant (FCA), showed greatly enhanced delayed-type hypersensitivity to intradermal injection of f-HANA antigen as compared with animals immunized with a simple mixture of f-HANA/B30-MDP or with f-HANA alone (data not shown). Regarding cytotoxic T-lymphocyte response of spleen cells from variously immunized mice to influenza-infected cells, virus-infected murine spleen cells (responder) that were co-cultured with stimulator cells, namely normal spleen cells infected with the virus and irradiated by X ray, caused a marked lysis of P815 target cells infected with live homologous virus. In sharp contrast, splenocytes from mice immunized with f-HANA alone were totally devoid of the activity. Under these assay conditions, the splenocytes from animals immunized with f-HANA/B30-MDP virosomes exhibited a definite cytotoxic activity, at least at an effector/target ratio of 20:1 (Fig. 4).

To see whether protection against viral infection is afforded by B30-MDP virosome vaccine, groups of BALB/c mice immunized with each of three kinds of A/Yamagata/120/86 vaccine were subjected to airborne exposure with mouse-adapted A/Yamagata virus two weeks after the vaccination. Protective effects of B30-MDP virosome vaccine was significant in terms of both the decrease in lung viral titer and the increase in survival rate: its efficacy was comparable to that of whole virus particle vaccine and better than that of free HANA vaccine (Table 3).

Study has also been done on a matter of practical concern to us all, namely the possible deleterious effects of storage on vaccine potency. Lyophilization of virosome

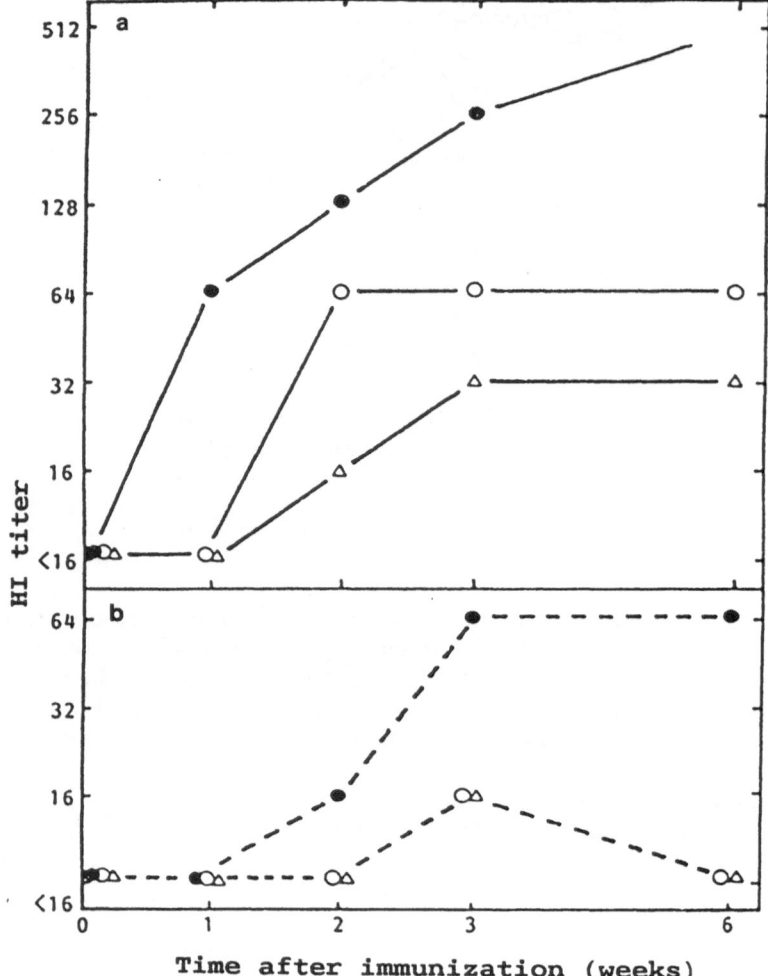

Fig. 2. Antibody production in mice immunized with f-HANA, f-HANA/B30-MDP (mixture), B30-MDP virosome. Groups of 5 ddy female mice (4-wks old) were immunized intraperitoneally by a single injection of each vaccine (0.5 ml). Dose/mouse: f-HANA, 0.4 μg; B30-MDP and cholesterol, 25 μg. f-HANA alone (△), f-HANA/B30-MDP (mix) (○), B30-MDP virosome (●). (a) Homologous virus, A/Bangkok/1/79(H3N2)), _____; (b) its variant, A/Philippines/2/82, - - -. (6)

vaccine in the presence of maltose or lactose did not damage the heightened immunogenicity of B30-MDP virosome vaccine (data not shown). Stability of a lyophilized virosome vaccine was then examined in terms of changes in virosome size and B30-MDP content. While the liquid virosomes were rather unstable, the lyophilized virosome vaccine was fully stable even under very hard conditions such as preservation

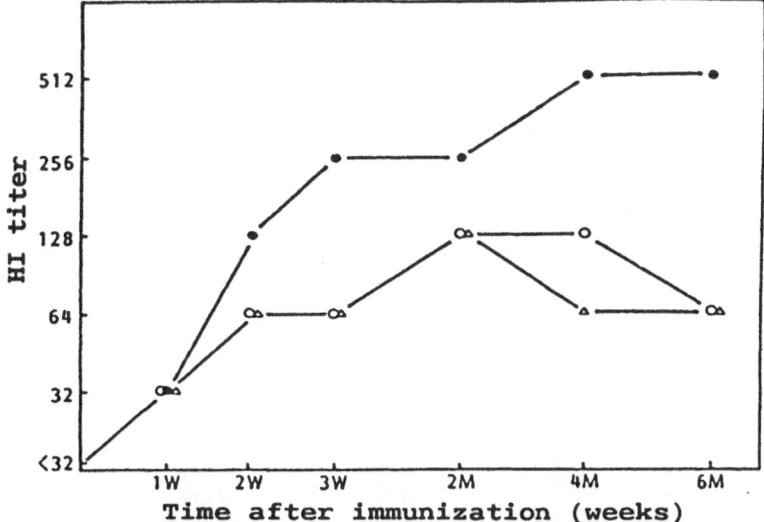

Fig. 3.   Comparison of duration of antibody production
in mice immunized with f-HANA) (▲), MDP
(mixture) (O), and B30-MDP virosome (●).
Dose/mouse: f-HANA, 3.3 μg; B30-MDP and
cholesterol, 33 μg. (6)

Immunogenicity and possible side-effects of f-HANA/B30-MDP virosome vaccine were
recently compared with those of a currently used HA vaccine with 77 healthy volunteers
(male adults) with informed consent (9). The volunteers were divided into eight groups.
Volunteers in each of seven test groups were vaccinated with one of the test virosome
vaccines and those of one reference group received the currently used HA vaccine. The
test or reference vaccine was subcutaneously injected into the upper arm once or twice
at an interval of four weeks. The results are summarized as follows:

Table 3. Protective Efficacy of Various Vaccines Prepared from A/Yamagata/120/86
(H1N1) in Mice against Airborne Challenge with Mouse-Adapted
A/Yamagata Virus (7).

Vaccine	Virus titer in lung days after challenge (PFU/ml/mice)	Survivors at indicated days(%)						
		4	5	6	7	8	9	10
Virosome	9.0×10^4	100	100	80	60	60	60	60
HA alone	9.5×10^5	100	100	80	40	40	20	0
Whole particle	1.2×10^6	100	100	100	60	60	60	60
Control	5.0×10^6	100	80	20	20	0	0	0

PFU, plaque-forming unit.

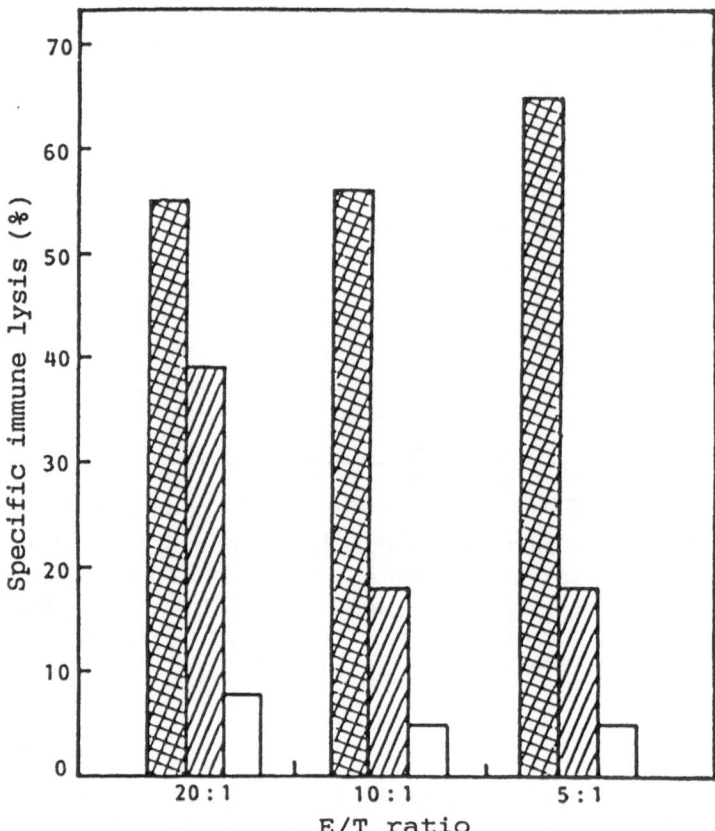

Fig. 4. Cytotoxic T-lymphocyte response of splenic cells from immunized mice to influenza virus-infected. Cytotoxic activity was expressed as man percentage for immune lysis of homologous virus-infected target cells. Live vaccine (□), B30-MDP virosome (□), f-HANA alone (□). (6).

(1) B30-MDP virosome vaccines containing f-HANA from A/Yamagata/120/86(H1N1) and A/Fukuoka/C29/85(H3N2), respectively, were highly immunogenic in terms of production of serum HI antibody even by a single vaccination. The potency of f-HANA/B30-MDP virosome vaccines was significantly higher than that of the currently used HA vaccine; (2) B/Nagasaki/1/87 f-HANA/B30-MDP virosome vaccine, however, was less immunogenic, and did not as effectively raised serum HI antibody, as was true of the currently used HA vaccine; (3) a side-effect that should not be overlooked was local redness and swelling accompanied by slight pain which frequently appeared on the day following the vaccination, and seemed dependent on the dose of B30-MDP. This local irritative reaction, nevertheless, faded within around five days, and did not seem to seriously disturb a practical use of f-HANA/B30-MDP virosome vaccine; and (4) immunization of the volunteers with the virosome vaccine did not cause either systemic reactions, including fever or abnormality in clinical and laboratory examinations significantly beyond those caused by the currently used HA vaccine, excepting the increase

Table 4. Stability of B30-MDP Virosome Vaccine - Comparison of Lyophilized Vaccine with Liquid Vaccine - (8)

Storage conditions	Mean diameter(nm)		Intact B30-MDP(%)	
(for 3 weeks at °C)	Liquid	Lyophilized	Liquid	Lyophilized
Before storage	226.5		100.0	100.0
5°	255.2	209.9	98.6	100.0
30°	336.1	224.9	90.0	99.0
40°	295.7	240.1	81.5	100.2
50°	660.7	267.4	62.6	100.3

of circulating leukocytes, especially neutrophils. Incidentally, the leucocytosis observed was due to one of the immunopharmacological activities inherent in MDP and its derivatives (1).

ADJUVANT EFFECTS OF B30-MDP AND/OR A SYNTHETIC LOW-TOXICITY LIPID A ANALOG, LA-15-PH ON IMMUNE RESPONSES TO *PLASMODIUM FALCIPARUM* MAJOR MEROZOITE SURFACE PROTEIN, GP195 (15-17)

The rationale for a strong adjuvant for subunit malaria vaccine is as follows: (a) low immunogenicity of synthetic and recombinant subunit malaria vaccine antigens in clinical trials using alum as adjuvant, and (b) vaccination studies in <u>Aotus</u> model with blood stage antigens suggesting that adjuvants of similar potency to FCA may be needed for protection. There are many studies substantiating this rationale. For instance, Siddiqui et al., demonstrated that vaccination of monkeys with *P. falciparum* major

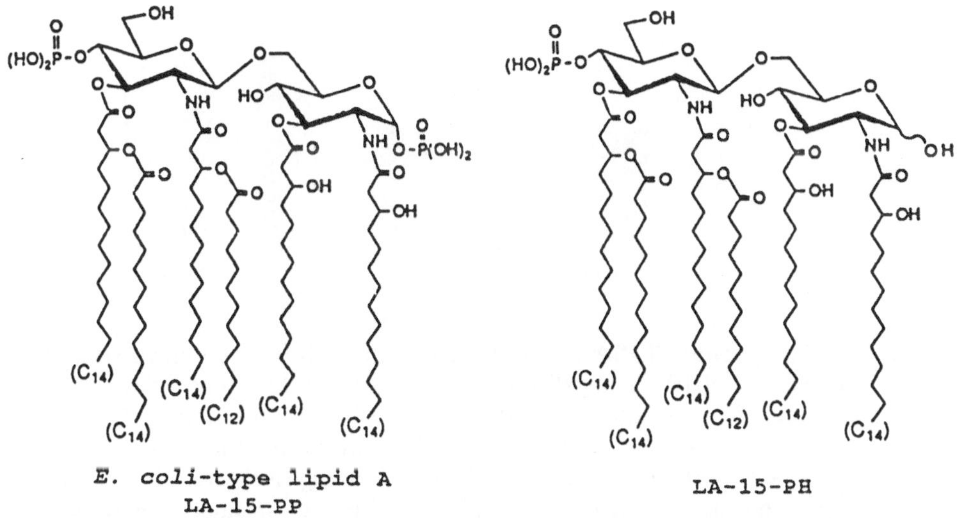

Fig. 5. Structure of *E. coli*-type lipid A (LA-15-PP) and its 4'monophospho derivative (LA-15-PH).

Table 5. Adjuvant Formulations for *P. falciparum* gp195 Immunogenicity Studies

Formulation: (i) FCA, (ii) B30-MDP in liposomes, (iii) LA-15-PH
   in liposomes, (iv) B30-MDP & LA-15-PH in liposomes, (v) liposomes, (vi) alum (Alhydrogel), (vi) saline.

Rabbits: Immunized 4x at 4 week intervals with 50µg gp195 in
   adjuvants.
Mice: Immunized 3x at 4 week intervals with 5µg gp195 in adjuvants.
Dosage: For B30-MDP and LA-15-PH: 200 µg for rabbits, 20 µg for
   mice.

merozoite surface protein, gp 195, in combination with FCA, could completely protect the monkeys from lethal infections (18-21). Our goal is to find a synthetic adjuvant that would replace FCA in potency without local and systemic side effects. B30-MDP and synthetic lipid A analog, LA-15-PH (22), the latter of which corresponds in chemical structure to monophosphoryl lipid A (MPL) described by late Professor Edgar Ribi (Fig. 5) were used here.

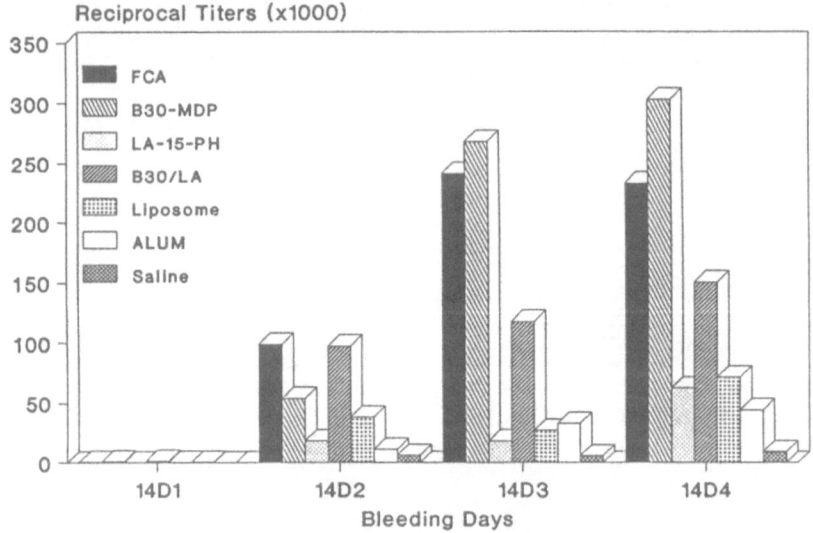

Fig. 6. Group-specific ELISA titers of sera from rabbits immunized with *P. falciparum* gp195 using different adjuvant formulations. Sera were obtained from 14 days after the primary (14D1), secondary (14D2), tertiary (14D3), and quaternary (14D4) immunizations. End-point titers are defined as serum dilutions having an OD of 0.2, and expressed as the geometric mean value ± standard deviation. Antibody titers of FCA, B30-MDP, and B30-MDP/LA-15-PH groups were significantly higher than the other adjuvant group ($p < 0.05$, one tail t-test) for 14D2, 14D3, and 14D4 bleedings. The differences among FCA, B30-MDP, and B30-MDP/LA-15-PH were not significant. (16).

We started the present study to determine if B30-MDP and/or LA-15-PH can induce immune responses to *P. falciparum* gp195 with as good efficacy as FCA. Adjuvant formulations for the gp195 immunogenicity studies in mice and rabbits are outlined in Table 5.

Fig. 6 shows antibody response in rabbits immunized with gp195 in different adjuvants. B30-MDP and B30-MDP/LA-15-PH were as effective as FCA in inducing high antibody levels to gp195, although the formulation of B30-MDP alone seemed a little more effective. LA-15-PH alone was not as good. In contrast, immunogenicity studies in mice with the same antigen/adjuvant formulations showed that B30-MDP, B30-MDP/LA-15-PH and LA-15-PH were more effective than FCA (Fig. 7). This indicates that the response to LA-15-PH may be species dependent.

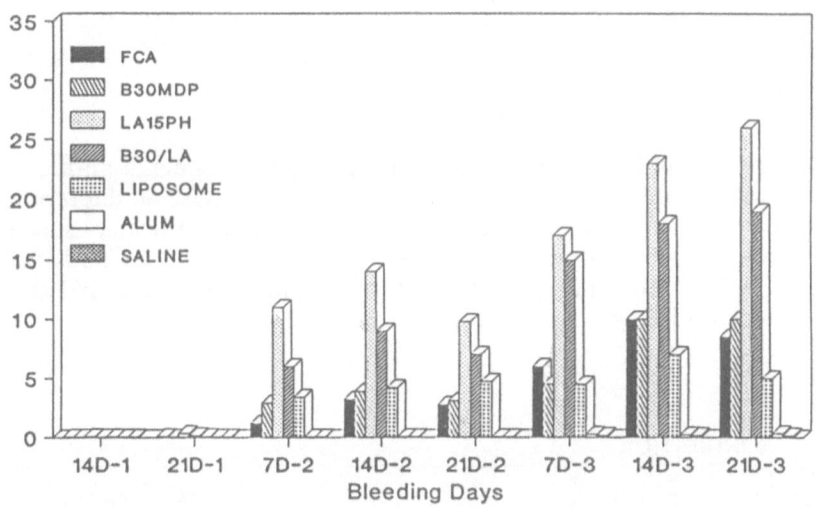

Fig. 7. ELISA antibodies to gp195 in $C_{57}BL/10$ mice immunized with *P. falciparum* gp195 in different adjuvant formulations. Data shown are assays done using pooled sera from weekly bleedings from 14 days after the primary immunization (14D-1) to 21 days after the tertiary immunization (21D-3). (15,17).

Fig. 8 presents the immunoglobulin isotype distribution of gp195-specific antibodies in mice. All synthetic adjuvants produced IgG1 and IgM antibodies. Moreover, formulations containing LA-15-PH, i.e., B30-MDP/LA-15-PH and LA-15-PH alone, induced a much higher IgG2b response than FCA or B30-MDP alone. Therefore, these formulations were able to induce a broader isotype distribution of anti-gp195 antibodies.

The biological activity of the anti-gp195 antibodies induced by different adjuvants was evaluated by an *in vitro* parasite growth inhibition assay, in considering that serum levels of the parasite growth inhibitory antibody well correlate with the extent of gp195 induced protective immunity against *P. falciparum* in <u>Aotus</u> monkeys. As shown in Fig. 9, FCA, B30-MDP/LA-15-PH and, to a lesser extent (two out of three rabbits) B30-MDP induced antibodies that significantly (>50%) inhibited parasite growth *in vitro*. Compound LA-15-PH alone was ineffective. These results are indirect evidence that B30-MDP and B30-MDP/LA-15-PH may be efficacious in potentiating protective immunity against *P. falciparum in vivo*.

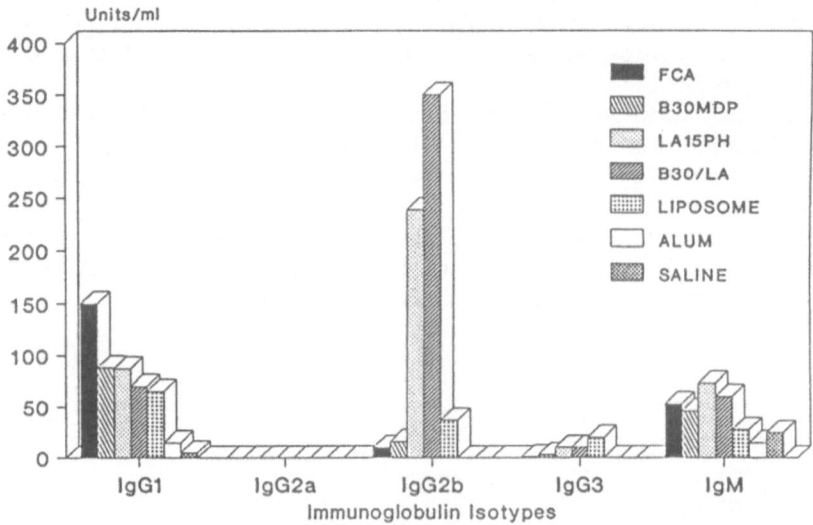

Fig. 8. Immunoglobulin isotype distribution of gp195 specific antibodies from $C_{57}BL/10$ mice immunized with *P. falciparum* gp195 in diferent adjuvant formulations. Data represent the results of pooled tertiary sera from six mice per adjuvant group. One unit per ml of gp195 antibody represents the ELISA absorbance equivalent to 1 µg/ml of a monoclonal antibody of the same isotype coated on plastic wells. (15)

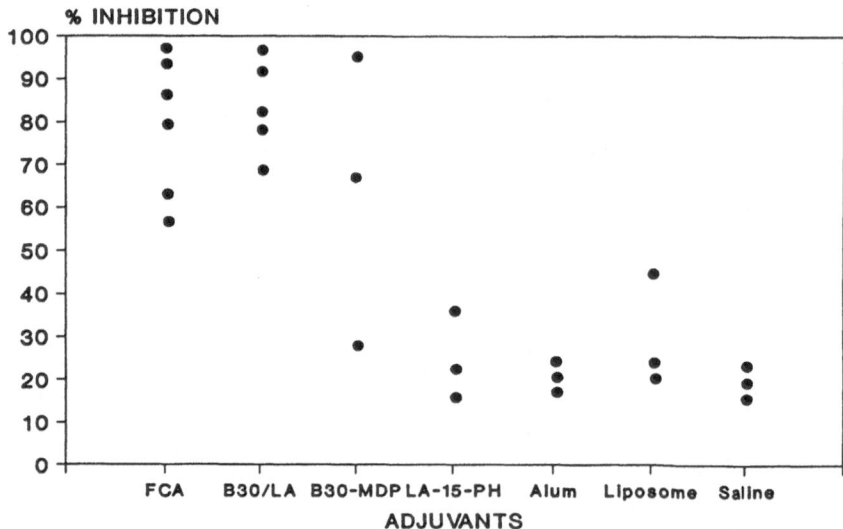

Fig. 9. The biological activity of anti-gp195 rabbit antibodies induced by different adjuvant formulations was evaluated by an *in vitro* parasite growth inhibition assay. The degree of inhibition is determined by the percentage of inhibition of growth over the 72 hr period. This is done by comparing the increase in percentage of parasitemia of cultures containing the pre-immune rabbit serum at 72 hr (P) with the increase in parasitemia from 0 hr in the cultures containing the respective immune serum at 72 hr (T)

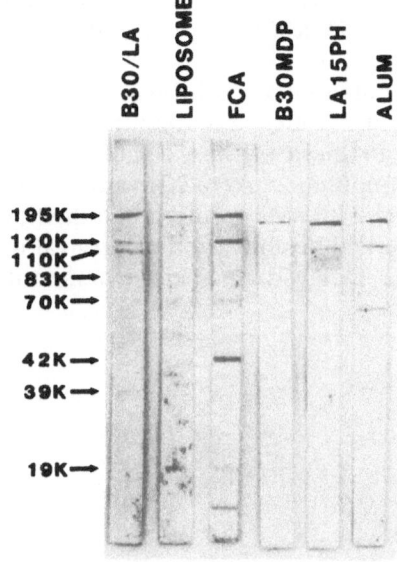

Fig. 10. Immunoblots of *P. falcipa-rum* whole parasite extracts using anti-gp195 sera from $C_{57}BL/10$ mice immunized with gp 195 in different adjuvant formulations. (17)

The specificity of the anti-gp195 antibodies induced by different adjuvants was also analyzed. The results in Fig. 10 show that the reactivity pattern with gp195 and its processed smaller fragments (gp195 is processed into a number of smaller fragments) were different. This suggests that although antibody titers to gp195 induced by the synthetic immunomodulators are similar, there are qualitative differences in these antibody populations. The fine specificities were also examined using the synthetic peptides VR-1 and VR-2, corresponding to two repeat sequence of gp195. As shown in Fig. 11, only FCA and B30-MDP/LA-15-PH induced anti-VR-1 antibodies, while reactivity to VR-2 was restricted to FCA. These data, together with those shown in the preceding paragraphs, indicate a definite influence of adjuvants on the specificity of immune responses.

To sum up: (a) Adjuvant formulations based on synthetic B30-MDP and LA-15-PH induced high levels of anti-gp195 immune responses comparable to those by FCA; (b) B30-MDP formulations are more efficacious than LA-15-PH in rabbits, while LA-15-PH is more effective than B30-MDP in mice; (c) immunogenicity studies of gp195 in mice suggested that adjuvants influence several parameters of the humoral responses in addition to antibody titer including isotypes and specificity; (iv) there is a lack of correlation among the parameters measured, which indicates that immunomodulators exert influence on these parameters independently; and (v) in rabbits, the B30-MDP/LA-15-PH and B30-MDP formulations induced biologically active antibodies as measured by an *in vitro* parasite growth inhibition assay. The degree of inhibition was similar to FCA. Thus, the first objective has been accomplished. Vaccination studies in <u>Aotus</u> monkeys using these adjuvants are to be conducted in the near future.

ADJUVANT EFFECTS OF B30-MDP IN INDUCTION OF SPECIFIC IMMUNITY AGAINST EXPERIMENTAL TUMORS IN GUINEA PIGS (23, 24)

Two types of experimental tumors, a solid-type L10 hepatocarcinoma and an acute B cell leukemia, EN-L2C that is much less immunogenic and much more malignant than L10 tumor were used in these studies. Table 6 indicates that two successive intradermal injections of a simple mixture of X-irradiated L10 or L2C tumor cells and 5 μg of B30-MDP at a one week interval resulted in induction of striking tumor-specific immunity as expressed by marked prolongation of survival time to challenging inoculation of the respective tumors. Delayed-type skin hypersensitivity was also elicited with

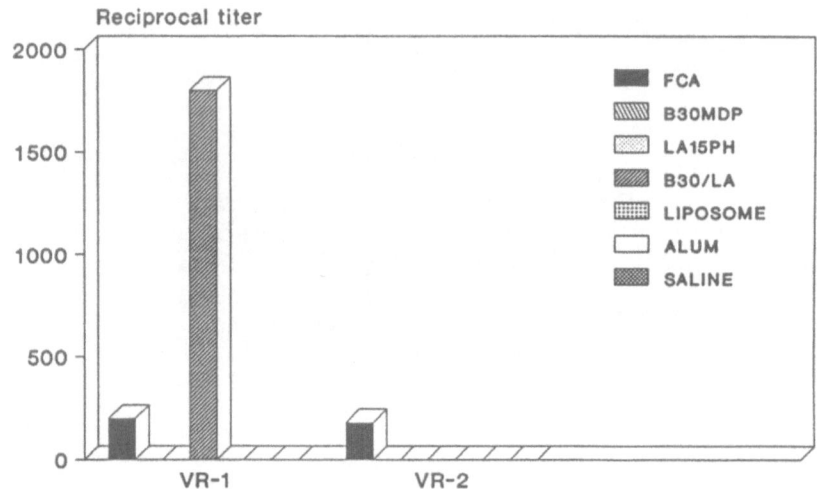

Fig. 11. Reactivity of sera from $C_{57}BL/10$ mice immunized with gp195 in different adjuvant formulations with synthetic peptides to variable repeat regions of gp195. VR-1 and VR-2 were coupled to bovine serum albumin (BSA) at a molar ratio of 100:1 (peptide:BSA). Pooled tertiary 21 day sera were used for this assay. (15,17)

only the tumor cells used for vaccination. Under the present assay conditions, MDP, a parent molecule of B30-MDP was totally ineffective.

The results so far obtained by Kataoka's group are summarized as follows: a) B30-MDP served as an effective adjuvant for induction of antitumor immunity in strain 2 guinea pigs against line 10 hepatocarcinoma and acute B cell leukemia (EN-L2C) by simple mixing with X-ray irradiated tumor cells in phosphate-buffered saline. In case of line 10 tumor, 3 M KC1 extract of the tumor (sTAA) could replace irradiated cells; (b) the immunity induced was tumor-specific and persisted for more than 100 days; (c) two 5 μg intradermal injections of B30-MDP with tumor antigen was highly effective in inducing antitumor immunity. This dose of B30-MDP with tumor antigens elicited only slight induration at the injection site. The induration disappeared quickly, within 2 wks;

Table 6. Induction of Specific Immunity against L10 and L2C Tumors in Guinea Pigs by A Mixture of X-ray-Irradiated Tumor Cells and B30-MDP (24)

A: Rejection of challenging tumors

Test animals	Survival time (days, mean±S.D.)	
(4/group)	L2C	L10
Control	16.5±1.3	63.3±3.5
L2C immune	>100	68.0±4.2
L10 immune	16.0±0.5	>100

B: Delayed-type skin reaction

Test animals	Size of reaction (mm, mean±S.D.)	
(4/group)	L2C	L10
Control	0	0
L2C immune	12.1±0.6	0
L10 immune	0	4.2±3.1

Groups of strain 2 guinea pigs were immunized by two successive id injections of X-irradiated L2C or L10 cells (1x10^6 each) mixed with 5 μg of B30-MDP at a one week interval. Seven days after the second injection, the animals were challenged with live 1x10^6 L10 cells or 2x10^5 L2C cells.

two 5 μg intradermal injections of B30-MDP with tumor antigen was highly effective in inducing antitumor immunity. This dose of B30-MDP with tumor antigens elicited only slight induration at the injection site. The induration disappeared quickly, within 2 wks; and (i.v.) the marked potency of a B30-MDP/tumor "vaccine" to induce specific anti-tumor immunity may allow a new approach to the immunotherapy of tumor.

BRM ACTIVITIES OF STREPTOCOCCAL LIPOTEICHOIC ACIDS (4)

The final topic of this paper concerns BRM activities of streptococcal lipoteichoic acids (LTA) that we have pursued for the past several years (25-27). The results so far obtained are summarized as follows: (a) *S. pyrogenes* LTA induced high levels of a cytotoxic factor against L-929 cells in the serum of *P. acnes*-primed mice; (b) the cytotoxic factor thus induced caused hemorrhagic necrosis of Meth A tumor carried by BALB/c mice. The factor inhibited the growth of a wide spectrum of tumor cell lines (mouse and human), and it discriminated between normal and tumor cells as did the LPS-induced TNF. These findings strongly suggest that the cytotoxic factor induced by LTA is a tumor necrosis factor (TNF); (c) the cytotoxic activity of LTA-induced TNF was completely neutralized by an antiserum raised against LPS-induced murine TNF; (d) the LTA-inducing ability of LTA resisted acid hydrolysis but was destroyed by mild alkaline treatment; (f) the LTA preparation caused a marked regression with hemorrhagic necrosis and complete cure of Meth A tumor established on BALB/c mice, in combination with MDP pretreatment; and (g) the LTA, unlike endotoxic LPS, was not lethally toxic in either *P. acnes*-primed or galactosamine-treated mice.

Induction by a nontoxic LTA of a TNF that seemed to be far less toxic as compared with a TNF induced by LPS so far as examined is a subject of great interest from the standpoint of utilizing streptococcal BRMs in medicine. However, until recently, progress of our study has been seriously hampered for some unidentified satisfactorily active in eliciting the serum TNF in properly primed mice. This difficulty reason by poor reproducibility to obtain *S. pyogenes* LTA preparations which are has been overcome by a recent study of Kato's group using LTA of *Streptococcus faecalis* ATCC 9790. That is to say, they demonstrated two molecular species of LTA that are different in chemical and biological activities (4).

Fig. 12 depicts an elution pattern of a crude *S. faecalis* LTA preparation by hydrophobic chromatography on an Octyl-Sepharose (CL-4B) column. Two LTA fractions, LTA-1 and LTA-2, were separated by differences in their content of acyl groups. Chemical analysis, together with the proposed structure by Fisher (28) suggests that the glycolipid moiety of LTA-1 and LTA-2 is diacylkojibiosyl-sn-glycerol and diacyl(phosphatidylkojibiosyl)glycerol, respectively.

Table 7 presents a representative assay result on TNF induction by LTA-1 and LTA-2 preparations. LTA-2 fraction is highly effective in TNF induction in the serum of *P.acnes*-primed mice, though its TNF-inducing ability was weaker than that of a reference *Salmonella enteritidis* LPS. Incidentally, none of the mice elicited with LTA-2 suffered from general prostration or showed unhealthy hair condition in spite of powerful production of TNF. This is in sharp contrast to the animals elicited with LPS. LTA-1 fraction, on the other hand, was scarcely active. This table also indicates that an acid-hydrolyzed LTA-2 exhibited a powerful activity to induce high levels of serum TNF, while an alkaline-treated LTA-2, namely, deacylated "LTA" did not show any activity. Though data are omitted, LTA-2 but not LTA-1 was capable of inducing serum TNF when used for elicitation in MDP-primed mice.

Marked difference was also noted between LTA-1 and LTA-2 with ability to induce interferons α/β and γ in *P. acnes*-primed mice (Table 8). Effects of acid- or alkaline-hydrolysis on the IFN-inducing ability of LTA-2 were essentially the same as those on TNF induction. Furthermore, Haruhiko Takada and Rieko Arakaki demonstrated that LTA-2, but not LTA-1, was capable of inducing high levels of serum IL-6 in MDP-primed mice, while both LTA-2 and LTA-1 and their glycolipid portions stimulated murine peritoneal macrophage cultures to induce thymocyte activating factor, TAF (both cell-free and cell-associated) and IL-6. They further showed that LPS low-responding C3H/HeJ mice were poor responders to induction of TNF and Il-6 by LTA-2 in combination with MDP priming (unpublished data).

Fig. 13 illustrates tumor regressive effects of LTA preparations on Meth A tumor established in BALB/c mice in combination with MDP pretreatment. Not only LTA-2 but also, unexpectedly, LTA-1 specimen caused marked tumor regression accompanied by extensive hemorrhagic necrosis. As shown in Table 9, a dose as small as 4µg either LTA-2 or LTA-1 brought about complete cure of Meth A tumor carried by BALB/c mice at a high rate. The antitumor effects of 4 µg of LTA preparations seemed to surpass those of 0.1 µg of a reference *E. coli* LPS. In this assay too, an acid-hydrolyzed LTA-2 was active, but alkaline-treated LTA-2 was inactive. Hydrolysis with acid or alkaline of LTA-2 may cause a split of the linkage between polyglycerophosphate backbone structure and glycolipid and a deacylation of a glycolipid moiety, respectively.

Experimental evidence so far obtained strongly suggests that a glycolipid fraction obtained by acid hydrolysis of LTA-2 has a structure responsible to produce cytokines such as TNF, IFN α/β and γ and IL-6 by *S. faecalis* lipoteichoic acids and to tumor regressive activity in combination with MDP pretreatment.

Nevertheless, the bioactive glycolipid has not yet been isolated in a satisfactorily purified state. So, purification and chemical characterization of a principle responsible for the BRM activities of LTA are now in progress with the object of synthesizing the bioactive principlein the laboratory of Professor Shoichi Kusumoto, Faculty of Science, Osaka University.

Table 7. TNF Production in Mice Primed with *P. acnes* by LTA Preparations Obtained from *S. faecalis* ATCC 9790 (4)

Test specimen	Dose (μg/mouse)	Cytotoxicity[a] (unit/ml)
LTA-1	500	60
	100	<20
	20	<20
LTA-2	500	8.0×10^6
	100	3.7×10^5
	20	4.0×10^4
Acid-hydrolysed LTA-2	500	1.3×10^8
	100	3.7×10^5
	20	1.9×10^4
Alkaline-hydrolysed LTA-2	500	<20
	100	<20
S. enteritidis LPS[b]	1	1.0×10^9
Control	None	<20

Female, 8 weeks old ICR mice (5 per group) were primed by ip injection of 1.5 mg (dry weight) of heat-killed P. acnes whole cells. One week later, the mice were elicited by iv injection of the indicated doses of test compounds. Serum specimens collected 1.5 h after the elicitation were pooled, and heated at 56°C for 30 min to reduce their nonspecific cytotoxic activity.
[a] Reciprocal of the highest dilution of pooled serum causing 50% lysis of L-929 cells.
[b] Bacto LPS-W.

Table 8. Induction of IFN-α/β and IFN-γ in Mice Primed with *P. acnes* by LTA Preparations Obtained from *S. faecalis* ATCC 9790 (4)

Test specimen	Dose (μg/mouse)	IFN-α/β (IU/ml)[a]	IFN-γ (IU/ml)[a]
LTA-1	500	<40	80
	100	<40	<40
LTA-2	500	320	640
	100	160	160
	20	<40	80
Acid-hydrolysed LTA-2	500	80	320
	100	<40	80
	20	<40	<40
Alkaline-hydrolysed LTA-2	500	<40	<40
E. coli LPS[b]	1	1280>	1280>
Control	None	<40	<40

Female, 8 weeks old ICR mice (3 per group) were primed by ip injection of 1.5 mg (dry weight) of heat-killed P. acnes whole cells. One week later, the mice were elicited by iv injection of the indicated doses of test compounds. Serum specimens were collected 1.5 h (for IFN-α/β) or 4 h (for IFN-γ) after the elicitation.
[a] The arithmetic mean of the reciprocal of the highest serum dilution causing 50% reduction of the cytopathic effect of vesicular stomatitis virus on L cells. The activity of IFN-γ was determined after the neutralization with anti-murine IFN-α/γ serum.
[b] Bacto LPS-W prepared from O111:B4.

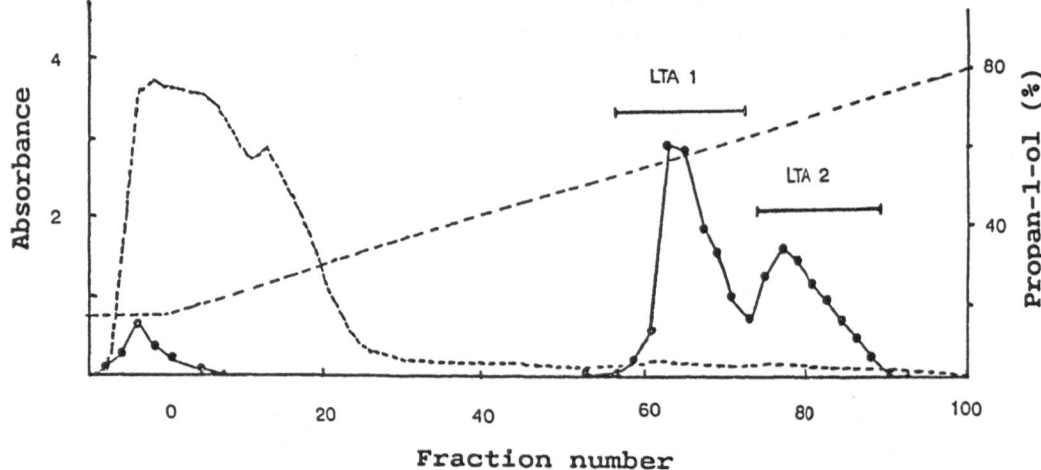

Fig. 12. Hydrophobic interaction chromatography of crude LTA from *S. faecalis* ATCC 9790 on an Octyl-Sepharose CL-4B column. The column was successively eluted with 15% propan-1-ol in 0.1 M sodium acetate buffer (Ph 4.5) and a linear gradient of 15-80% propan-1-ol in the same buffer. Fractions were collected and monitored by their phosphorus content (●) and absorbance at 260 nm (----). (4)

CONCLUDING REMARKS

1. There are a number of biomedically active compounds (BRMs) of bacterial origin that modulate the host functions , particularly natural and acquired defense mechanisms in various ways. Most of them are located in cell surface layers of parasitic bacteria, whether pathogenic or indigenous, probably reflecting the fact that these surface layers are a place of intimate contact with the front line of the host defense systems.

2. The fact worthy of note is that bacterial cell surface BRMs do not exert so strong or direct a cytotoxicity on host cells as typical exotoxins do: even endotoxic lipopolysaccharides and cord factors are far less lethally toxic than exotoxins in terms of specific activity on the basis of weight.

3. These facts lead us to speculate that development of both the immunomodulating activities of bacterial BRMs and the reactivity of the host against them have evolved through selection and mutation based on the host-parasite interactions from time immemorial.

4. As a medical bacteriologist and immunopharmacologist, I have had much interest in the pathological roles of bacterial cell surface BRMs, namely in their involvement with pathogenesis of bacterial infections (exogenous or endogenous) on one hand, and also in possible utilization of these BRMs to enhance of host defense functions against "not-self" agents on the other, although the present paper is exclusively concerned with the latter interest.

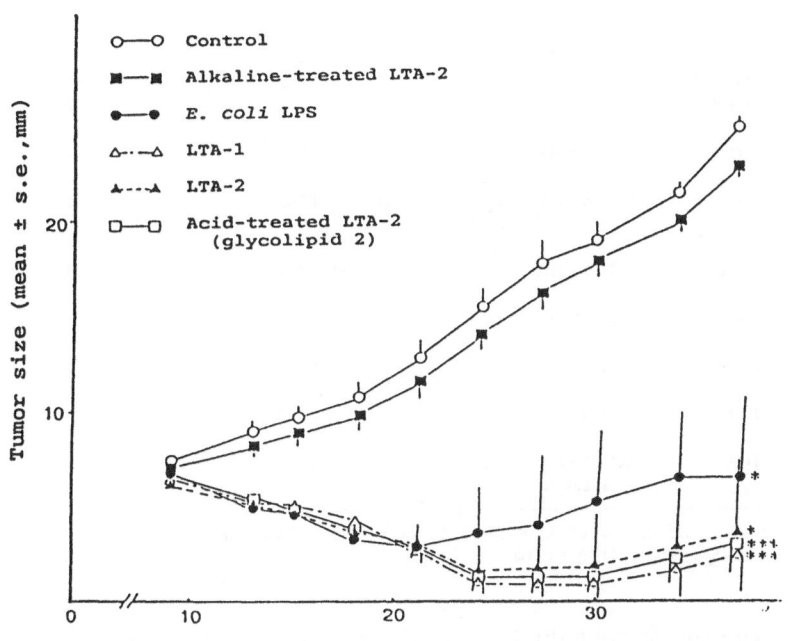

Days after tumor inoculation

Fig. 13. Regressive effects of various LTA preparations of Meth A fibrosarcoma (solid type) established in BALB/c mice. Groups of five BALB/c mice bearing Meth A tumor (7-8 mm in average diameter) were pretreated by the i.v. injection of MDP (100 μg/mouse). Four hr later, the MDP-treated groups received the i.v. injection of test LTA preparations (100 μg/mouse) or reference *E. coli* (0111:B4) LPS (0.1 μg). The group receiving neither MDP nor a test sample served as the control. Significant difference from the control on day 37: (*)P<0.05; (***)P-<0.001. (4)

Table 9. Induction of Hemorrhagic Necrosis and Complete Cure of Meth A Fibrosarcoma Established in BALB/c Mice by Administration of LTA with or without MDP Pretreatment (4)

Group	MDP treatment	Test specimen	Dose (μg/mouse)	Hemorrhagic necrosis and complete cure[a]
1	Yes	E. coli LPS[b]	0.1	3
2	Yes	LTA-1	100	4
3	Yes		20	3
4	Yes		4	3
5	Yes	LTA-2	100	4
6	Yes		20	5
7	Yes		4	4
8	Yes	Acid-hydrolysed LTA-2	100	4
9	Yes	Alkaline-hydrolysed LTA-2	100	0
10	Yes	Saline	None	0
11	No	E. coli LPS[b]	0.1	0
12	No	LTA-1	100	0
13	No	LTA-2	100	0
14	No	Acid-hydrolysed LTA-2	100	0

[a]Number of positive out of a total of five mice.
[b]Bacto LPS-W from 0111:B4.

5. Bacterial BRMs are a double-edged sword realized by shifts of delicate balance between beneficial and harmful effects on the host, as is the case with inflammatory reactions. Elucidation of a bioactive center of bacterial BRMs and subsequent chemical synthesis of the active principle and its analogs or derivatives provide us the means for creation of novel compounds in which the balance between "good" and "bad" effects shifts in favor of the host. There is good prospect of success in winning safe and useful remedies mimicking bacterial cell surface BRMs.

ACKNOWLEDGEMENTS

The author is grateful to Dr. Kuniaki Nerome and Dr. Akira Oya, Department of Virology and Rickettsiology, National Institute of Health, Tokyo; Mr. Kunio Ohkuma, the Chemo-Sero-Therapeutic Institute, Kumamoto; Dr. Masahide Kaji, Jr., the 1st Department of Internal Medicine, Kurume University School of Medicine, Kurume; Dr. George S. N. Hui and Dr. Wasim A. Siddiqui, Department of Tropical Medicine and Medical Microbiology, University of Hawaii, Honolulu; Dr. Tetsuo Kataoka and Dr. Tohru Tokunaga, Department of Cellular Immunology, National Institute of Health, Tokyo; Dr. Keijiro Kato, Dr. Osamu Tsutsui and Mr. Susumu Kokeguchi, School of Dentistry, Okayama University, Okayama, and Dr. Haruhiko Takada and Mrs. Rieko Arakaki, Faculty of Dentistry, Kagoshima University, Kagoshima, Dr. Shoichi Kusumoto, Faculty of Science, Osaka University for generous help in preparation of this paper, particularly for permission to reproduce the figures and tables printed in their original articles and also to quote data from the papers in press or submission and those unpublished.

REFERENCES

1. H. Takada and S. Kotani, Immunopharmacological activities of muramylpeptides, in: "Immunology of the Bacterial Cell Envelope", D. E. S. Stewart-Tull and M. Davis, ed., John Wiley and Sons, Chichester, pp. 119-152 (1985).

2. H. Takada and S. Kotani, Structural requirements of lipid A for endotoxicity and other biological activities, CRC Crit. Rev. Microbiology, 16:477-523 (1989).

3. Y. Kawai and K. Akagawa, Macrophage activation by an ornithine-containing lipid or a serine-containing lipid, Infect. Immun., 57:2086-2091 (1989).

4. O. Tsutsui, S. Kokeguchi, T. Matsumura, and K. Kato, Relationship of the chemical structure and immunobiological activities of lipoteichoic acid from Streptococcus faecalis (Enterococcus hiraeATCC 9790), FEMS Microbiol. Immunol. 76:211-218 (1991).

5. G. Lemaire, J. P. Tenu, and J. F. Petit, Natural and synthetic trehalose diesters as immunomodulators, Med. Res. Rev., 6:243-274 (1986).

6. K. Nerome, Y. Yoshioka, M. Ishida, K. Okuma, T. Oka, T. Kataoka, A. Inoue, and A. Oya, Development of a new type of influenza subunit vaccine made by muramyldipeptide-liposome: enhancement of humoral and cellular immune responses, Vaccine 8:503-509 (1990).

7. K. Ohkuma, T. Honda, K. Tadakuma, S. Goto, T. Oka, and M. Sakoh, Immunological studies on influenza MDP-virosome vaccine. [II], Clinic and Virus, 18:91-95 (1990). [in Japanese]

8. K. Ohkuma, T. Honda, T. Oka, and M. Sakoh, Pharmaceutical studies on influenza MDP-virosome vaccine, Clinic and Virus, 18:79-85 (1990). [in Japanese]

9. M. Kaji, Jr., Y,. Kaji, M. Kaji, K. Ohkuma, T. Honda, T. Oka, M. Sakoh, S. Nakamura, K. Kurachi, and M. Sentoku, The phase I clinical tests of influenza MDP-virosome vaccine (KD-5382), Vaccine (in press).

10. H. Fukumi, Production and potency standardization of haemagglutinin subunit influenza vaccine, International Symposium Influenza Vaccine for Men and Horses, Ser. Immunobiol. Stand., 20:99 (1973).

11. M. Tsujimoto, S. Kotani, F. Kinoshita, S. Kanoh, T. Shiba, and S. Kusumoto, Adjuvant activity of 6-O-acyl-muramyldipeptides to enhance primary cellular and humoral immune responses in guinea pigs: Adaptability to various vehicles and pyrogenicity, Infect. Immun., 53:511-516 (1986).

12. M. Tsujimoto, S. Kotani, T. Shiba, and S. Kusumoto, Adjuvant activity of 6-O-acyl-muramyldipeptides to enhance primary cellular and humoral immune responses in guinea pigs: Dose-response and local reactions observed with selected compounds, Infect. Immun. 53:517-521 (1986).

13. M. Tsujimoto,M., S. Kotani, T. Okunaga, T. Kubo, H. Takada, T. Kubo, T. Shiba, S. Kusumoto, T. Takahashi, Y. Goto, and F. Kinoshita. Enhancement of humoral immune responses against viral vaccine by a nonpyrogenic 6-O-acyl muramyldipeptide and synthetic low toxic analog of lipid A. Vaccine, 7:39-48 (1989).

14. M. A. Wells, F. A. Ennis, and P. Albrecht, Recovery from a viral respiratory infection. II. Passive transfer of immune spleen cells to mice with influenza pneumonia, J. Immunol. 126:1042-1046 (1981)

15. G. S. N. Hui, S. P. Chang, L. Q. Tam, A. Kato, S. E. Case, C. Hashiro, W. A., Siddiqui, S. Kotani, T. Shiba, and S. Kusumoto, Characterization of antibody responses induced by different synthetic adjuvants to the Plasmodium falciparum major merozoite surface precursor protein, gp195, in: "Vaccine 90", T. M. Chanock, R. A. Lerner, F. Brown, H.. Ginsberg, ed., Cold Spring Laboratory, New York, pp.477-484 (1990).

16. G. S. N. Hui, L.Q.Tam, S.P.Chang, S.E.Case, C.Hashiro, W.A. Siddiqui, T.Shiba, S.Kusumoto, and S.Kotani, Synthetic low toxicity muramyl dipeptide and monophosphoryl lipid A replace Freund complete adjuvant in inducing growth-inhibitory antibodies to the *Plasmodium falciparum* major merozoite surface protein, gp195, Infect. Immun., 59:1585-1591 (1991).

17. G. S. N. Hui, S. P. Chang, H. Gibson, A. Hashimoto, C. Hashiro, P. J. Barr, and S. Kotani, Influence of adjuvants on the antibody specificity to the *Plasmodium falciparum* major merozoite surface protein, gp195, J. Immunol.(in press).

18. W. A. Siddiqui, L. Q. Tam, S. C. Kan, K. Kramer, S. E. Case, K. L. Palmer, K. M. Yamaga, and G. S. N. Hui, Induction of protective immunity to monoclonal antibody-defined *Plasmodium falciparum* antigens requires strong adjuvant in Aotus monkeys, Infect. Immun., 52:314-318 (1986).

19. W. A. Siddiqui, L. Q. Tam, K. J. Kramer, G. S. N. Hui, S. E. Case, K. M. Yamaga, S. P. Chang, E. B. T. Chan, and S. C. Kan, Merozoite surface coat precursor protein completely protects Aotus monkeys against *Plasmodium falciparum* malaria, Proc. Natl. Acad. Sci. USA, 84:3014-3018 (1987).

20. W. A. Siddiqui, Role of adjuvants in malaria vaccine, in: "Immunotherapeutic Prospects of Infectious Diseases", K. N. Masihi and W. Lange ed., Springer-Verlag, Berlin, pp.325-335 (1990).

21. W. A. Siddiqui, Where are we in the quest for vaccines for malaria?, Drugs 41:1-10 (1991).

22. S. Kotani, H. Takada, M. Tsujimoto, T. Ogawa, I. Takahashi, T. Ikeda, K. Otsuka, H. Shimauchi, N. Kasai, J. Mashimo, S. Nagao, A. Tanaka, S. Tanaka, K. Harada, K. Nagaki, H. Kitamura, T. Shiba, S. Kusumoto, M. Imoto, and H. Yoshimura, Synthetic lipid A with endotoxic and related biological activities comparable to those of natural lipid A from *Escherichia coli* Re-mutant, Infect. Immun., 49:225-237 (1985).

23. T. Kataoka and T. Tokunaga, A synthetic adjuvant effective in inducing antitumor immunity, Jpn. J. Cancer Res., 79:817-820 (1988).

24. T. Kataoka, M. Kinomoto, M. Takegawa, and T. Tokunaga, Effect of a synthetic adjuvant for inducing anti-tumor immunity, Vaccine, 9:300-302 (1991).

25. A. Yamamoto, H. Usami, M. Nagamuta, S. Sugawara, S. Hamada, T. Yamamoto, K. Kato, S. Kokeguchi, and S. Kotani, The use of lipoteichoic acid (LTA) from *Streptococcus pyogenes* to induce a serum factor causing tumor necrosis, Br. J. Cancer, 51:739-742 (1985).

26. H. Usami, A. Yamamoto, Y. Sugawara, S. Hamada, T. Yamamoto, K. Kato, S. Kokeguchi, H. Takada, and S. Kotani, A nontoxic tumour necrosis factor induced by streptococcal lipoteichoic acids, Br. J. Cancer 56:797-799 (1987).

27. H. Usami, A. Yamamoto, W. Yamashita, Y. Sugawara, S. Hamada, T. Yamamoto, K. Kato, S. Kokeguchi, H. Ohokuni, and S. Kotani, Antitumor effects of streptococcal lipoteichoic acids on Meth A fibrosarcoma, Br. J. Cancer 57:70-73 (1988).

28. W. Fisher, Physiology of lipoteichoic acids in bacteria, Adv. Microb. Physiol., 29:233-302 (1988).

DETERMINATION OF THE ANTIINFECTIOUS ACTIVITY OF RU 41740

(BIOSTIM) AS AN EXAMPLE OF AN IMMUNOMODULATOR

I. Ounis

CASSENNE Laboratories, Tour Roussel Hoechst
1 Terrasse Bellini, 92080 Puteaux, France

SUMMARY

Evaluation of the anti-infectious activity of an immunomodulator performed either *in vitro* or *in vivo* in animals as in humans must answer three questions :

- what are the targets ?
- what models should be used to study the mechanism of action?
- what methodology should be selected for the assessment of therapeutic benefit?

In the case of RU 41470 (Biostim), an immunomodulator with a known structure and of biological origin affects immunocompetent cells and two essential mediators: Il1 and CSF.

Because of multiple interactions between anti-infectious, anti-inflammatory and anti-allergic responses, as well as the pleiotropism of mediators, there exists no absolute predictive index of activity *in vivo* and, independently of models of immune deficiency, experimental infections are a particularly useful pharmacological model for study of the anti-infectious activity of any immunomodulator. In this model, RU 41470 tested by oral, intraperitoneal and aerosol administration, proved to be active regardless of the type of infectious agent for extracellular bacteria, intracellular bacteria, viruses or yeasts. Because of special local features of anti-infectious defences (pulmonary, cutaneous), the target organ must be identified when studying mechanism of action.

RU 41740 stimulates the metabolic activities of alveolar macrophage and the target organ is the respiratory tract. From a clinical pharmacology standpoint, stimulation of different immune components has been investigated with RU 41470 at different dosages, using double-blind versus placebo designs. Target pathology, regardless of severity, includes a risk of infection and the existence of an immunological deficiency. Chronic bronchitis is a reference pathology since patients are subject to episodes of infection, resulting in acute decompensation and contributing to worsening of the ventilatory obstructive disorder. Clinical efficacy in terms of anti-infectious prophylaxis must be evaluated by a strict methodological approach: randomized double-blind placebo-controlled trials with prolonged follow-up.

RU 41740 is effective in prophylaxis of respiratory infections in chronic bronchitis (reduction in the number of respiratory infections, their duration and in antibiotic

Microbial Infections, Edited by H. Friedman *et al.*
Plenum Press, New York, 1992

consumption) and in prophylaxis of respiratory tract infections in children over one year old.

Clinicians faced with the perplexity of the mechanism of action of immunomodulators and their number are preoccupied above all by the response which such an anti-infectious immunomodulator can offer in a context of clinical reality. Evaluation of the anti-infectious activity of an immunomodulator takes place *in vitro* as well as *in vivo* in the animal then in man and must answer three questions:

- what are the sites of impact?
- what models should be used to study mechanism of action?
- what methodology should be selected for the assessment of therapeutic benefit?

The validity of this evaluation depends on purification of the compound studied, guarantee of the reproducibility of the biological activities studied and of the possibility to study the distribution.

During the development of Biostim (RU 41470), CASSENNE Laboratories were confronted by the specific difficulties which arise in the development of any stimulant immunomodulator.

RU 41470 is an immunomodulator consisting of two highly purified compounds obtained from *Klebsiella pneumoniae* (13). It belongs to the group immunomodulators of biological (bacterial or fungal) origin, together with lentinan and MDP. Immunomodulators have been classified in terms of their origin but a classification based upon their immunological activities is useful for comparative study of the mechanism of action of different compounds (15-35). Purification methods have included spectrometry, nuclear magnetic resonance, gas chromatography and high performance liquid chromatography.

It has neither mutagenic nor teratogenic actions and its animal toxicity per os is very slight. Distribution has been studied in the animal. RU 41470 passes through the intestinal wall and evidence has been found of early preferential distribution to the lymphatic system (19).

What Are the Sites of Impact *In Vitro*?

A large number of difficulties arise when attempting to identify the sites of impact of a compound with an immunological action. Immunocompetent cells are numerous and are multipotential in terms of activities, (example: the macrophage). In addition, these activities differ according to the tissue site (circulating monocyte, macrophage or alveolar macrophage). Impacts on immunocompetent cells may vary in relation to the pharmaceutical form of the compound (nanoparticles, liposomes).

A large number of mediators are involved in anti-infectious defences (Il1 - TNF - Il2 - IFN γ) and their activities have been more precisely identified following the development of recombinant mediators. However evaluation of their actions is made difficult since they are sometimes superimposable (Il1 - Il6) or poorly determined (macrophage growth factors).

Mediators are responsible for interactions of the immune system and stimulate the expression of histocompatibility receptors.

The immunological sites of impact of RU 41740 are known.

RU 41740 acts upon all the immunocompetent cells involved in anti-infectious defenses (monocytes, macrophages, T and B lymphocytes, polymorphonuclears, NK cells) (11-34-37).

Biostim stimulates the monocytes and alveolar macrophages to secrete interleukin 1 (21) as well as CSF activity (2). These two mediators play a central role in the initiation of the immune response.

Unfortunately, in the area of immunology, determination of sites of impact is not sufficient to deduce the mechanism of action of the compound studied. Anti-infectious, anti-inflammatory and anti-allergic responses are linked and neuromediators and mediators of inflammation may interact in the anti-infectious reaction.
In view of the complexity of immunological mechanisms, consistency between different experimental results is essential in the development process of an immunomodulator.

Even when sites of impact are known, no absolute predictive index of *in vivo* activity exists.

What Models Should Be Used to Study Mechanism Of Action?

Acceptability, in particular from an immunological standpoint, must be evaluated at all phases during the *in vivo* study of an immunomodulator. While a certain degree of toxicity may be tolerable in serious situations where there is a major risk of infection, this is no longer the case in benign disorders and this may lead to the interruption of development at any stage.

The development of any anti-infectious immunomodulator includes two essential stages: demonstration of immune stimulation and study of anti-infectious activity. The target organ of the compound must also be identified at the same time.

Stimulation of the immune system is easier to demonstrate in models of immune deficiency. The latter may be induced, or not, as is the case during the neonatal period and in athymic animals. Induced immune deficiency may be iatrogenic (anti-CD4, anti-CD8, corticosteroids, anti-inflammatory drugs) or secondary to a burn, to malnutrition.

RU 41740 has been studied in these situations of immune deficiency. It stimulates the humoral response (14)(immune deficiency by malnutrition) as well as phagocytosis (10)(immune deficiency induced by corticosteroids). The migratory capacity of polymorphonuclear neutrophils specifically depressed by niflumic acid is restored in the presence of RU 41740.

The impacts of RU 41740 on the immune system enable it to stimulate all the components of immunity: phagocytosis, cell-mediated immunity, humoral immunity.

Experimental infections are a particularly useful model to study the anti-infectious activity of an immunomodulator. It is essential to demonstrate anti-infectious activity *in vivo* because of the lack of any immunological criterion with an absolute predictive value.

The first aim of this integrated model is to test the anti-infectious defence system overall. This can be studied in the presence or absence of immune deficiency.

Antibiotics are commonly used in infections and the second aim of experimental infections is to study both the acceptability of combined prescription as well as any possible synergistic action between the immunomodulator and the antibiotic.

The methodology involved in an experimental infection is to provoke a lethal infection, whether acute, continuous for a period of several days or chronic. Lethal infection is a situation in which anti-infectious defenses are strongly solicited but the same players are involved in a benign infection. Immunological abnormalities accompanying a chronic and acute infection may differ. Thus Bamberger (5) showed the existence of a decrease in the bactericidal function of neutrophils in the presence of a chronic Staphylococcus aureus abscess while nothing similar occurred in the case of acute abscesses. Various routes of administration must be used (IV, IP, PO) and the different dosages tested enable the determination of an active dose.

Evaluation criteria are :

- mean survival rate during the days following the experimental infection;
- infectious invasion with count of infectious agents in the kidneys, lungs and lymphatic organs (spleen, nodes).

RU 41740 has been tested by oral, intraperitoneal and aerosol administration and is active regardless of the infectious agent responsible (21-25), whether bacteria growing by extracellular proliferation (*Staphylococcus aureus*, *Streptococcus pneumoniae*) or bacteria growing by intracellular proliferation (*Listeria monocytogenes*, *Salmonella*

Table 1. RU 41740 - Experimental Infections

	SPECIES	ACTIVE DOSES/DAY
Bacterial infections :		
- <u>Extracellular organisms</u>		
* Escherichia coli	Mice	30-100 mg/kg
* Pseudomonas aeruginosa	Mice	30-100 mg/kg
* Klebsiella pneumoniae	Mice	20 mg/kg
* Streptococcus pneumoniae	Mice	30-100 mg/kg
- <u>Intracellular organisms</u>		
* Listeria monocytogenes	Pig	20-200 mg/kg
* Erysipelothrix rhusiopathiae	Mice	1-10 mg/kg
Viral infection :		
* Influenza	Mice	30-100 mg/kg
Fungal infection :		
* Candida albicans	Mice	10 mg/kg

typhimurium, etc). This same activity has been shown in viral infections (33)(influenza virus, encephalomyocarditis virus) and fungal infections (Candida albicans) (Table 1).

Animals were given a single dose on the same day as that of infection or repeated doses for up to seven days before infection.

The target organ must be identified because of the special·local features of anti-infectious defences (pulmonary defences, cutaneous defences).

The target organ of RU 41740 is the respiratory tract. RU 41740 stimulates the metabolic activities of the alveolar macrophage: increased ability to secrete interleukin 1 (22), stimulation of the secretion of enzymes involved in the regulation of mucus (18)(hyaluronidase, β glucuronidase), stimulation of phosphomethyltransferase activity (31) and inhibition of sialidase activity (26), indicative of an increase in cell reactivity.

What Methodology Should Be Selected for The Assessment of Therapeutic Benefit?

Rationale of Clinical Pharmacological Studies. When seeking a possible anti-infectious action, as in all other areas, clinical pharmacological studies must cover as a priority study of different dosages and, following animal pharmacological studies, confirmation of activity *in vivo* in man. Activity of the compound in the various components of the immune process must be investigated.

In the case of RU 41740, doses tested varied from 2 to 32 mg/day. Populations studied using double-blind versus placebo designs were varied but all had immune deficiency. This involved cell-mediated immunity (anergic patients with lymphoma in remission (27) or patients with a carcinoma (6-23)). Humoral-mediated immunity was studied in patients given influenza vaccine in combination with treatment (32). Study populations for the evaluation of phagocytosis included chronic bronchitis (7) and elderly individuals in poor general condition (1). Parameters for study of phagocytosis were phagocyte index, percentage of isolated phagocytic cells, fungicidal action on Candida and oxygen free radical metabolism evaluated by chemoluminescence.

Selecting A Target Pathology. A target pathology should be evaluated both in terms of clinical pharmacology, in order to demonstrate prophylactic efficacy in a study model, as well as clinically in its indications and using appropriate methodology.

How Should Target Pathology Be Determined? Target pathology for an anti-infectious immunomodulator involves, regardless of its severity, a risk of infection and the presence of immunological deficiency. A distinction is drawn between pathological states in which the situation is dominated by infection, with varying degrees of underlying immune disorders, and others where it is the immunological deficiency situation which predominates (Table 2).

When infection predominates, the context may be that of an acute, chronic or recurrent infection.

An example of acute infection is measles in an endemic context which is accompanied by transitory immune deficiency which may nevertheless have very serious consequences.

In the case of chronic infections accompanied by immune abnormalities, e.g. HIV infection, tuberculosis, leprosy, onchocerciasis, therapeutic evaluation comes up against methodological problems. Cell-mediated immunity is particularly affected in these circumstances. The natural history of the disease involving a prolonged course, any therapeutic benefit is more difficult to interpret. In addition, such patients are usually on multiple medications which complicates evaluation of the acceptability of the studied drug.

Recurrent infections such as urinary tract infections, chronic bronchitis, cystic fibrosis and recurrent respiratory tract infections in the child are situations in which the absence of an event is of interest: e.g., absence of an episode of infection in the case of recurrent respiratory tract infections. Difficulties in evaluation are linked to the variability of criteria indicating the onset of infection and to limited epidemiological data.

Cystic fibrosis (28) and chronic bronchitis are chronic disorders characterized by recurrent respiratory tract infections which lead to impairment of respiratory function. Certain immunological abnormalities have been described in chronic bronchitis, in particular abnormalities of chemotaxis and of phagocytosis (16-30). Abnormalities of mucus are responsible for chronic bacterial colonization and recurrent respiratory infections. In cystic fibrosis, the secondary inflammatory reaction results in destruction of lung parenchyma.

Table 2. Target Pathology for an Anti-Infectious Immunomodulator

1 **INFECTION** **SITUATION**	ACUTE INFECTION AND IMMUNE ABNORMALITIES - measles CHRONIC INFECTION AND IMMUNE ABNORMALITIES - HIV infection - tuberculosis - leprosy - onchocerciosis - malaria etc. RECURRENT INFECTIONS - recurrent urinary tract infections - recurrent respiratory tract infections in the child - chronic bronchitis
2 **IMMUNE DEFICIENCY** **SITUATION**	- elderly - surgical stress - burns - aplasia - primary immune deficiency
3 **CHRONIC AND RECURRENT** **INFECTION WITHOUT IMMUNE** **DEFICIT TARGET**	CYSTIC FIBROSIS (bacterial colonization, chronic inflammation, lung destruction) ANTI-INFECTIOUS PROPHYLAXIS MODEL

The risk of infection is major in such disorders and cystic fibrosis in particular is a model of anti-infectious prophylaxis bearing in mind the seriousness of the disease.

In a clinical situation in which immune deficiency predominates, this may involve a primary immune deficiency or a secondary (age, surgical stress, burns etc.) immune deficiency.

Methodology of Efficacy Trials in Anti-infectious Prophylaxis

Clinical efficacy in anti-infectious prophylaxis must be evaluated by a strict methodological approach: randomized double-blind placebo-controlled trials, with prolonged follow-up.

Evaluation criteria, whether direct or indirect, in these therapeutic trials must be appropriate to the context of clinical practice (4-9). They must also ensure the demonstration of therapeutic benefit (Table 3).

It is easier to link therapeutic benefit to direct criteria such as reduction in the number of infections, free interval between infections, duration of infections or the duration of antibiotic therapy. Assessment by the investigator as well as the duration of infections reflect their severity. The latter can also be evaluated specifically, e.g. using a semi-quantitative scale.

Survival, disease progression in terms of respiratory function and limitation of costs resulting from respiratory infections are indirect criteria.

Table 3.   Clinical Evaluation Criteria for an Anti-Infectious Immunomodulator

- SUITABLE FOR CLINICAL PRACTICE

- RELEVANT

- INDIRECT

- survival

- incremental impairment of respiratory function

- costs

- DIRECT

- number of infections

- number of patients without infection

- duration of infections

- severity

- duration of antibiotic treatment

- investigator's assessment

In the case of RU 41740, prophylaxis against recurrent respiratory tract infections has been confirmed in chronic bronchitis, even at the advanced stage of chronic cor pulmonale (3-12-36). Chronic bronchitis is a reference pathology. There are more than two million chronic bronchitis sufferers in France (29), with twenty thousand deaths per year imputable to the disease, acute decompensation resulting from infection being a major cause of mortality. Chronic bronchitis patients are subject to infection, this very probably contributing to worsening of the obstructive ventilatory disorder.

Efficacy results have concerned reduction in the number of respiratory tract infections, their duration and antibiotic consumption. Hugonot (24) found evidence, in a double-blind placebo-controlled trial lasting one year, of the prophylactic action of RU 41740 against respiratory tract infections in elderly institutionalized individualso, in a precarious general condition because of metabolic disease, cardiac failure or chronic bronchitis, for example.

Because of its good safety/acceptability, on the basis of clinical parameters as well as various immunological parameters, RU 41740 has been studied in children over 1 year old. Studies used the same methodology and the same results were obtained concerning efficacy (8).

It is important in the case of an immunomodulator to confirm that clinical efficacy is accompanied in parallel by the same immunological activity found in basic studies.

Grassi showed in 36 chronic bronchitis patients with abnormalities of phagocytosis (phagocyte index, fungicidal action on Candida) that RU 41470 had an immunological action on these parameters and at the same time was effective in terms of anti-infectious prophylaxis (reduction in the number of infections, duration of respiratory infections)(17).

Evaluation of Immunological Aceptability

The clinical development of an anti-infectious immunomodulator must include evaluation of the benefit/risk ratio and efficacy results must be viewed in the context of acceptability results. Certain risks must be analyzed in particular : allergic risk, auto-immune risk.

In the case of RU 41740, acceptability was evaluated in the animal then in human both in clinical trials and in specific pharmacological studies.

In the animal, Touraine studied the NZB/NZW mice with a spontaneous disorder similar to human lupus erythematosus. Treatment influenced neither the course of proteinuria nor survival.

No change was found in parameters measured both in the adult with chronic bronchitis and in the child: rheumatoid factor, anti-DNA antibodies, anti-erythrocyte antibodies, circulating immune complexes, IgE. Safety in double-blind placebo-controlled trials was good even in allergic individuals.

The development programme of RU 41740 has provided some answers to the three questions which arise to assess the value of any anti infectious immunomodulator compound.

Its immunological targets are known. RU 41740 stimulates all the components of the immune process, as reflected by the protective action demonstrated in the model of experimental infections and which occurs regardless of the type of infectious agent.

RU 41740 is indicated in prophylaxis of respiratory infections in chronic bronchitis and of recurrent respiratory tract infections in the child over 1 year old.

RU 41740 remains a research compound since its immunological impacts continue to be elucidated. New target disorders are being investigated in the context of anti-infectious prophylaxis and the therapeutic benefit offered to the patient is thus more clearly defined.

REFERENCES

1. A. Alfetra, A. Amoroso, G. M. Ferri, P. Basso, G. Guide, L. Bonomo, *In vivo* effects of RU41740 in aged humans: evaluation of some immunological parameters, Immunol. Clin. 7:39-45 (1988).

2. J. P. Andreux, M. Renard, and M. H. Andreux, Modulation of murine hemopoiesis by repeated injection of glycoprotein extract from *klebsiella pneumoniae*, Int. J. Immunopharmacol. 8:147-154 (1986).

3. D. Anthoine, B. Blaive, G. Cabanieu, J. Chretien, A. Danrigal, C. H. Ducreuzet, D. Dusser, P. Leophonte, J. Migueres, J. F. Muir, and C. H. Ollagnier, Etude en double aveugle du Biostim dans la prevention des surinfections des patients atteints de bronchopathie chronique, Rev. Pneumol. Clin. 41 3:213-217 (1985).

4. M. D. Anthonisen, M. D. Manfreda, C. P. W. Warren, E. S. Hershfield, G. K. M. Harding, and N. A. Nelson, Antibiotic therapy in exacerbations of chronic obstructive pulmonary disease, Ann. Intern. Med. 106:193-204 (1987).

5. D. M. Bamberger and B. L. Herndon, Bactericidal capacity of neutrophils in rabbits with experimental acute and chronic abscesses, J. Infect. Dis. 162:186-192, 1990.

6. H. Blomgren, H. Viland, B. Cedermark, T. Theve, and U. Ohman, Oral treatment with RU41740 (Biostim) in patients with advanced colorectal cancer: influence on the blood lymphocyte population, Int. J. Immunopharmac. 2:71-76, 1989.

7. J. Bonde, R. Dahl, R. Edeltein, A. Kok-Jensen, L. Lazer, L. PUnakivi, A. Seppala, U. Soes-Petersen, and K. Viskum, The effect of RU 41740 an immune modulating compound in the prevention of acute exacerbations in patients with chronic bronchitis, Eur. J. Respir. Dis. 69:235-241 (1986).

8. P. Bonfils, Prophylaxie des infections des voies respiratoires chez l'enfant par le RU41740, Presse Med., 17:1430-1432 (1988).

9. C. Brambilla, Infections respiratoires recidivantes: definition des populations a risque et criteres d'evaluation dans les etudes d'efficacite de la prophylaxie, M/S, suppl. n°7:19-24 (1990).

10. F. Capsoni, F. Minonzio, E. Venegoni, A. M. Ongari, G. Guide and C. Zanussi, *In vitro* modulation of polymorphonuclear leukocyte functions by a glycoprotein extract from *klebsiella pneumoniae* (RU 41740): correction of corticosteroid-induced defects and phagocytosis and adherence, Int. J. Immunotherapy, IV 2: 107-115 (1988).

11. F. Capsoni, F. Minonzio, E. Venegoni, A. M. Ongari, P. L. Meroni, G. Guide, and C. Zanussi, *In vitro* and *ex vivo* effect of RU41740 on human polymorpho-nuclear leukeocyte function, Int. J. Immuno. 10:121-133 (1988).

12. P. Carles, F. Fournial, R. Bollinelli, Interet du Biostim dans la prevention des episodes de surinfection chez les insuffisans respiratoires par bronchite chronique ou dilation des bronches, Gaz Med. France, 88 25:1-3 (1981).

13. B. Chaumet, M. and Ch. Boissier, Structure et distribution du RU 41740, Presse Med. 17:1423-1425 (1988).

14. N. V. Christou, I. Zakaluzny, J. C. Marshall, and C. W. Nohr, The effect of the immunomodulator RU 41740 (Biostim) on the specific and nonspecific immunosuppression induced by thermal injury of protein deprivation, Arch. Surg. 123:207-210 (1988).

15. J. P. Devlin and K. D. Hargrave, The design and synthesis of immune regulatory agents: targets and approaches. Tetrahedron 14:4327-4369 (1989).

16. A. Fietta, C. Bersani, V. De Rose, A. Grassi, P. Mangiarotti, M. Uccelli, and C. Grassi, Evaluation of systemic host defense mechanisms in chronic bronchitis. Respiration 53:37-43 (1988).

17. A. Fietta, C. Bersani, V. De Rose, P. Mangiarotti, C. Merlini, M. Uccelli, G. Guidi, G. Gialdroni Grassi, Double-blind trial RU 41740 vs. placebo: immunological and clinical effects in a group of patients with chronic bronchitis. Respiration 54:145-152 (1988)

18. Fiszer-Szafarz, M. Rommain, C. Brossard, and P. Smets, Hyaluronic acid-degrading enzymes in rat alveolar macrophagesand in alveolar fluid: stimula-tion of enzyme activity after oral treatment with the immunomodulator RU 41740. Biology of the Cell 63: 355-360 (1988).

19. M. Fortier, A. Bonfils, R. Zalisz, M. Heyman, and P. Smets, Pharmacokinetic profile of RU 41740, a bacterial immunomodulator, in mice, rats, and monkeys. Internatl. J. of Phamaceutics, 52:27-36 (1989).

20. P. Gepner, Immunotolerance du RU 41740. Presse Med. 17:1458-1460 (1988).

21. C. Griscelli, B. Rospierre, J. Montreuil, B. Fournet, G. Bruvier, J. M. Lang, C. Marchiani, R. Zalisz, and R. Edelstein, Pouvoir immunomodulateur de glycoproteines isolees de *Klebsiella pneumoniae*. In: Immunomodulation by microbial products and related compounds. Y. Yamamura and S. Kotani, ed., Proc. of an Internatl. Symp., Osaka, July 27-29, 1981, Amsterdam, Oxford, Princeton, Excerpta Medica 563:261-265 (1982).

22. M. Guenounou, F. Vacheron, R. Zalisz, P. Smets, and J. Agneray, Immunological activities of RU 41740, a glcoprotein extract from Klebsieela pneumoniae. I. Activation of murine B cells and induction of interleukin-1 production by macrophages. Ann. Immunol. (Inst. Pasteur) 135D:59-69 (1984).

23. J. Herman, M. C. Kew, A. and A. R. Rabson, The effect of RU 41740 (Biostim) on the production of interleukin 1 by monocytes and enriched large granular lymphocytes in normals and patients with hepatocellular carcinoma. Cancer. Immunol. Immunother. 21a;26-30 (1986).

24. R. Fugonot, L. M. Gutierrez, and L. Hugonot, Action preventive des immunomodulateurs sur les infections respiratoires des sujets ages. Presse Med. 17:1445-1449 (1988).

25. V. Joly, Action d'un immunomodulateur, le RU 41740, sur les infections experimentales. Presse Med. 17:1430-1432 (1988).

26. C. Lambre, J. C. Almayrac, A. Boussairi, A. Greffard, H. De Cremous, K. Atassi, and J. Bignon, Retentissement de l'administration d'un extrait de paroi de klebsiella pneumoniae sur l'immunite locale pulminaire. Rev. Mal. Resp. 6:R154-R194 (1989).

27. J. M. Lang, J. A. Gastaut, J. J. Sotto, J. Troncy, C. Marchiani, Enhancement of delayed cutaneous hypersensitivity by oral administration of RU 41740 (Biostim) in lymphoma patients. A randomized double blind multicentric trial. Int. J. Immunopharma. 8:687-690 (1986).

28. G. Lenoir, Mucoviscidose. Encycl. Med. Chir (Paris-France) Poumons 6039 A[95] 9 188: 1-22.

29. F. Neukirch et al, La bronchite chronique. Rev. Mal. Resp. 5:331-346 (1988).

30. H. Nielsen, and J. Bonde, Association of defective monocyte chemotaxis with recurrent acute exacerbations in chronic obstructive lung disease. Eur. J. Respir. Dis. 68:200-206 (1986).

31. Y. Pacheco, P. Fonlupt, M. Dubois, N. Biot, M. Perrin-Fayolle, and H. Pacheco, La methylation des phospholipides membranes du macrophage alveolaire dans la sarcoidose pulmonaire et les fibroses pulmonaires interstitielles diffuses. Pathol. Bio.. 34, 10:1067-1073 (1986).

32. M. L. Profeta, P. L. Meroni, G. Guidi, R. Palmieri, G. Palladino, V. Cantone, and C. Zanussi, Influenza vaccination with adjuvant RU 41740 in the elderly. Lancet, April 25, 8539 1:973 (1987).

33. A. Rudent, R. Zalisz, A. M. Quero, and P. Smets, Enhanced resistance of mice against influenza virus infection after local administration of glycoprotein extracts from klebsieela pneumoniae. In: J. Immunopharmacol. 7, 4, 525-531 (1989).

34. S. Sozzani, W. Luini, L. Braceschi, and F. Spreafico, The effect ofBiostim (RU 41740) on natural killer activity in different mouse organes. Int. J. Immunopharmacol, 8 (8):845-853 (1986).

35. F. Spreafico, Problems and challenges in the use of immunomodulating agents. Int. Arch. Allergy Appl. Immunol. 76, suppl. 1:108-111 (1985).

36. J. R. Viallat, C. Boutin, P. Farisse, and D. Constantini, Etude en double aveugle d'un immunomodulateur d'origin bacterienne (Biostim) dans la prevention des episodes infectieux chez le bronchitique chronique. Poumon Coeur 39:53-57 (1983).

37. C. Wood, and G. Moller, Influence of RU 41740, a glcoprotein extract from klebsielle pneumoniae, on the murine immune system. KK. II. RU 41740 facilitates the responses to Con A in otherwise unresponsive T. enriched cells. J. Immunol. 125 (1):131-136 (1985).

MDP DERIVATIVES AND RESISTANCE TO BACTERIAL INFECTIONS

IN MICE

M. A. Parant[1], F.J.Parant[1], C. Le Contel[1], P. Lefrancier[2] and
L. Chedid[2]

[1] Lab. Immunopharmacology, CNRS, Paris, France
[2] VACSYN France SA, Paris

INTRODUCTION

Among the numerous biological response modifiers (BRMs) capable of increasing nonspecific immunity to infections, MDP (or muramyl dipeptide) derivatives represent remarkable tools for experimental studies and some of them appear to be suitable for human use because of immunostimulating properties can be separated from some unwanted side effects.

A approach combining immunostimulating agents with chemotherapeutic drugs can be considered in various types of infection in view of lowering the dose of drugs having significant toxic or immunosuppressive effects. Moreover, chemotherapy of infections can be unsatisfactory in immunocompromised hosts, i.e. in the absence of adequate cellular defences. The concept of host defence stimulation comes under renewed and increasing consideration due to the increasing number of problem infections in patients with qualitative and quantitative defects.

In many experimental models administration of cytokines led to heightened resistance to an infectious challenge. However, treatment with these endogenous mediators was often accompanied by considerable side effects. In addition, a clear understanding of cytokine effects on microbial growth would seem to be a prerequisite for their clinical application in infectious diseases. Some endogenous soluble factors may enhance microorganism proliferation, acting as growth factors (1,2). Moreover, all cytokines interact with each other in a network-like fashion, making unclear whether a biological response is caused subsequent to the application of the exogenous cytokine or by stimulation of the endogenous cytokine network. In contrast, administration of a cytokine may result in the down-regulation of another cytokine production. It is therefore unclear how to determine an optimum biological modifying dose of a cytokine and there are many documented examples of bell-shaped dose-response relationships (3). Recombinant proteins appear to be more immunogenic that natural cytokines and to elicit neutralizing antibodies in patients (4).

The possibility that could be considered for nontoxic BRMs such as MDP derivatives is related either to their capacity to directly stimulate phagocytic cells, to

Microbial Infections, Edited by H. Friedman *et al.*
Plenum Press, New York, 1992

induce the production of cytokines at the proper site and at appropriate level, and/or to the synergistic effect they may display with cytokines. Such combined treatment may warrant more extensive preclinical evaluation.

MAIN FEATURES OF MDP ADMINISTRATION

MDP and its derivatives were used in numerous experimental models as adjuvants of specific immune responses and it has been shown that due to the size of such glycopeptides, they were non-immunogenic by themselves unless they were coupled to a carrier (reviewed in 5).

Only some of adjuvant active compounds displayed an enhancement of nonspecific immunity in animal models (6). MDPs are to be given before the infectious challenge or at most a couple of hours later. When administered prior to infection, MDPs are less effective than endotoxin, but in contrast to the toxic bacterial component, they do not cause a negative phase of increased susceptibility when given simultaneously with the organism. Another striking difference with LPS is the capacity of some MDP deriva-tives to protect animal after oral administration to adult and also to newborn mice (7) which are sensitive to infections, and cannot be stimulated by endotoxin whereas MDPs were shown to be very potent immuno-stimulants by parenteral route in animals only 2 to 8 days old . It should be noted that recombinant TNF like endotoxin was also only effective after three weeks of age. A high degree of sensitivity to infection was establish-ed when the spleen had been removed at one week of age even if animals were challenged with *K. pneumoniae* five to seven weeks later. MDP as well as endotoxin was able to protect the sensitized animals against the challenge.

Another interesting observation was that conjugation procedures with an ap-propriate carrier, natural or synthetic, immunogenic or non-immunogenic, were used for enhancing antiinfectious activities of muramyl dipeptides (8). This type of investigation has a potential for developing antibacterial and antiviral agents endowed with both specific and nonspecific activities. After conjugation, the potency of MDPs to enhance resistance to infection was greater, but some side effects such as fever were more pronounced (8).

Antibiotic-immunostimulant combination against bacterial or fungal infections has a therapeutic usefulness, it has been shown that the median effective dose of antibiotic can be lowered by association with MDPs (9). Several examples have been reported indicating that a small dose of an antimicrobial agent potentiated the effect of MDP, particularly in immunodepressed animals (7).

LIPOPHILIC VERSUS HYDROPHILIC MDP DERIVATIVES

Lipophilic MDP derivatives stably incorporated within liposomes displayed increased antitumor effects (10,11). Even when administered as a suspension in saline, lipophilic MDPs have often been shown to be more effective against infections than the hydrophilic analog (7). The greater potency of lipophilic MDPs was demonstrated against a larger spectrum of bacterial challenges and they were particularly effective in immunodepressed animals (reviewed in 7).

In our experiments, a lipophilic MDP derivative, MDP-dipalmitoyl-glycerol or MDP-GDP (11), administered in saline produced constantly a more protective effect against *Klebsiella pneumoniae*, *Streptococcus pneumoniae* or *Pseudomonas aeruginosa* than that obtained with its hydrophilic analog. Moreover, a smaller dose of MDP-GDP was found effective after incorporation into liposomes. In the assay reported in Fig. 1, a dose about 1,000-fold smaller than usual produced a definitive survival in almost 80% of the mice infected with *K. pneumoniae*. An intravenous injection of *Candida albicans* cells produced a very severe disease which may kill all mice after a delay of 2-3 weeks.

of the mice infected with *K. pneumoniae*. An intravenous injection of *Candida albicans* cells produced a very severe disease which may kill all mice after a delay of 2-3 weeks. Hydrophilic MDP was found almost ineffective against a systemic infection with *C. albicans* but several injections of MDP-GDP before the challenge produced a significant

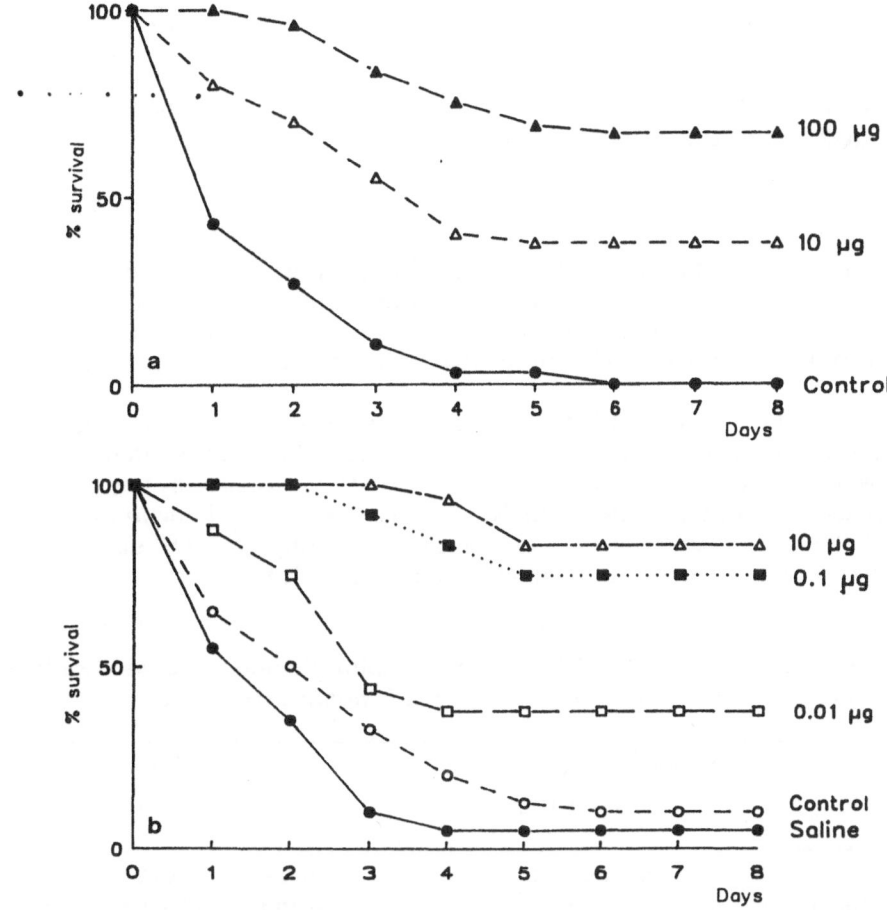

Fig 1.   Protective effect of MDP-GDP in saline or in liposomes in mice infected with 10^4 *Klebsiella pneumoniae*. Pretreatment was given intravenously one day before the intravenous challenge.
A - MDP-GDP in saline B - MDP-GDP in liposome

delay in mortality as shown in Table 1 that reports survival rate after 10 days. All the animals were not definitively protected, even when MDP-GDP had been incorporated into liposomes (Table 1). Association of MDP-GDP to a sub-optimal dose of amphotericin B significantly improved the survival rate.

Table 1. Protective Effect of MDP-GDP Incorporated Into Liposomes against *Candida albicans* Infection

Treatment i.v.	Survival on day +10
Saline Controls	0/16
Liposome controls	0/16
MDP-GDP in saline 100 µg	12/16**
MDP-GDP in liposomes 10 µg	10/16**
30 µg	12/16**

a dose per day from day -4 to day -1 before the intravenous challenge with 2×10^6 C.albicans organisms

b ** = $\underline{p} < 0.01$

The same compound MDP-GDP was used to prevent mortality in mice infected with Influenza APR8 virus by the intranasal route. Whereas hydrophilic MDPs were devoid of any protective effect when given intravenously for 4 days before the challenge, the same protocol with MDP-GDP produced a high percentage (more then 75%) of survivors.

A side effect of the lipophilic MDP derivative is its pyrogenicity. In rabbits, MDP-GDP induced a fever response with a dose (2 µg/kg) 12-fold lower than hydrophilic MDP. However, prevention of the rise in body temperature caused by any of pyrogenic MDPs was achieved with indomethacin. Regarding resistance to infections, it has been shown that the antipyretic drug did not reduce the potency of MDP derivatives, and even enhanced their effect in immunodepressed mice (8,12).

Table 2. Protective Effect of MDP-GDP in Cyclophosphamide-treated Mice Infected with *Mycobacterium fortuitum* or *Pseudomonas aeruginosa*

Treatment i.v. on day -5	Infectious Challenge i.v. on day 0	No of survivors on: day +3	day +5	day +12
Controls	M.fortuitum	8/24	0/24	
MDP-GDP 300 µg	"	24/24	23/24	22/24**
MDP(D,D)-GDP 300 µg	"	20/24	13/24	6/24*
Controls	P.aeruginosa	2/24	2/24	2/24
MDP-GDP 300 µg	"	14/24	12/24	12/24**
MDP(D,D)-GDP 300 µg	"	6/24	3/24	3/24

a All mice received 6 mg Cyclophosphamide i.p. on day -4 before the challenge with 7×10^6 M.fortuitum or 6×10^5 P.aeruginosa organisms.

b * = $\underline{P} < 0.05$, ** = $\underline{p} < 0.01$.

IMMUNOCOMPROMISED HOSTS

Immunosuppressive drugs or radiation therapy are known to impair natural resistance to infection. Such therapy results in neutropenia and myelosuppression which are frequently associated with opportunistic infections. Cyclophosphamide administration in mice was used as a model to increase the susceptibility to various organisms. Thus, immunosuppressed mice were killed with 10^5 *C. albicans* organisms instead of 2×10^6 in normal mice and well protected by MDP-GDP.

An increase in the number of both typical and nontuberculous mycobacterial infections has been reported in HIV patients [13,14]. Since disseminated infection due to *Mycobacterium fortuitum* has been reported in immunocompromised hosts [14], and since the organisms is poorly pathogenic in normal mice, some assays were carried out to evaluate the mouse resistance after cyclophosphamide (6 mg) administration. Under these conditions, mice infected with 7×10^6 organisms died within 5 days (Table 2). A single injection of MDP-GDP one day before cyclophosphamide administration produced a highly significant protective effect and a definitive survival in most of the animals. In comparison, another lipophilic derivative prepared with the MDP stereoisomer was less effective (Table 2). Mice sensitized to Pseudomonas infection were markedly protected by pretreatment with MDP-GDP under the same conditions.

In additional assays, it was shown that the lipophilic MDP was able to enhance the recovery of neutrophil number decreased subsequent to cyclophosphamide treatment and to produce a rapid bone marrow recovery. Moreover, pretreatment with MDP-GDP did not impair efficacy of the alkylating agent in mice inoculated with L1210 leukemia cells (Table 3).

POSSIBLE ROLE OF CYTOKINES

In 1987, we reported the fist evidence that a recombinant cytokine, TNF alpha, was capable of stimulating nonspecific resistance to infection [15]. Soon after, IL-1 was

Table 3. MDP-GDP Administration did not Modify the Efficacy of Cyclophosphamide on Mouse L1210 Leukemia

Treatment after inoculation of L1210 cells on day +1 (mg i.v.)	on day +2 (i.p.)	MST	30-day survivors
Saline	none	12.8	0/20
MDP-GDP 0.3	none	13.0	0/20
MDP-GDP 1	none	12.9	0/20
Saline	Cy	> 30	20/20
MDP-GDP 0.3	Cy	> 30	20/20
MDP-GDP 1	Cy	> 30	20/20

a (Balb/c x DBA2) Fl mice received 300 L1210 cells i.p. on day 0
b Mean survival time
c Cy = cyclophosphamide 6 mg.

179

also shown to display a beneficial role in antibacterial resistance (16). In both cases, the most effective time of administration was either concomitantly or a short time before or after the bacterial challenge. The effect of an LPS contamination in the cytokine preparations was ruled out since both cytokines were effective in LPS-nonresponsive mice (15,16). The mechanisms involved in the protective effect of both cytokines appeared different (17).

Because muramyl peptides are known to directly activate monocytes/macrophages and polymorphonuclear cells, we examined whether they also displayed an indirect effect in stimulating the production of cytokines. In vitro incubation of human monocytes, rabbit or mouse macrophages with adjuvant active muramyl peptides produced the release of IL-1, TNF and IL-6. However, in mice receiving an MDP derivative, the presence of cytokine in the blood was not easily detectable. MDP-GDP is the only one to induce a low level of serum TNF and IL-6. Then, we provided evidence that some MDP derivatives were able to prime animals for an enhanced cytokine production when given a few hours before a challenge with LPS or bacteria (18). The difference between hydrophilic MDP and the lipophilic derivative was related to the duration of the MDP-GDP effect, 2 days after treatment the TNF level produced following LPS injection was still higher (Fig. 2). Comparison between several MDP derivatives led to the conclusion that only those displaying a protective effect in infected mice were capable of preparing TNF and IL-1 producing cells to release huge amounts of cytokine at the early beginning of infection (Table 4).

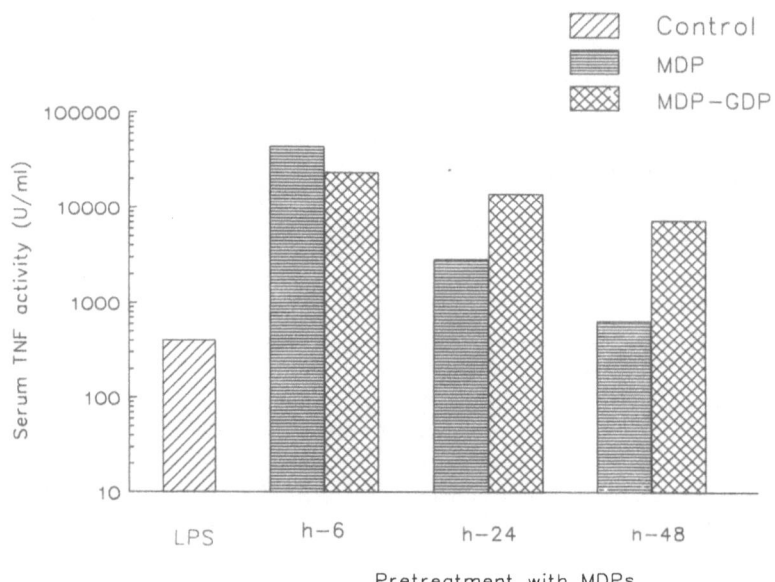

Fig 2.   Influence of pretreatment with MDP or MDP-GDP on LPS-induced TNF production in mice. All mice received an intravenous injection of 25 μg LPS and were bled 2 hrs later. Intravenous administration of MDP or MDP-GDP (300 μg) was made at various time intervals prior to LPS. TNF activity in serum was evaluated by the bioassay on L929 cell line.

Table 4. Protective Effect against *K. pneumoniae* Infection and priming Effect for Induction of TNF Production in Mice of Several Adjuvant MDPs

Pretreatment	K.pneumoniae infection	TNF in serum of LPS-treated mice
MDP = AcMur-L-Ala-D-iGln	+ +	+ + +
Murabutide = MDP(NH2)-OnBu	+ +	+ + +
MDP-GDP	+ +	+ + +
MDP(D,D)-GDP	+ + +	+ + +
MDP-A--L	+ + + +	+ + + +
Murametide = MDP(NH2)-OMe	−	−
MDP(Gly) -OnBu	−	−
MDP(L-Val) -OnBu	−	−
MDP(L-Thr)	−	−
MDP(D-Nle) -OnBu	−	−
MDP(D-NLe) -OMe	−	−

Beside the enhanced production of cytokines, MDP derivatives have been described to potentiate macrophage functions (reviewed in 19). Purified polymorphonuclear cells (PMN) responded to muramyl peptides by a marked increase of the oxidative burst in response to FMLP and of the killing of *C. albicans* cells (20). This is a direct effect since it could be observed in the absence of other cell populations. However, TNF was able to stimulate PMN functions, and when combined with MDP, the cytokine displayed a strong capacity to enhance the bactericidal potency of PMN (20).

INFLUENCE OF A DISPERSING AGENT ON MDP-GDP STIMULATING EFFECTS

In the above-mentioned experiments, lipophilic compounds were suspended in saline by sonication or by using a grinder. Before being injected, the preparation has to be stirred up. In the following assays, MDP-GDP was associated with a trehalose derivative, di-octanoyl-D-trehalose sulfate or DOTS, allowing to obtain a stable homogenous dispersion in the aqueous medium. After lyophilisation, an identical dispersion was obtained by adding water.

Several comparative experiments have already been carried out. It has been shown that the protective effect of MDP-GDP against a *K. pneumoniae* infection was not modified in any aspect by injecting the compound dispersed by DOTS in comparison with the compound suspended in saline. DOTS solution alone was used in all *in vitro* and *in vivo* assays and was regularly found negative, comparable to medium or to saline. Large doses of DOTS were injected intravenously to rabbits without inducing fever. *In vitro*, MDP-GDP dispersed with DOTS was found slightly more active than the compound suspended in saline, as measured by TNF and IL-6 production in monocyte cultures, or by certain PMN functions such as oxidative response and inhibition of *C. albicans* growth. The *in vivo* release of serum TNF was enhanced by pretreatment of mice with MDP-GDP 6 hrs before injection of LPS. The priming effect was stronger when mice were pretreated with the combination of MDP-GDP/DOTS. Therefore, it does not seem that the formulation used to administer the lipophilic derivative impair its immunostimulating properties.

The most surprising result obtained with MDP-GDP and DOTS was related to the toxic synergism with LPS. It has been reported by Ribi (21) that guinea pigs were rendered more susceptible to the toxic effect of LPS by the simultaneous administration of MDP. We confirmed a similar but less marked toxic synergism in mice and showed that there are large differences between MDP derivatives in their capacity to increase LPS toxicity (23). Although TNF production is often linked with LPS toxicity, it has been shown that there was no correlation between the toxicity of LPS in animals treated with MDPs and the priming effect for an enhanced TNF level in response to LPS (18). MDP-GDP in saline caused both an increase in LPS toxicity and a priming effect for TNF release. However, when MDP-GDP was suspended in DOTS, its effect on LPS toxicity was decreased (Table 5) whereas its capacity to enhance LPS-induced TNF production in the blood was increased. It was shown in further assays that DOTS injected simultaneously with LPS did not modify the lethal effect of the bacterial component.

Results in this report underline the role of formulation in treatment with an immunostimulant in mice subsequently challenged with a pathogenic organisms. In most cases, enhanced effectiveness of MDP molecules is accompanied by stronger side effects such as pyrogenicity and increase in LPS toxicity. However, when MDP-GDP was combined with DOTS, the effect of the lipophilic derivative on LPS toxicity was decreased whereas immunostimulating properties did not seem to be impaired. In addition, the presentation of the lipophilic compound as a lyophilized product, easily reconstituted with an aqueous medium, allows to perform comparative assays with an homogenous and stable preparation.

REFERENCES

1. M. Denis, D.Campbell, and E.O.Gregg, Interleukin-2 and granulocyte-macrophage colony-stimulating factor stimulate growth of a virulent strain of *Escherichia coli*, Infect. Immun. 59:1853 (1991).

2. M. Denis, Growth of *Mycobacterium avium* in human monocytes: identification of cytokines which reduce and enhance intracellular microbial growth. Eur. J. Immunol. 21:391 (1991).

3. C. Huber, and M. Herold, The importance of patient monitoring in clinical cytokine trials: use of serum markers to define biologically active doses. Cancer Surveys 8:809 (1989).

4. G. Antonelli, M. Currenti, O. Turriziani, and F. Dianzani. Neutralizing antibodies to interferon-alpha: Relative frequency in patients treated with different interferon preparations. J. Infect. Dis. 163:882 (1991).

5. F. Audibert, C. Leclerc, and L. Chedid. Muramyl peptides as immunopharmacological response modifiers, p. 307, in: "Biological Response Modifiers", P.F.Torrence, ed., Academic Press, Orlando (1985).

6. M. A. Parant, Muramyl peptides as enhancers of host resistance to bacterial infections, p. 235, In: "Immunopharmacology of Infectious Diseases: Vaccine Adjuvants and Modulators of Nonspecific Resistance", J.A. Madje, ed., Alan Liss, Inc., New York (1987).

7. M. A. Parant, and L. Chedid, Stimulation of nonspecific resistance to infections by synthetic immunoregulatory agents, Infection, 12:230 (1984).

8. M. Parant, G. Riveau, F. Parant, C. A. Dinarello, S. M. Wolff, and L. Chedid, Effect of indomethacin on increased resistance to bacterial infection and on febrile response induced by muramyl dipeptide, J. Infect. Dis., 142:708 (1980).

9. F. M. Dietrich, W. Sackman, O. Zak, and P. Dukor, Synthetic muramyl dipeptide immunostimulants; protective effects and incrased efficacy of antibiotics in experimental bacterial and fungal infections in mice, p. 1730, In: "Current Chemotherapy and Infectious Diseases", J.D. Nelson, G. Grossi, eds., Am. Soc. Microbiol. (1981).

10. E. S. Kleinerman, K. L. Erickson, A. J. Schroit, W. E. Fogler, and I. J. Fidler, Activation of tumoricidal properties in human blood monocytes by liposomes containing lipophilic muramyl tripeptide, Cancer Res., 43:2010 (1983).

11. N. C. Phillips, M. L. Moras, L. Chedid, P. Lefrancier, and J. M. Bernard, Activation of alveolar macrophage tumoricidal activity and eradication of experimental metastases by freeze dried liposomes containing a new lipophilic muramyl dipeptide derivative, Cancer Res., 45:125 (1985).

12. M. Parant, G. Riveau, and L. Chedid, Influence of indomethacin on various immunopharmacological effects of muramyl dipeptide (MDP), Agents Actions, 7:33 (1980).

13. A. J. Crowle, D. L. Cohn, and P. Poche, Defects in sera from acquired immunode-ficiency syndrome (AIDS) patients and from non-AIDS patients with *Mycobacterium avium* infection which decreases macrophage resistance to M.avium. Infect. Immun., 57:1445 (1989).

14. J. B. Sack, Disseminated infection due to *Mycobacterium fortuitum* in a patient with AIDS, Rev. Infect. Dis., 12:961 1990).

15. M. Parant, F. Parant, M-A. Vinit, and L. Chedid, Action protectrice du tumor necrosis factor (TNF) obtenu par recombinaison génétique contre l'infection expérimentale bactérienne et fongique, C. R. Acad. Sci. Paris, 304 Série III:1 (1987).

16. C. J. Czuprynski, and J. F. Brown, Recombinant murine interleukin 1 alpha enhancement of nonspecific antibacterial resistance, Infect. Immun., 55: 2061 (1987).

17. M. Parant, Possible mechanisms in endotoxin-induced modulation of nonspecific immunity, in: "Cellular and Molecular Aspects of Endotoxin Reactions" p.395, A.Nowotny, J.J.Spitzer and E.J.Ziegler, eds., Elsevier Science Publishers, Amsterdam (1990).

18. M. Parant, F. Parant, M-A. Vinit, C. Jupin, Y. Noso, and L. Chedid, Priming effect of muramyl peptides for induction by lipopolysaccharide of tumor necrosis factor production in mice, J. Leukocyte Biol., 47:164 (1990).

19. C. Leclerc, and L. Chedid, Macrophage activation by synthetic muramyl peptides p.1, in: "Lymphokines", E.Pick, ed., Academic Press Inc., London (1982).

20. C. Jupin, M. Parant, and L. Chedid, Effect of muramyl peptides and tumor necrosis factor on oxidative responses of human blood phagocytes, Immunol. Letters, 22:187 (1989).

21. E. E. Ribi, J. L. Cantrell, K. B. Von Eschen, and S. M. Schwartzman, Enhance-ment of endotoxin shock by N-acetylmuramyl-L-alanyl-(L-seryl)-D-isoglutamine (muramyl dipeptide), Cancer Res., 39:4756 (1979).

22. N. C. Phillips, C. Tsoukas, and L. Chedid, Abrogation of azidothymidine-induced bone marrow toxicity by free and liposomal muramyl dipeptide, in: "Immuno-therapeutic Prospects of Infectious Diseases" p. 135, K.N.Masihi and W.Lange, eds., Springer-Verlag, Berlin, (1990).

23. L. A. Chedid, M. A. Parant, F. M. Audibert, G. J. Riveau, F. J. Parant, E. Lederer, J.P. Choay, and P. L. Lefrancier, Biological activity of a new synthetic muramyl peptide adjuvant devoid of pyrogenicity, Infect. Immun., 35:417 (1982).

THERAPEUTIC AND PROPHYLACTIC EFFECTS OF ROMURTIDE

AGAINST EXPERIMENTAL ANIMAL INFECTIONS

Yoshihito Niki and Osamu Tatara*

Department of Primary Health Care and Preventive Medicine
and Department of Internal Medicine*
Kawasaki Medical School
Kurashiki City, Japan

INTRODUCTION

Infectious complications are common and often life-threatening problems in immunocompromised patients. Despite remarkable developments of new anti-microbial agents and life-supporting treatments in the past decade, certain infections remain difficult to control when the immune condition of a patient is severely depressed. In these patients without adequate cellular and/or humoral defenses, even potent antimicrobial agents may become less effective. And patients who survive the first episode of infection are at high risk for recurrence of subsequent periods of immuno-suppression.

Immunopotentiating substances and biological response modifiers, such as muramyl dipeptide (MDP) and various kinds of cytokines have been increasingly recognized to be of major importance for treating infections. MDP is a minimal structural requirement for adjuvant activities of the bacterial cell wall, and the compound and its analogs have been revealed to enhance the resistance to infection (1). In the present report, we evaluated the biological activity of Romurtide, a promising synthetic analog of MDP, in our animal models of Klebsiella and Aspergillus pulmonary infection.

Effects of Romurtide on *K. pneumoniae* Pneumonia in Mice

We investigated the synergistic effects of Romurtide and Cefmenoxime, a standard third generation cephalosporin, in the treatment of experimental *Klebsiella pneumonia* in mice. Mice were infected with 10^4 cfu of *K. pneumonia* by inhalation of an aerosolized bacterial suspension. Quantitative cultures of lungs of mice showed rapid proliferation of the organism 2 to 3 days after the infection. About 95% of the untreated animals died of a systemic Klebsiella infection within one week. When mice were treated with a subcutaneous injection of Cefmenoxime twice daily for 3 days, starting 1 day after the infection, a dose-dependent increase of survival rates was observed.

Microbial Infections, Edited by H. Friedman *et al.*
Plenum Press, New York, 1992

To clarify the synergistic effect of Romurtide, mice were treated with Cefmenoxime at a daily dose of 40 mg/kg (lower than the ED_{50}). Romurtide was injected subcutaneously at a dose of 100 µg/animal on 3 different schedules. When Romurtide was administered one day prior to the infection, a significantly higher survival rate of 80% was observed on day 7. Even with a single dose of 100 µg of Romurtide administered one day after the infection, the survival rate observed was higher than that of mice treated with Cefmenoxime alone; still, more than 50% of the mice in this group died within a week. One additional injection of the compound at 3 days after the infection, however, improved the survival rate to 60%. Quantitative cultures of *K. pneumoniae* in lungs of animals treated with Cefmenoxime alone showed a retardation of the proliferation compared to those of control mice. Most animals, however, showed higher counts at days three and four. In the lungs of mice treated with the combination chemotherapy, the growth of the bacteria was apparently suppressed, and rarely exceeded the initial counts of 10^4 cfu (colony forming unit) even 4 days after the infection. A histopathological study of lungs of these animals also showed evidence of a protective activity of Romurtide. In a lung section of a mouse treated with a prophylactic dose of Romurtide one day prior to infection and sacrificed 6 hr after infection, we can see an early response of macrophages and neutrophils into alveolar spaces, while no inflammatory cell infiltration was observed in the sections of lungs from control animals at this point. An accumulation of inflammatory cells in alveolar spaces became apparent 18 - 24 hr after the infection. A prophylactic dose of 100 µg of Romurtide affected the functions of the alveolar macrophages. Mean total cell count in alveolar lavage fluid from animals with Romurtide prophylaxis was significantly higher than that from PBS treated control animals at 24 hr after the infection; however, no significant difference was observed at 2 and 3 days after the infection.

The chemotactic and phagocytic activities of alveolar macrophages harvested from mice 1 to 4 days after a single dose of Romurtide were also examined. As shown in Table 1, the migration of macrophages through a filter was enhanced significantly at 2 to 4 days after Romurtide injection. The phagocytic index of these macrophages was also significantly higher than that of PBS treated control at 24 hr after the infection. It is suggested that the enhancement of host resistance to experimental *Klebsiella pneumonia* by Romurtide is attributed in part to an augmentation of the functions of macrophages and neutrophils.

Effects of Romurtide on *A. fumigatus* Lung Infection in Rats

Recently, not only bacterial but fungal infection has become another problem in immunocompromised patients. Pulmonary aspergillosis, in particular, is a potentially

Table 1. Stimulation of Alveolar Macrophage Functions in Mice Treated with 100 µg/mouse Romurtide

Function	Treatment	Days after single dose treatment			
		1	2	3	4
Chemotaxis	PBS	95.8±19.2	45.6±28.3	37.2±17.1	69.2±25.1
	ROM	64.6±18.7	160.3±15.0 *	91.0±39.6	135.2±48.6 *
Phagocytosis	PBS	25.9± 7.2	27.7± 4.0	25.3± 7.1	27.5± 5.8
	ROM	45.5± 0.9 *	36.3± 5.5	36.5±11.3	32.3± 7.5

Chemotaxis: number of chemotactic cells/HPF
Phagocytosis: % of macrophages with Latex particles
*: $P < 0.05$

life-threatening complication among neutropenic patients with leukemia or lymphoma and bone marrow transplant recipients. In spite of efforts to improve diagnosis and treatments over the last decade, early and specific diagnosis of this infection is still difficult, and treatment may be toxic and often fails.

We have three different models of pulmonary aspergillosis: an acute model, a model of chronic infection, and a model where recurrent infection is induced by immunosuppression. The prophylactic and therapeutic effects of Romurtide were evaluated in these models. In the acute model, male SD rats were treated for two weeks with three doses per week of 100 mg/kg of cortisone acetate while being maintained on a low protein (8%) diet and receiving tetracycline via their drinking water. Animals were infected by surgically exposing their trachea and injecting a suspension of 10^6 conidia of *A. fumigatus*. Animals develop a rapidly progressive aspergillus infection that remains confined to the lungs. Mortality in this model was 100% by 7 to 10 days post infection. In this acute model, we administered 3 to 7 doses of 500 or 1,000 µg of Romurtide beginning one day prior to infection. Doses were administered by subcutaneous injection or by gastric lavage. In all experiments, the survival of untreated control was 0% by day 8. In some experiments, Romurtide appeared to delay mortality, however, as sole therapy, the drug did not have a statistically significant effect on mortality in the acute model of pulmonary aspergillosis.

We have used this acute model to evaluate the effectiveness of various antifungal agents, including aerosol amphotericin B treatment and prophylaxis (2-4). The aerosol route of amphotericin B is more efficient than the intravenous route in delivering the drug to the lung; it also should be less toxic since it limits accumulation in other organs. We also investigated the activity of Romurtide in the acute model, when combined with aerosol amphotericin B. Three daily doses of Romurtide were administered by subcutaneous injection beginning one day prior to infection; a single dose of 1.6 mg/kg of aerosol amphotericin B prophylaxis was administered 2 days prior to infection. In this experiment, the survival rate was not significantly different for animals that received both aerosol amphotericin B and Romurtide and animals that received aerosol amphotericin B alone (Figure 1). We also compared the effects of the compound when

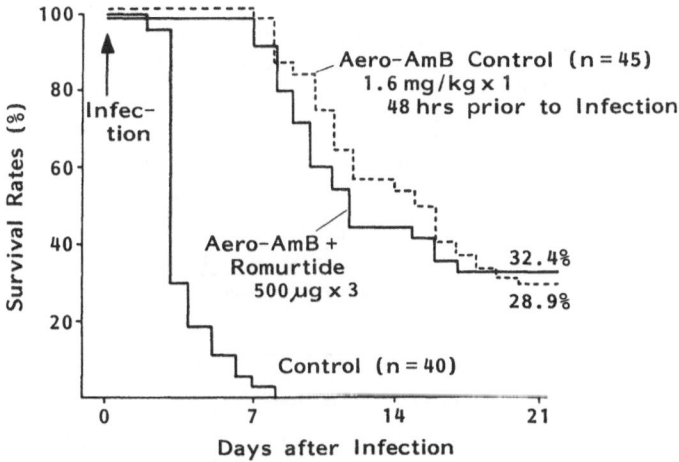

Romurtide; 500 µg s.c., day -1, 0, +1

Fig. 1. Combination therapy with amphotericin B and Romurtide against acute Rat Pulmonary Aspergillosis.

combined with intraperitoneal doses of amphotericin B. The doses of amphotericin B were highly effective, however, and the combination with Romurtide did not improve survival over that of amphotericin B alone.

We were concerned that our inability to demonstrate the therapeutic effect of Romurtide might result from the fact that the high inoculum, acute model produced a very severe, rapidly progressive infection. Because the infection is quickly fatal, subtle enhancements of non-specific immunity are unlikely to significantly alter the outcome. So next we investigated the effects of Romurtide in models of more indolent or chronic infection. To produce a model of chronic infection, we used a lower inoculum, 10^5 conidia, and discontinued steroids and low protein diet at one week post infection. In this model, Romurtide, when given alone at high doses, did not prevent mortality; however, it significantly delayed mortality compared to controls. By day 21, the survival rate among controls was 23.5% and was 43% among animals that had received 9 doses of Romurtide. Figure 2 shows the effects of aerosol amphotericin B alone or combined with Romurtide in the chronic model. Romurtide, given daily for three days beginning one day prior to infection did not enhance the activity of amphotericin B. However, we did observe a reduction in mortality when amphotericin B treated animals were treated with three doses per week of Romurtide beginning on the first day post infection.

It is well recognized that immunosuppressed patients who develop aspergillosis may experience a recurrence of the infection during a subsequent period of immunosuppression. In our chronic model, by the end of week four post infection, animals had regained more than 60% of the weight lost during the steroid induction period and appeared to be healthy. However, at necropsy, they had more than 10^4 cfu of *A. fumigatus* in their lungs. Histopathological sections of lungs from these animals showed many well developed focal lesions which were encapsulated by macrophages, lymphocytes and multi-nucleated giant cells. In the center of the lesions we could see branching septate hyphae and necrosis in the silver stained section. We adapted this chronic model to investigate the question of recurrence. Animals received a prophylactic dose of aerosol amphotericin B, steroids were discontinued one week post

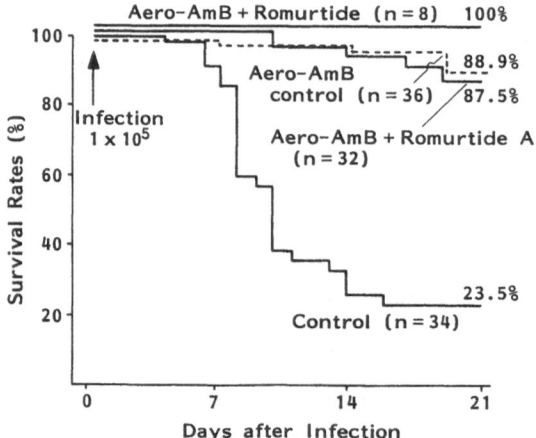

Aero-AmB; 1.6 mg/kg x 1 48 hrs prior to Infection
Romurtide A; 500 µg x 3 day −1, 0, +1
Romurtide B; 500 µg x 3/week day +1 ∿ +20

Fig. 2. Combination therapy with amphotericin B and Romurtide against chronic Rat Pulmonary Aspergillosis.

infection and then restarted four weeks post infection. During the second period of steroid treatments, animals also were treated with Ceftriaxone to prevent bacterial infections. When steroids were discontinued, there was a slow but steady clearance of the fungus from the lungs, and when steroids are restarted the organism began to proliferate. The infection had spread rapidly throughout the entire lung and the mortality rate was 90% by the seventh week of the second course of steroids. We studied the activity of Romurtide in this model. Figure 3 shows the results of quantitative cultures during weeks 4 through 7 post infection. As mentioned previously, when steroids were discontinued, as shown on the left, there was slow clearance of aspergillus from the lungs; the rate of clearance was similar for the control and Romurtide treated animals. When steroids were restarted, as shown on the right, there was a proliferation of aspergillus in the lungs of control rats. In contrast, despite the reinstitution of steroid immunosuppression, aspergillus was cleared from the lungs when animals were treated with Romurtide. Total mortality rate of animals at the seventh week of steroid treatment was significantly higher in the control group. In this model of recurrent aspergillosis, a second course of steroids caused reactivation and progression of the infection, while treatment with Romurtide appeared to prevent reactivation and to promote clearance of the fungus.

DISCUSSION

Romurtide was selected as a compound showing higher biological activity but less pyrogenicity than other compounds through studies on the structure-activity relationship. The profile and underlying mechanisms of the stimulating effects of Romurtide on resistance to infection in animals have been investigated and reported by many investigators. The drug has been confirmed to enhance several immune-functions as shown in Table 2. It could stimulate not only animal macrophages but also human macrophages to induce interleukins, CSFs and complement component and so on.

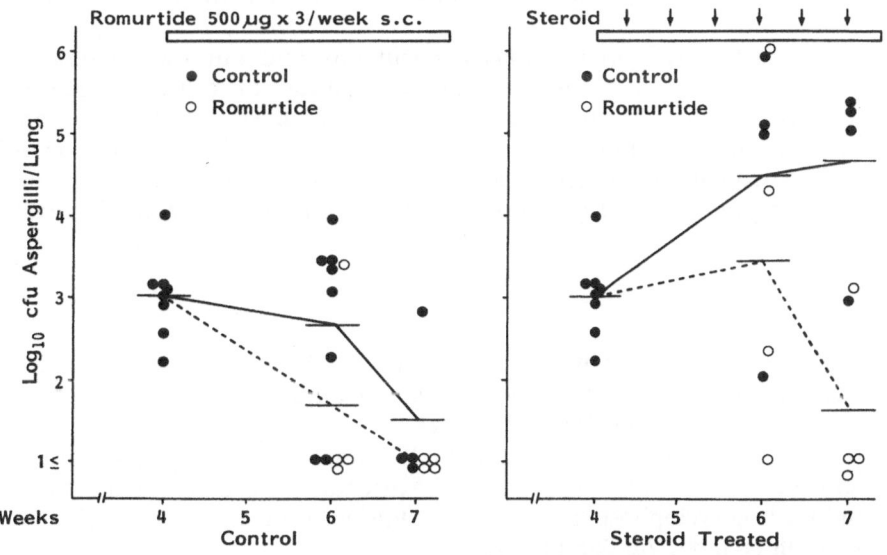

Fig. 3. Effects of Romurtide in control and steroid retreated rats. Quantitative Culture of Aspergilli in Rat Lung.

Table 2. Factors and Cytokines Induced by Romurtide

IL-1	Monocyte/Macrophage	in vitro	mouse, human
IL-6	Monocyte/Macrophage	in vitro	monkey, human
G-CSF	Monocyte/Macrophage	in vitro	mouse, human
M-CSF	Monocyte/Macrophage	in vitro	mouse, human
TNF	Monocyte/Macrophage	in vitro	mouse, human
IFN-γ	Serum	in vivo	mouse
PGE$_2$	Monocyte/Macrophage	in vitro	mouse, human
Complement components	Serum	in vivo	mouse, human
	Macrophage	in vitro	mouse
Fibronectin	Serum	in vivo	mouse

Otani T., et al.

Repeated subcutaneous injection with 1 mg/animal of Romurtide increased significantly the peripheral counts of WBCs and platelets in monkeys. Increase of the WBC count was due to a marked increase of neutrophils and monocytes. Osada and co-workers reported that the compound activates chemotactic, phagocytic and killing activity of neutrophils, in addition to increasing the peripheral count. On the basis of this immune potentiating activity, Romurtide is supposed to have the ability to stimulate host resistance mechanisms.

Otani et al. investigated the protective effects of Romurtide against systemic infection with *Pseudomonas aeruginosa*, *Staphylococcus aureus* and *Candida albicans* in mice (1). They reported that administration of only one dose of the compound at 100 µg/mouse one day prior to the infection led to significant increase in the survival rates of animals in all organisms tested. Timing of the administration of the compound greatly influenced its protective activity. In an *Escherichia coli* infection model, the activity was greatest when Romurtide was administered one day before infection. They concluded that in a systemic infection model the organisms disseminated rapidly from the injection site, so that it may not be possible by simultaneous treatment with Romurtide to stimulate the immune systems soon enough to control the infection. They also showed that the therapeutic effects of various antimicrobial agents are potentiated by combined use of Romurtide. Combined chemotherapy with Romurtide and antibiotics reduced the ED_{50} values of antibiotics to one-half to one-fifth those of solo therapy.

In conclusion, based on previous experiments and our own, Romurtide can effectively prevent the establishment and reactivation of a variety of infectious microorganisms in normal and immunosuppressed hosts. The mechanism of enhancement of host resistance may at least involve qualitative and quantitative activation of neutrophils and macrophages. The clinical usefulness of Romurtide in the treatment and prophylaxis of serious opportunistic infections deserves further evaluation.

REFERENCES

1. T. Otani, K. Katami, T. Une, Y. Osada, and H. Ogawa, Restoration by MDP-Lys (L18) of resistance to *Pseudomonas pneumonia* in immunosuppressed guinea pigs, Microb. Immun. 28:1077 (1984).
2. R. Nakajima, Y. Ishida, F. Yamaguchi, T. Otani, Y. Ono, M. Nomura, T. Une, and Y. Osada, Beneficial effect of Muroctasin on experimental leukopenia induced by cyclophosphamide or irradiation in mice, Arzneim. Forsch./Drug Res. 38 (II), Nr. 7a: 986 (1988).

3. T. Otani, Y. Une, and Y. Osada, Stimulation of non-specific resistance to infection by Muroctasin, Arzneim. Forsch./Drug Res. 38 (II), Nr. 7a:969 (1988).
4. Y. Niki, E. M. Bernard, H. J. Schmitt, W. P. Tong, F. F. Edwards, and D. Armstrong, Pharmacokinetics of aerosol amphotericin B in rats, Antimicrob. Agents Chemother. 34 (1):29 (1990).
5. Y. Niki, E. M. Bernard, F. F. Edwards, H. J. Schmitt, B. Yu, and D. Armstrong, Model of recurrent pulmonary aspergillosis in rats, J. Clin. Microb. 29 (7): in press (1991).

ENHANCEMENT OF HOST RESISTANCE TO OPPORTUNISTIC

INFECTIONS BY UBENIMEX (BESTATIN)

Masaaki Ishizuka[1], Fuminori Abe[2], Shigeru Abe[3], Katsuhisa Uchida[3], Tatsuo Ikeda[3], Noriko Ito[3], Kayo Aoyagi[3], Hideyo Yamaguchi[3] and Tomio Takeuchi[1]

Institute for Chemotherapy[1], Microbial Chemistry Research Foundation, Shizuoka; Research Laboratories[2], Nippon Kayaku Company, Tokyo; Teikyo University School of Medicine[3], Tokyo, Japan

SUMMARY

Ubenimex is a low molecular biological response modifier (BRM) which has been demonstrated to have antitumor and immunomodulatory activities. In this study, the effect of ubenimex on infection with *Candida albicans* was investigated in normal and immunosuppressed mice, and it showed a prophylactic effect. In normal mice, ubenimex prolonged survival time in a dose-dependent manner. In immunosuppressed mice treated with cyclophosphamide 4 days before infection, ubenimex at 5 mg/kg for 5 days significantly increased the number of survivors. Significant improvement in peritoneal leukocyte counts and in function of neutrophils including phagocytosis and release of activated oxygen was observed in ubenimex-treated mice. These results indicate that ubenimex is a potent BRM for prevention against opportunistic infections.

INTRODUCTION

Ubenimex was found by Umezawa et al. to be an inhibitor of aminopeptidases in culture filtrates of *Streptomyces olivoreticuli* (1) and it has now been synthesized (2). We have undertaken a variety of immunopharmacological studies on ubenimex and demonstrated that it acts on macrophages, T cells and bone marrow cells and has antitumor activity (3-12). It is thought that ubenimex enhances the production and/or induction of various cytokines from immunocompetent cells and modulates the function of antitumor effectors such as macrophages, cytotoxic T cells and natural killer (NK) cells (13-19). Since ubenimex is a small molecular weight substance, its pharmacological behavior has been examined in detail. A study on microdistribution demonstrated that it is prolifically distributed on macrophages in lymphoid organs and in a solid tumor (20,21). Biochemical and histochemical studies proved that one of the binding sites is leucine aminopeptidase which is located on the cell surface (23). It is well absorbed by oral administration (21), has extremely low toxicity (24) and can be safely administered over long periods (25). Well controlled clinical trials have evaluated the efficacy and toxicity of ubenimex (26-36). In one of these studies, prolongation of the duration of

Microbial Infections, Edited by H. Friedman *et al.*
Plenum Press, New York, 1992

remission and survival time were observed in adult non-lymphocytic leukemia patients who reached a complete remission by induction and consolidation chemotherapy combined with ubenimex (28). In the course of clinical study, it was pointed out that ubenimex might be effective in preventing secondary infection. Several preclinical studies of ubenimex in experimental infections have been reported. Harada et al. reported that ubenimex enhanced resistance to *Listeria monocytogenes* infection in mice (37). Dicknite et al. demonstrated that it reduced bacterial persistence in experimental chronic *Salmonella typhimurium* infection in mice (38), while Tanaka et al. observed that it has a prophylactic effect on experimental pyelonephritis induced by *Pseudomonas aeruginosa* (39). Against virus infections, ubenimex enhanced antibody formation to *Herpes simplex* (40) and improved CD4$^+$ lymphocyte counts in AIDS and AIDS-related complex patients (41). In this paper, we report the effect of ubenimex on *Candida albicans* infection in mice.

MATERIALS AND METHODS

Specific pathogen free, inbred CD1 (4 week, female) or C3H/He (7 week, male) mice were purchased from Charles River Japan, Atsugi. Ubenimex was prepared by Nippon Kayaku Co., Ltd., Tokyo. Cyclophosphamide was purchased from Shionogi Co., Ltd, Osaka. Ubenimex and cyclophosphamide were dissolved in physiological saline and filtered through a Millipore filter (0.22 μm). *Candida albicans* TIMM No. 1768 and No. 0239 were maintained at the Research Center for Medical Mycology, Teikyo University. The cultural condition and the preparation of *Candida albicans* suspension for infection were performed as described previously (42). Peritoneal leukocyte counts and the function of neutrophils including phagocytosis and chemiluminescence response were examined as described by Lehrer et al. (43).

RESULTS AND DISCUSSION

First, prophylactic effect of ubenimex on *Candida albicans* infection was examined in normal mice. CD1 mice were infected i.v. with 1 x 10^6 cells of *Candida albicans*.

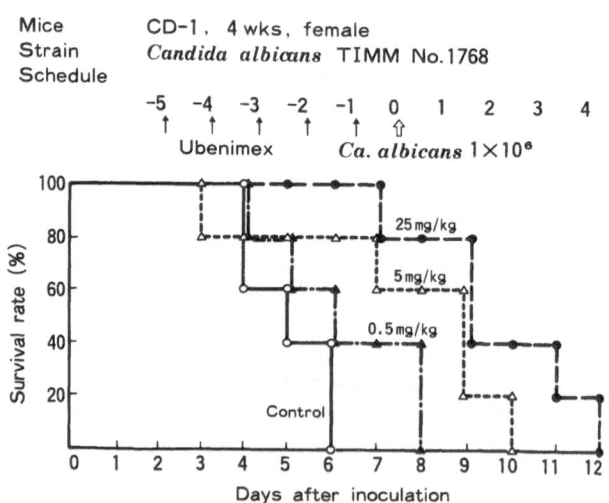

Fig. 1. Prophylactic treatment with ubenimex in normal mice.

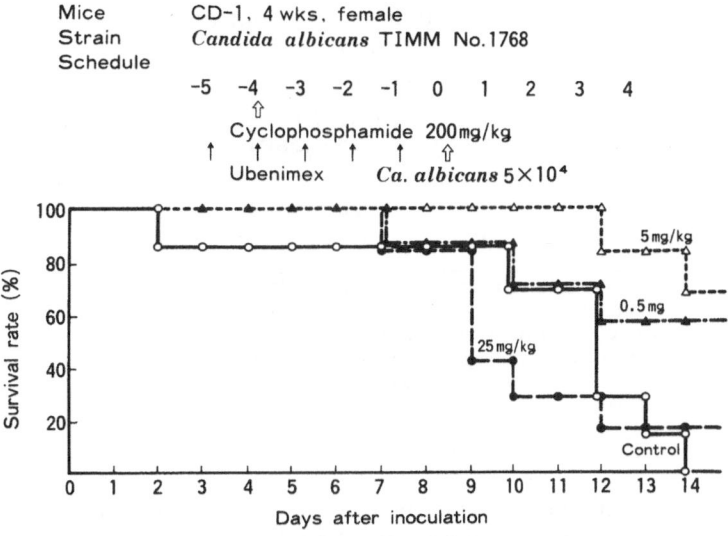

Fig. 2. Prophylactic treatment with ubenimex in immunosuppressed mice.

Ubenimex was administered daily for 5 days starting 5 days prior to the infection. As shown in Figure 1, ubenimex prolonged the survival time of mice in a dose-dependent manner, whereas all non-treated mice died within 6 days after being infected. The prophylactic effect on *Candida albicans* was examined in immunosuppressed mice which were treated with cyclophosphamide. Cyclophosphamide was i.p. injected 4 days before infection at a single dose of 200 mg/kg. The immunosuppressed mice died 14 days following inoculation of 5 x 10^4 cells of *Candida albicans*. Five days before the infection, mice were given ubenimex daily for 5 days and the number of infected survivors was determined. As shown in Figure 2, treatment with ubenimex prolonged the mouse survival. The number surviving more than 14 days was as follows: 4 out of 7 mice at 0.5 mg/kg, 4 out of 6 mice at 5 mg/kg and 1 out of 7 mice at 25 mg/kg. Survivors' kidneys were removed to examine their Candida count. Vertical sections of the left side kidneys were rubbed on Candida GS agar and cultured for 48 hr at 37°C under aerobic conditions. No Candida colony was observed in 2 and 1 out of 4 mice at doses of 0.5 and 5 mg/kg, respectively.

These results indicate that ubenimex has prophylactic activity against Candida infection in normal and immunosuppressed mice. It has been shown that neutrophils have an important role in opportunistic infection including Candida infection. Thus, the effect of ubenimex on neutrophils in cyclophosphamide-treated mice was evaluated. Mice were given cyclophosphamide (200 mg/kg, i.p.) once and given 5 mg/kg of ubenimex daily for 4 days; 24 hr after the last administration, the total number of peritoneal exudate cells (PEC) and the number of neutrophils in PEC was determined. Though cyclophosphamide reduced the total number of PEC to 63% that of normal mice, ubenimex treatment kept the reduction to 46%. Moreover, the number of neutrophils was reduced about 43% in the cyclophosphamide-treated group, but only 25% in the ubenimex-treated group. The effect on phagocytosis of *Candida albicans* and production of superoxide anion by neutrophils was also investigated. Mice were given ubenimex daily for 4 days; one day thereafter neutrophils in casein-elicited PEC of C3H/He mice were collected by Ficoll-sedimentation method and these cells were mixed with *Candida albicans* TIMM No. 0239 opsonized by mouse serum. After incubation for 30 min at 37°C, the number of phagocytic cells was determined. As

Table 1. Effect of Ubenimex on Neutrophil Phagocytosis of *Candida albicans*

Treatment	Dose (mg/kg)	No. of phagocytic cells (mean±SD)	(%)
None		18.9±3.2	100
Ubenimex	25	28.0±7.5*	148

C3H/He mice were given ubenimex daily for 4 days (n=6). Formalin-killed *C. albicans* (2×10^7) and neutrophils (4×10^6) were incubated, and phagocytosis was determined by trypan blue and eosin Y staining.
*$p < 0.05$

shown in Table 1, ubenimex significantly increased the number of phagocytic neutrophils. The effect of ubenimex on chemiluminescence response of PEC was examined in cyclophosphamide treated mice. As shown in Table 2, compared with the cyclophosphamide treated group, ubenimex at 0.5 to 25 mg/kg enhanced the release of superoxide anion from PEC. These results indicate that the administration of ubenimex is effective in increasing the number of neutrophils and also in activating their functions.

We thus demonstrated that a low molecular immunomodulator, ubenimex enhanced host resistance to the infection of *Candida albicans*. This may be due to the activation of neutrophils by ubenimex. As reported previously, ubenimex activates macrophages to be tumoricidal, and augments interleukin 1 production, release of activated oxygens, prostaglandins and lysosomal enzymes (15, 18, 44, 45). In this study, it was shown that ubenimex additionally activates neutrophils.

We recently reported that the treatment of mice with ubenimex increases regulation of the frequency and absolute numbers of colony forming unit granulocyte-macrophage (CFU-GM) (46). As a result, there is an increase in bone marrow cellularity and in the absolute neutrophil counts in cyclophosphamide myelosuppressed mice. In human, Jarstrand et al. demonstrated that ubenimex increased granulocyte phagocytosis when orally administered (47). These results are consistent with the results described above. Ohmori et al. reported that ubenimex significantly enhanced GM-CSF

Table 2. Enhancement of Chemiluminescence Response of Peritoneal Exudate Cells (PEC) in Cyclophosphamide (CY)-treated Mice by Ubenimex

Treatment	Dose (mg/kg)	Counts per minute (mean+SD)	(%)
Saline		2053±223	100
Ubenimex	0.5	2544±476	124
	5	3026±328*	147
	25	3059± 77**	149

C3H/He mice were given CY (200 mg/kg) once and given ubenimex daily for 4 days (n=3). PEC (3×10^5 cells/100µl), picibanil (1mg/ml, 25µl) and luminol (2×10^{-6}M, 100µl) were incubated. Chemiluminescence was mesured by PicoLite 6100 (Packard).
*$p < 0.05$, **$p < 0.01$

production of human peripheral mononuclear cells (48). Shibuya et al. reported that ubenimex modulates high affinity receptor for GM-CSF on leukemic cell line TF-1 (49). Augmentation on CFU-GM production by ubenimex was more obvious in mice in which immune function had been suppressed by irradiation or cyclophosphamide treatment. These effects on bone marrow cells and neutrophils might be one reason ubenimex shows protective activity against infections. It is believed that ubenimex has therapeutic advantages as an anti-infective agent to opportunistic infections in myelosuppressed patients.

REFERENCES

1. H. Umezawa, T. Aoyagi, H. Suda, M. Hamada, and T. Takeuchi, Bestatin, an inhibitor of aminopeptidase B, produced by actinomycetes, J. Antibiot. (Tokyo) 29:97 (1976).
2. R. Nishizawa, T. Saino, M. Suzuki, T. Fujii, T. Aoyagi, and H. Umezawa, A facile synthesis of bestatin, J. Antibiot. (Tokyo) 36:698 (1983).
3. H. Umezawa, M. Ishizuka, T. Aoyagi, and T. Takeuchi, Enhancement of delayed-type hypersensitivity by bestatin, an inhibitor of aminopeptidase B and leucine aminopeptidase, J. Antibiot. (Tokyo) 29:857 (1976).
4. M. Ishizuka, T. Masuda, N. Kanbayashi, S. Fukazawa, T. Takeuchi, T. Aoyagi, and H. Umezawa, Effect of bestatin on mouse immune system and experimental murine tumors, J. Antibiot. (Tokyo) 33:642 (1980).
5. M. Ishizuka, H. Saito, Y. Sugiyama, T. Takeuchi, and H. Umezawa, Effect of bestatin on the mouse immune system and experimental murine tumors, J. Antibiot. (Tokyo) 33:653 (1980).
6. F. Abe, K. Shibuya, M. Uchida, K. Takahashi, H. Horinishi, A. Matsuda, M. Ishizuka, T. Takauchi, and H. Umezawa, Effect of bestatin on syngeneic tumors in mice, Gann 75:89 (1984).
7. F. Abe, K. Shibuya, J. Ashizawa, K. Takahashi, H. Horinishi, A. Matsuda, M. Isizuka, T. Takeuchi, and H. Umezawa, Enhancement of antitumor effect of cytotoxic agents by bestatin. J. Antibiot. (Tokyo) 38:411 (1985).
8. F. Abe, M. Hayashi, H. Horinishi, A. Matsuda, M. Ishizuka, and H. Umezawa, Enhancement of graft-versus-host reaction and delayed cutaneous hypersensitivity in mice by ubenimex, J. Antibiot. (Tokyo) 39:1172 (1986).
9. K. Ebihara, F. Abe, T. Yamashita, K. Shibuya, E. Hayashi, K. Takahashi, H. Horinishi, M. Enomoto, M. Ishizuka, and H. Umezawa, The effect of ubenimex on N-methyl-N'-nitro-N'nitrosoguanidine-induced stomach tumors in rats, J. Antibiot. (Tokyo) 39:966 (1986).
10. J. E. Talmadge, M. Koyama, A. Matsuda, C. Long, and F. Abe, Immunotherapeutic properties of bestatin: Mechanism of activity, in: "Recent Results of Bestatin," H. Umezawa, ed., Jpn. Antibiot. Res. Assoc., Tokyo (1987).
11. K. Nemoto, F. Abe, K. Karakawa, H. Horinishi, and H. Umezawa, Enhanced of colony formation of mouse bone marrow cells by ubenimex, J. Antibiot. (Tokyo) 40:894 (1987).
12. F. Abe, M. Schneider, P. L. Black, and J. E. Talmadge, Chemoimmunotherapy with cyclophosphamide and bestatin in experimental metastasis in mice, Cancer Immunol. Immunother. 29:231 (1989).
13. K. Saito, K. Takegoshi, T. Aoyagi, H. Umezawa, and Y. Nagai, Stimulatory effect of bestatin, a new specific inhibitor of aminopeptidase, on the blastogenesis of guinea pig lymphocytes, Cell. Immunol. 40:247 (1978).

14. I. Florentin, N. Kiger, M. Bruley-Rosset, J. Schulz, and G. Mathe, Effect of seven immunomodulators on different types of immune responses in mice, in: "Human Lymphocyte Differentiation," B. Serrou and C. Rosenfeld, eds., North Holland Biomedical Press, Amsterdam (1978).

15. H. U. Schorlemmer, K. Bosslet, and H. H. Sedlacek, The ability of the immuno-modulating dipeptide bestatin to activate cytotoxic mononuclear phagocytes, Cancer Res. 43:4148 (1983).

16. B. E. Dunlap, S. A. Dunlap, and D. H. Rich, Effect of bestatin on in vitro responses of murine lymphocytes to T-cell stimuli, Scand. J. Immunol. 20:237 (1984).

17. T. Noma, B. Klein, D. Cupissol, J. Yata, and B. Serrou, Increased sensitivity of IL-2-dependent cultured T cells and enhancement of in vitro IL-2 production by human lymphocytes treated with bestatin, Int. J. Immunopharmacol. 6:87 (1984).

18. J. E. Talmadge, B. F. Lenz, R. Pennington, C. Long, H. Phillips, M. Schneider, and H. Tribble, Immunomodulatory and therapeutic properties of bestatin in mice, Cancer Res. 46:4505 (1986).

19. F. Abe, A. Fujii, K. Yoshimura, E. Yugeta, M. Ishizuka, and T. Takeuchi, Immunomodulatory and Therapeutic characteristics of bestatin (ubenimex), Int. J. Immunotherapy 6:203 (1990).

20. H. Miyazaki, Pharmacokinetics and metabolism of bestatin in humans and rats by gas chromatography-mass spectrometry, in: "Small Molecular Immunomodifiers of Microbial Origin," H. Umezawa, ed., Pergamon Press, New York (1981).

21. M. Koyama, Pharmacokinetics and biotrans-formation of bestatin, in: "Recent Results of bestatin 1986," H. Umezawa, ed., Jpn. Antibiot. Res. Assoc., Tokyo (1987).

22. T. Yamashita, J. Ito, F. Abe, K. Takahashi, T. Takeuchi, and M. Enomoto, Autoradiographic study of tissue distribution of [3H]-ubenimex in IMC carcinoma-bearing mice, Int. J. Immunopharmacol. 12:755 (1990).

23. W. E. G. Muller, D. Schuster, R. K. Zahn, A. Maidhof, G. Leyhausen, D. Falke, R. Koren, and H. Umezawa, Properties and specificity of binding sites for the immunomodulator bestatin on the surface of mammalian cells, Int. J. Immunopharmacol. 4:393 (1982).

24. K. Ito, J. Hamada, Y. Irie, T. Hagiwara, Y. Sakai, M. Hayashi, T. Sakakibara, M. Suzuki, Y. Irie, M. Horiguchi, M. Kurata, T. Machida, M. Tsubosaki, A. Matsuda, and N. Konoha, Toxicological studies on bestatin III, Chronic toxicity test and recovery study in beagle dogs, Jpn. J. Antibiot. 36:3053 (1983).

25. T. Ichikawa, Clinical studies by bestatin on genitourinary cancer, Jpn. J. Antibiot. 38:166 (1985).

26. H. Majima, Phase I clinical study of bestatin, in: "Recent results of bestatin 1985," H. Umezawa, ed., Jpn. Antibiot. Res. Assoc., Tokyo (1985).

27. H. Blomgren, F. Edsmyr, P. L. Esposti, and I. Naslund, Immunological and hematological monitoring in bladder cancer patients receiving adjuvant bestatin treatment following radiation therapy. A prospective randomized trial, Biomed. Pharmacother. 38:143 (1984).

28. K. Ota, S. Kurita, K. Yamada, T. Masaoka, Y. Uzuka, and N. Ogawa, Immuno-therapy with bestatin for acute nonlymphocyticleukemia in adults, Cancer Immunol. Immunother. 23:5 (1986).

29. K. Ishihara, Phase III controlled studies of bestatin iRn malignant tumors of the skin, in: "Recent results of bestatin 1985," H. Umezawa, ed., Jpn. Antibiot. Res. Assoc., Tokyo (1985).

30. M. Takada, and M. Fukuoka, Controlled study of ubenimex on inoperable lung cancer, in: "Bestatin Workshop," Proc. 16th Int. Congr. Chemotherapy, Jerusalem (1989).

31. T. Yasuo, S. Ohshima, N. Nakano, and Y. Kotake, A randomized clinical trial of ubenimex in resected lung cancer, in: "Bestatin Workshop," Proc. 16th Int. Congr. Chemotherapy, Jerusalem (1989).

32. C. Mouritzen, Bestatin as adjuvant treatment inoperated stage I and II non-small cell lung cancer, in: "Recent results of bestatin 1986," H. Umezawa, ed., Jpn. Antibiot. Res. Assoc., Tokyo (1986).

33. K. Yoshinaka, T. Tanaka, S. Sakagami, T. Saeki, M. Nishiyama, T. Toge, M. Niimoto, and T. Hattori, Prospective randomized controlled study of bestatin in gastric cancer surgery, Jpn. J. Cancer Chemother. 15:493, (1988).

34. Y. Kuramoto, Clinical research of prevention of recurrence of superficial bladder cancer. Cooperative study of clinical efficacy of bleomycin intravesical instillation and bestatin, Hinyokiyo. 31:1861 (1985).

35. Y. Inuyama, H. Miyake, M. Horiuchi, and C. Taketa, Adjuvant therapy with bestatin for squamous cell carcinoma of the head and neak, in: "Resent results of bestatin 1985," H. Umezawa, ed., Jpn. Antibiot. Assoc., Tokyo, (1986).

36. K. Isono, Effect of bestatin on primary tumor and prognosis in patients with esophageal carcinoma, in: "Recent results of bestatin 1985," H. Umezawa, ed., Jpn. Antibiot. Assoc., Tokyo (1986).

37. Y. Harada, A. Kajiki, K. Higuchi, T. Ishibashi, and M. Takemoto, The mode of immunopotentiating action of bestatin: enhanced resistance of *Listeria monocytogenes* infection, J. Antibiot. (Tokyo) 36:1411 (1983).

38. G. Dickneite, F. Kaspereit, and H. H. Sedlacek, Stimulation of cell-mediated immunity by bestatin correlates with reduction of bacterial persistence in experimental chronic *Salmonella typhimurium* infection, Infect. Immun. 44:168 (1984).

39. N. Tanaka, Y. Kumamoto, T. Hirose, and A. Yokoo, Study of the prophylactic effect of ubenimex on experimental pyelonephritis induced *Pseudomonas aeruginosa* in neutropenic mice, J. Jpn. Assoc. Infec. Diseases 63:748 (1989).

40. A. Knoblich, W. G. Muller, V. Harle-Grupp, and D. Falke, Enhancement of antibody formation against Herpes simplex virus in mice by the T cell mitogen bestatin, J. Gen. Virol. 65:1675 (1984).

41. G. Mathe, H. Umezawa, L. Misset, S. Brienza, C. Canon, M. Musset, and P. Reizenstein, Immunomodulating properties of bestatin in cancer patients, a phase II trial, Biomed. Pharmacother. 40:379 (1986).

42. K. Uchida, Experimental Candidasis, Jpn. J. Med. Mycol. 29:169 (1988).

43. R. I. Lehrer, L. G. Ferrari, J. P-Delafield, and T. Sorrell, Fungicidal activity of rabbit alveolar and peritoneal macrophages against *Candida albicans*, Infect. Immun. 28:1001 (1980).

44. H. U. Schorlemmer, K. Bosslet, G. Dickneite, G. Luben, and H. H. Sedlacek, Studies on the mechanism of the immunomodulator bestatin in various screening test systems, Behring Inst. Mitt. 74:157 (1984).

45. K. Shibuya, E. Hayashi, F. Abe, K. Takahashi, H. Horinishi, M. Ishizuka, T. Takeuchi, and H. Unezawa, Enhancement of interleukin-1 and interleukin-2 release by ubenimex, J. Antbiot. (Tokyo) 40:363 (1987).

46. F. Abe, A. Matsuda, M. Schneider, and J. E. Talmadge, Effects of bestatin on myelopoietic stem cells in normal and cyclophosphamide treated mice, Cancer Immunol. Immunother. 32:75 (1990).

47. C. Jarstrand, and H. Blomgren, Increased granulocyte phagocytosis after oral administration of bestatin, a new immunomodulator, J. Clin. Lab. Immunol. 7:115 (1982).

48. F. Omori, S. Okamura, K. Haga, H. Baba, C. Kawasaki, T. Tanaka, K. Sugimachi, and Y. Niho, Ubenimex stimulates production of colony-stimulating factor from human peripheral mononuclear cells *in vitro*, Biotherapy 4:787 (1990).

49. K. Shibuya, S. Chiba, M. Hino, T. Kitamura, K. Miyazono, and F. Takaku, Effect of ubenimex on proliferation and differentiation of human bone marrow cells and leukemic cell lines, and the possible mechanisms of its action, in: "Experimental Hematology Today-1989," N. C. Gorin and L. Doray, eds., Springer-Verlag, New York (1990).

POTENTIATION OF HOST RESISTANCE AGAINST MICROBIAL

INFECTIONS BY LENTINAN AND ITS RELATED POLYSACCHARIDES

Yutaro Kaneko[1] and Goro Chihara[2]

[1]Ajinomoto Co., Inc., Tokyo, Japan
[2]Biotechnology Research Center
Teikyo University, Kawasaki, 216 Japan

INTRODUCTION

Cytotoxic anticancer agents are generally accompanied by severe side effects in a host, and reduce resistance against cancer and infectious diseases, especially by destroying lymphoid cells and bone marrow cells. As a result, many cancer patients die of various kinds of pneumonitis, septicemia, uremia or other secondary diseases. There are, however, reliable clinical evidence to support the existence of intrinsic resistance against cancer and infectious disease in the human body. An increase in this resistance may be one of the most important facets in the development of pharmaceutical therapy against such diseases.

In Oriental medicine as practiced in Asian countries, the fundamental principle is to regulate homeostasis of the entire body and bring the diseased person to his normal state, rather than to attack the disease directly. Using this concept, we re-examined the antitumor effects of many folk remedies which had long been believed to be effective in cancer patients, and we isolated lentinan having marked antitumor activity from *Lentinus edodes* (Berk.) Sing., an edible mushroom (1).

Lentinan is a fully purified $(1 \rightarrow 3)$-β-D-glucan with $(1 \rightarrow 6)$-β-D-glucopyranoside branches, and its physico-chemical properties are strictly characterized (2). This polysaccharide consists of a triple helix structure, which is important in its biological activities (3-5). The chemical properties of lentinan so far characterized are found to be essential for its immunopharmacological study and clinical use. Lentinan has no toxic side effects in animal models and its LD_{50} is over 2000 mg/kg i.p. The antitumor activity of lentinan was originally confirmed by using sarcoma 180 transplanted in CD-1 mice (2). However, it exhibited a marked antitumor effect not only on allogeneic hosts, but also on various kinds of syngeneic and autochthonous hosts (6-7), and prevented cancer metastasis (8-9) and methylcholanthrene-induced chemical carcinogenesis (7). A five year randomized follow-up control study on lentinan in phase III with advanced or recurrent stomach or colorectal cancer showed that the percent survival ratios of the stomach cancer patients in the lentinan treated group were about 19.5%, 10.4% and 6.5% after 1, 2 and 3 year treatment, while those in the control group were 2.9%, 2.9% and 0%, respectively (10). Lentinan therapy, therefore, showed very good results in prolongation of life span. This article, however, concerns the effects of lentinan and its related polysaccharides on various kinds of microbial infectious diseases including AIDS,

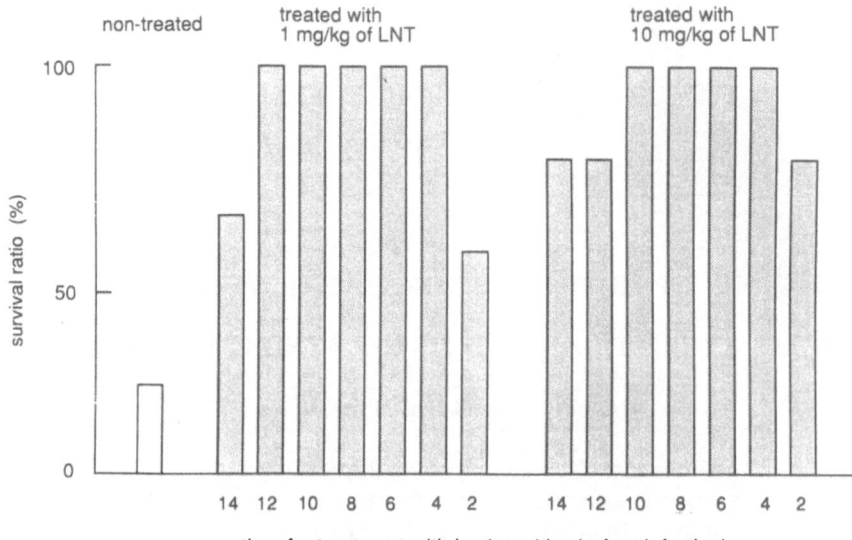

Fig. 1.   The anti-infection activity of lentinan on *Listeria monocytogenes*.
ICR mice were infected by the intravenous route with 2.5 x 10⁶ of
Listeria monocytogenes (12), and lentinan was intravenously
administered 2 to 14 days prior to the infection, at doses of 1 or
10 mg/kg. The survival mice were counted 14 days after the
infection.

their mode of action mechanism as host defense potentiators (HDPs) (11), and their
future possibilities.

Effects of Lentinan or Yeast Glucan on Bacterial Infections

Antibacterial activities of lentinan were examined using Listeria monocytogenes
(12) or *Mycobacteria tuberculoses* (13). ICR mice were infected by the intravenous route
with *Listeria monocytogenes*, and lentinan was intravenously administered 2 to 14 days
prior to the infection, at doses of 1 or 10 mg/kg. The significant anti-infection activity
was observed when lentinan was injected 4 to 10 days before infection (Fig. 1). The
same activity was observed in cyclophosphamide treated mice (Fig. 2). The preventive
effect of lentinan against *Listeria monocytogenes* infection was reduced by carrageenan
administration (Fig. 3) which is known to be a macrophage inhibitor. Similar
observations using zymosan were reported by Matsuo et al. (14).

Lentinan and yeast glucan showed a relapse-preventive effect in mice experimentally infected with tuberculosis (13). ddY Mice were infected with *Mycobacteria
tuberculoses* and then treated by intensive chemotherapy with a three-drug combination,
for instance, (streptomycin + isoniazid + rifampicin or ethanbutol + isoniazid +
rifampicin) for 5 months. After termination of chemotherapy, mice were divided into
three groups of those treated with lentinan, with yeast glucan and non-treated mice as
a control. Lentinan or yeast glucan was administered for 4 wks and again for 4 weeks
after a one month interval. The growth of latent tubercle bacilli in the lung and spleen
was examined by cultivation of tissue homogenates during the period of glucan
treatment and the succeeding 5 months.

The results indicated that lentinan was highly effective in inhibiting the reappearance of tubercle bacilli in the treatment regimen with chemotherapy, for example, as shown in Fig. 4.

Yeast glucan was also shown to be effective in animal models in other types of bacterial infections such as *Escherichia coli* (15), *Staphylococcus aureus* (16), and *Pseudomonas aeruginosa* (17).

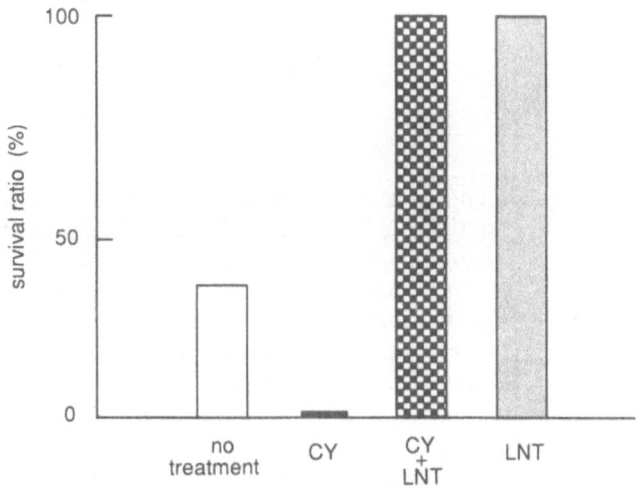

Fig. 2. The anti-infection activity of lentinan on *Listeria monocytogenes* in cyclophosphamide treated mice (12). ICR mice were infected by the intravenously route with 2.5 x 10^6 of *Listeria monocytogenes*. Cyclophosphamide was intraperitoneally administered 1, 3, 5 days prior to the infection, at dose of 50 mg/kg, and lentinan was intravenously administered 7 days prior to the infection, at dose of 3 mg/kg. The survival mice were counted 14 days after the infection.

Antiviral Activity

Antiviral activities of polysaccharides, lentinan or yeast glucan, were observed in the assay systems employing adenovirus type 12 (Ad- 12) (18), VSV-encephalitis virus (19), Abelson virus (19), Hepatitis virus MHV-A59 (20), Herpes simplex I & II (17) and human immunodeficiency virus (HIV) (21). Strain difference between C3H/He and C57BL/6 mice in tumorigenesis of Ad- 12 was observed, that is, C3H/He mice were susceptible to Ad-12, but C57BL/6 was entirely refractory.

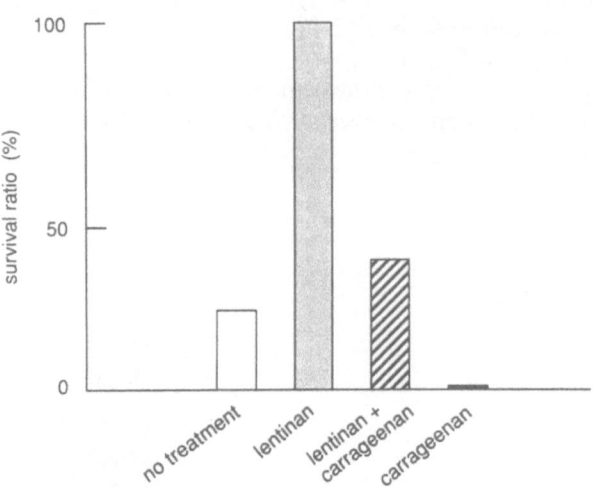

Fig. 3. The carrageenan reduction on the antiinfection
activity of Lentinan. ICR mice were treated with
Listeria monocytogenes and lentinan (12) in the
same manner as Fig. 2. Carrageenan was intraperitoneally administered 24 hrs before the
infection. The survival mice were counted 14
days after the infection.

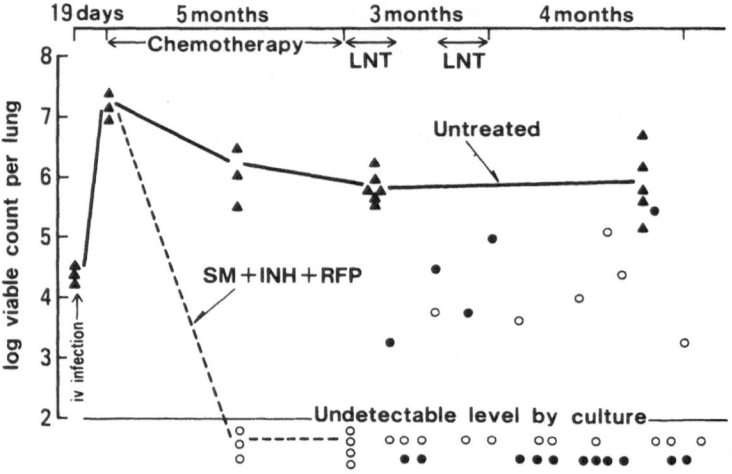

Fig. 4. The relapse-limiting effect of lentinan (LNT) administered after termination of chemotherapy with SM + INH
+ RFP in experimentally infected tuberculous mice, as
revealed in lung viable counts (13). Note: ● LNT-treat-
ed, ○ LNT-untreated individual animal, ▲ Control
animals. SM: streptomycin, INH: isoniazid, RFP:
rifampicin.

The natural cytotoxicity of normal spleen cells from C3H/He mice against Ad- 12 transformed cells was entirely inert, while the cytotoxic activity of normal spleen cells from C5BL/6 was significantly high. Therefore, such strain difference in tumorigenesis might be attributed to the normal spleen cell cytotoxicity (Fig. 5). When lentinan was intraperitoneally administered to C3H/He mice at doses of 10 mg/kg 14, 16 and 18 days before the viral infection, a significant inhibition of tumor incidence was obtained (18) and the natural cytotoxicity of normal spleen cells from lentinan injected mice was

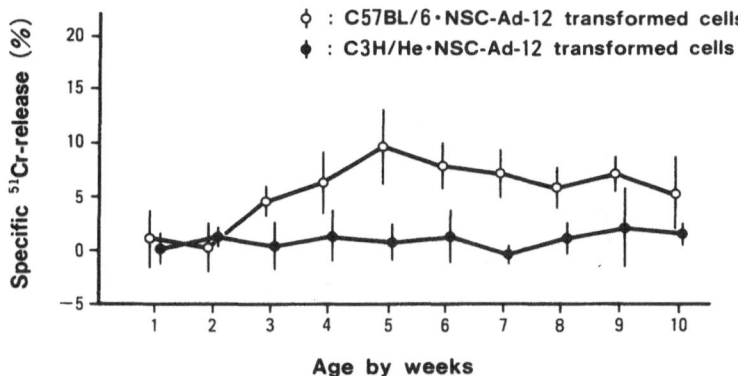

Fig. 5.   Kinetics of development of cytotoxic normal spleen cells (NSC) in C3H/He and C57BL/6 mice (18).   At weekly intervals after birth, spleens of C3H/He and C57BL/6 mice were pooled from 5 to 10 animals depending on the tissue size, respectively.   Splenic lymphocytes (NSC) were assayed for their cytotoxic activity in the ^{51}Cr-release test on syngeneic Ad12-transformed cells. Results given by the NSC of C3H/He (●) and C57BL/6 (○) mice were calculated as specific ^{51}Cr-release (%) and are expressed as the mean (●, ○) + standard error of 7 replicates specimen. Extents of statistical significance (Student's t test) for the difference in the NSC activities between the 2 mouse strains at respective ages were as follows: $p < 0.01$ at 5 wks; $p < 0.05$ at 4, 6, 7 and 8 wks; and $p < 0.1$ at 3, 9 and 10 wks after birth.

greatly augmented (Fig. 6). Though it is not confirmed whether or not lentinan induces viricidal activities against Ad- 12, it is clear that lentinan stimulates host cells to inhibit tumorigenesis originating from viral infection.

A protective activity augmented by lentinan against vesicular stomatitis virus (VSV) infection was shown using (BALB/c x C57BL/6)Fl mice. Female mice were intranasally infected with 1.2 x 105 pfu of VSV-Indian strain in 0.05 ml. Lentinan was intraperitoneally injected 3 days before infection and consecutively administered four days post infection, and complete protection was obtained as shown in Fig. 7.

The potentiating effect of lentinan on Abelson virus-induced immunity against syngeneic Abelson tumor was also demonstrated in BALB/c female mice. Lentinan (40 μg) was intraperitoneally injected on day 20 to 29 post subcutaneous Abelson virus immunization, and Abelson tumor was challenged on day 32. Augmentations of reduction in tumor incidence and tumor volume were observed 14 days after tumor challenge. The mechanisms of action of lentinan was not determined in this experiment, but it was suggested that it induced an antitumor activity through host mediation.

With respect to HIV treatment, only the reverse transcriptase inhibitor, AZT, has been approved as a drug. Nevertheless, AZT treatment has often had to be abandoned since it generally has toxicity detrimental to AIDS patients, while, the replication

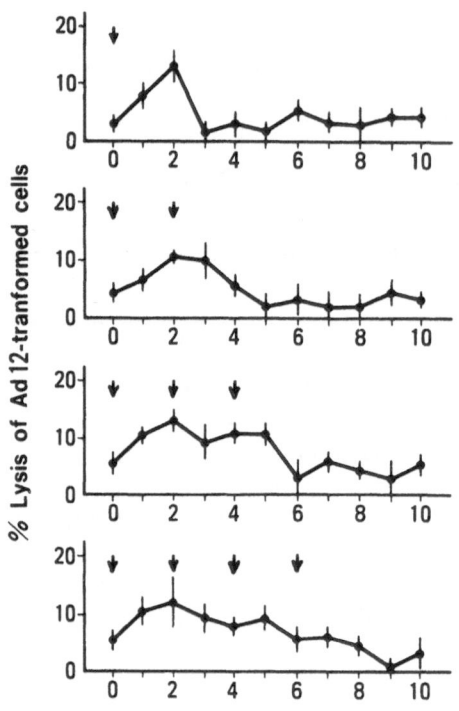

Days after Lentinan administration

Fig. 6.　Induction of cytotoxic normal spleen cells (NSC) in C3H/He mice. Groups of CH/He mice, 4 to 6 wks of age, were treated with Lentinan (10 mg/kg) intraperitoneally 1 to 4 times at 1-day intervals as indicated by the arrows in the figure, respectively. Daily from 0 to 10 days after the initial treatment, splenic lymphocytes (NSC) were pooled from 5 animals and assayed for their cytotoxic activity in the ^{51}Cr-release test on syngeneic Ad 12-transformed cells. Results were calculated as specific ^{51}Cr-release (%) and are expressed as the mean (●) + standard error of 7 replicates per specimen. Significant elevation (p < 0.05 by Student's t test) of the NSC activity as compared with that (3.4 + 3.2% for specific ^{51}Cr-release) in the untreated animals was observed on day 3 in the top graph; on days 2 and 3 in the second graph; and on days 1, 2, 3, 4 and 5 in the third and bottom graphs, respectively, after the initial treatment with Lentinan.

of HIV instantly appears after the discontinuance of AZT as shown in Fig. 8 (21). Lentinan, when used in combination with AZT, suppressed the appearance of HIV antigen expression on the cell surface more strongly than did AZT alone after removing AZT (Fig. 9)(21). This combination will reduce the dosage of AZT and prove more effective in the treatment of AIDS patients.

Different aspects of polysaccharides in anti-HIV activity have been found in those derivatives as sulfated polysaccharides, that is, dextran sulfate (22), mannan sulfate (23), pentosan sulfate (24), lentinan sulfate (25) or curdlan sulfate (Fig. 10)(26). These polysaccharide sulfates showed a complete inhibition of viral replication including cell-fusion due to syncytia formation *in vitro* which was not inhibited by reverse transcriptase inhibitors. The disadvantage of polysaccharide sulfate is that it generally possesses strong anticoagulant activity. Curdlan sulfate is evaluated as having the most potential as a candidate for HIV treatment because of its high effectiveness and low toxicity. Thus chemical modification of polysaccharides appears to be very important in the introduction of new treatment methods in HIV therapy. Furthermore, the combination of lentinan, curdlan sulfate and reverse transcriptase inhibitors in the treatment of HIV infected patients will provide an opportunity to treat not only AIDS patients, but asymptomatic HIV carriers to prevent the development of symptoms, because any attempt in therapy for patients with AIDS or HIV-carriers must include both the inhibition of viral replication and the enhancement of host immune system.

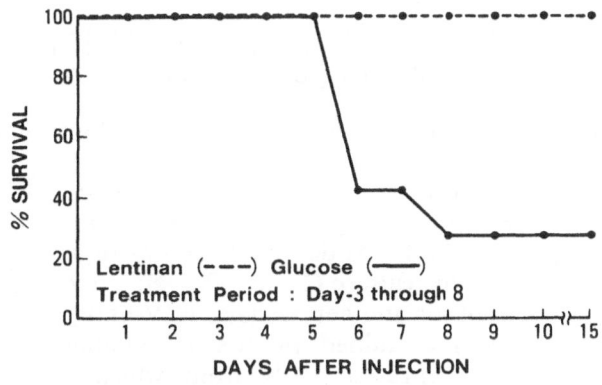

Fig. 7. Effect of Lentinan on VSV-encephalitis in (BALB/c x C57BL/6)F1 mice (19).

Antiparasitic Activity

A host resistance against malaria infection was found in the stimulation of cell-mediated immunity (27). Lentinan has been investigated as an antiparasitic agent taking into consideration that it is a T-cell oriented adjuvant (28-29). The first report was on observation of the potentiation of Schistosome granuloma formation by lentinan in mice (30). Since granulomatous tissue response to parasitic infection is understood to be a reaction of host defense and host morbidity (31-32), lentinan is categorized as an immune potentiator.

CBA/J mice were sensitized intraperitoneally with *Schistosoma mansoni* eggs, and production of synchronized hypersensitivity granulomas in their lung was done by intravenous egg challenge on day 7. Lentinan was injected intraperitoneally on days -7, 0, and +4 in relation to egg challenge at a dose of 250 µg/mouse, and the lung granulomas were measured 8 days after challenge. As shown in Table 1, described by Byram (30) lentinan significantly potentiated lung granulomas in comparison with the control mice. Similar observations were noted with lentinan in unsensitized mice. In

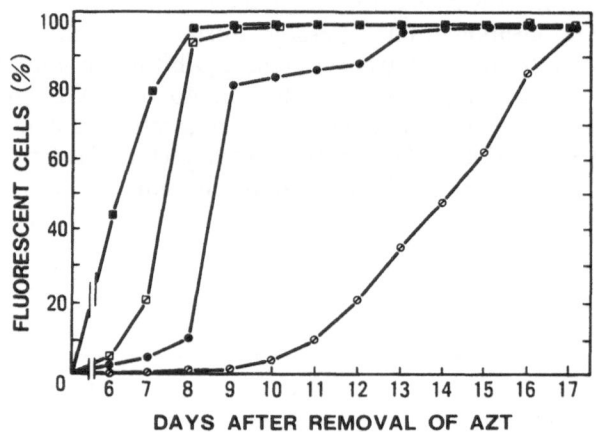

Fig. 8. Effect of elimination of AZT from culture fluid on the appearance of antigen-positive cells (21). MT-4 cells were infected with HIV (M.O.I. = 0.01) and cultured for 10 days in the presence of 1 (□) or 5 (O) µM AZT. At day 11, AZT was removed from a part of the culture (1 µM AZT, ■; 5 µM AZT, ●) and kept for another 12 days. HIV antigen expression was examined by the IF method. Closed symbols represent cells from which AZT was removed at day 11. Open symbols represent cells incubated with AZT throughout the experiment.

contrast, nude mice showed no susceptibility to lentinan administration in the induction of hypersensitive granulomas. The histopathologic picture of lentinan-potentiated granulomas was very characteristic: numerous large, pale stained macrophages, reduced and redistributed eosinophil populations, and frequent, extensive central necrosis which was uncommon in unpotentiated schistosome foci. Liver granulomas in cercaris-induced *Schistosome mansoni* infection were also augmented by lentinan, as were pulmonary granulomas in *Schistosome japonicum* using CBA/J mice.

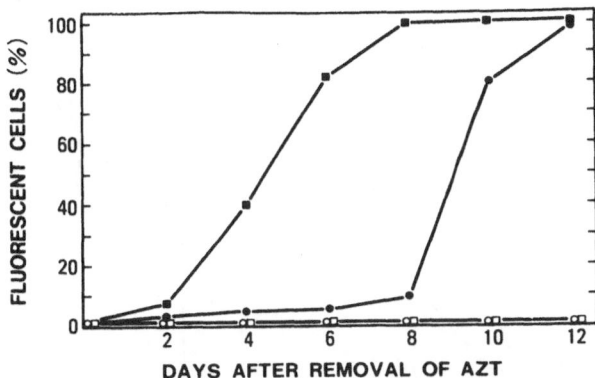

Fig. 9. Restoration of the activity of AZT by lentinan (21). MT-4 cells were infected with HIV and cultured for 10 days as in Fig. 8. After infection, cells were maintained with 1 μM AZT ■; 1 μM AZT + 100 μg/ml lentinan □, 5 μM AZT ●, or 5 μM AZT + 100 μg/ml lentinan 0 for 10 days. At day 11, AZT was removed from the culture medium and expression of HIV antigens was examined as above. Day 0 corresponds to 11 days after infection with HIV.

Table 1.   Lentinan Potentiation of Pulmonary Granulomas Following the Intravenous Injection of *Schistosoma mansoni* Eggs into CBA/J mice (30)

	No. of granulomas measured/ No. of mice	Mean lesion diameter ±SE (μg)	P value
Sensitized mice - eggs injected intraperitoneally on Day - 7 then intravenously on Day 0			
Lungs taken on Day +8			
Control	432/8	179.8 ± 2.9	
Lentinan*	263/8	361.1 ± 10.4	<0.001
Unsensitized mice - eggs injected intravenously on Day 0			
Lungs taken on Day +8			
Control	124/6	164.7 ± 5.2	
Lentinan**	143/7	283.2 ± 12.2	<0.001
Lungs taken on Day +16			
Control	156/6	230.9 ± 8.3	
Lentinan**	147/7	319.4 ±17.9	<0.001
Lungs taken on Day +16			
Control	70/6	187.7 ± 12.3	
Lentinan***	90/6	294.4 ± 21.5	<0.001

   *   Lentinan dose = 250 μg(Days -7, 0, +4)
  **   Lentinan dose = 250 μg(Days -14, -7, 0, +4)
 *** Lentinan dose = 250 μg(Day 0)

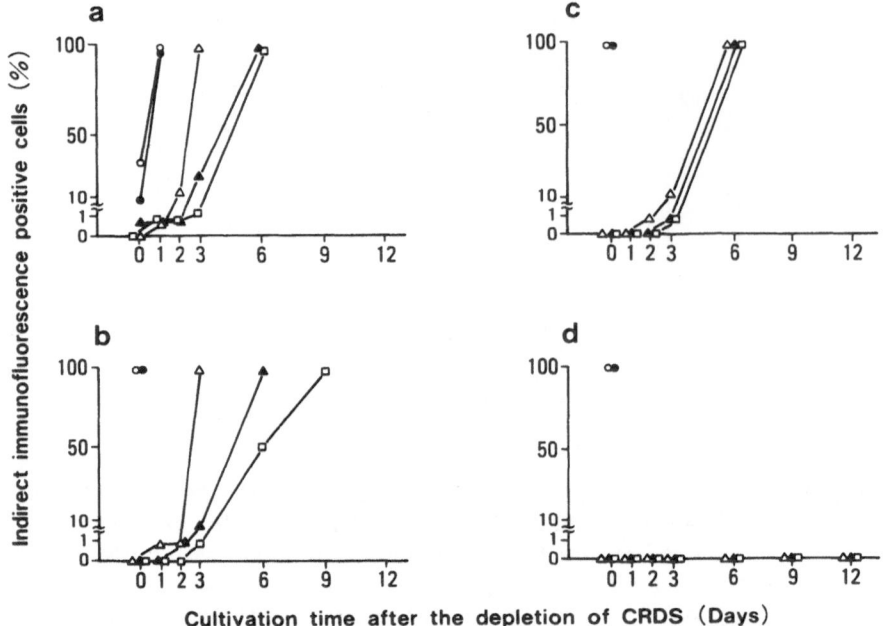

Fig. 10. Inhibitory effect of CRDS (M.W. = (7.9 + 0.6) x 10^4 daltons. S.C. = 15.2
+ 0.3%) against HIV-1 in CRDS depleted cell culture following co-
cultivation of MT-4. MT-4/HIV-1 and different concentrations of CRDS
[(O)0 µg/ml, (●)0.5 µg/ml, (∆)5 µg/ml, (▲)50 µg/ml and (□) 500 µg/ml]
for different co-cultivation periods [(a) 24 hr, (b) 48 hr, (c) 72 hr or (d) 168
hr]. MT-4 cells were suspended at a concentration of 12 x 10^5 cells/ml and
CRDS solution were prepared at concentration of 1.0, 10, 100, 1000 µg/ml.
MT-4 cell suspension (0.25 ml) and CRDS solution (0.5 ml) were mixed in
microtiter wells. HIV-1 producing MT-4 cell suspension (0.25 ml) with cell
density of 6 x 10^5 cells/ml was added to these microtiter wells ((MT-
4/HIV-1)/MT-4=0.5). After incubation of 24 hr, 48 hr, 72 hr or 168 hr at
37°C and 0.5% (v/v) CO atmosphere, each cultured cell suspension was
centrifuged to separate cells from supernatant, and separated cells were
suspended in the CRDS-free culture medium. These cell suspensions were
centrifuged again and the separated cells were suspended in the CRDS-free
culture medium for further cultivation. On day 0, 1, 2, 3, 6, 9, 12 after
cultivation without CRDS, HIV-1 antigen expression was examined (26).

The effect of lentinan on the resistance of mice to *Mesocestoides corti* was reported
by White et al. in 1988 (32). Lentinan was intraperitoneally injected to CBA/H mice
before and after infection with tetrathyridia of *Mesocestoides corti* and the number of
tetrathyridia in peritoneal cavity and liver at 40 days post infection was counted. As
shown in Table 2, multiple injections of lentinan for prophylaxis or therapy resulted in
a distinct reduction in the number of parasites. The liver granulomas were larger than
those in controls and encapsulated parasites were dead or dying. Similar histopathologic changes to those in *Schistosome mansoni* were observed in the lentinan treated

Table 2. The Effect of Prophylactic and Therapeutic Doses of Lentinan on *M. corti* Infections in CBA/H mice (32)

Group	No. per group	Lentinan Dose (μg)	Time administered*	Mean No. of tetrahydria at 40 days p.i. (mean ± SEM) Peritoneal cavity	P	Liver	P	Total	P	Mean multiplication rate	% of mean total from control
1	11	-	-	1259±235		502±136		1761±314		35	
2	10	1000	-28	747±107	NS	455± 69	NS	1202±147	NS	24	68
3	10	250 500 1000	-28, -21 -14, -7 -1	444± 97	0.05	352± 54	NS	796±150	0.05	16	45
4	10	250 500 1000	0, +7 +14, +21 +28	298± 70	0.01	200± 40	NS	498±105	0.01	10	28

211

mice. Combining those two findings in assessing its antiparasitic activity, it is most likely that lentinan is a potent immune stimulating agent which induces cell-mediated cytotoxic activities through macrophages and/or T-cells. Another type of polysaccharide, yeast glucan was reported also to be effective against *Plasmodium berghei* (33) and *Leishmani donovani* (34).

Antifungal Activity

Glucan, isolated from the yeast cell wall of *Saccharomyces cervisiae*, was examined to evaluate its ability to prevent *Candida albicans* sepsis with or without operation (35). Glucan (0.45 mg/mouse) was intravenously injected to male C57BL16J mice on day 10, 7, 4 or 1 prior to midline laparotomy and an intravenous challenge with *C. albicans (3 x 10^6)* was made 2 hrs post-operation. Glucan effectiveness was also evaluated under the same conditions of administration and challenge without operation.

The control mice with or without operation showed a 20% or 47% survival respectively, at 10 days post-challenge. On the contrary, the glucan-treated mice showed a respective 73% (p<0.05) or 100% survival. Macrophage phagocytosis was significantly enhanced by glucan administration and histopathological observations showed that glucan markedly inhibited the renal pathology associated with *C. albicans* challenge. These data suggest that glucan stimulated macrophage function and reduced renal dysfunction caused by *C. albicans* to induce the increased survival.

Possible Mode of Action

It is most likely that lentinan stimulates both specific and non-specific reactions of host defense mechanisms to induce the augmentation of antitumor, antibacterial, antiviral, antiparasitic or antifungal activity, because lentinan has shown T-cell dependent, B-cell dependent, NK-cell dependent and macrophage dependent immune reactions in animal
models (36).

One of the most characteristic activities of lentinan is T-cell oriented adjuvant, which augments helper-T cell activity (28, 29, 37). T-cell augmentation of lentinan is also understood by the results that nude mice treated with the substance were incapable of antiparasitic activity in *Schistosoma mansoni* infection (30), and that neonatal thymectomized mice did not show any antitumor activity with lentinan treatment (37); the administration of antilymphocyte serum decreased the antitumor effect (38).

Another important function of lentinan is macrophage stimulation. Lentinan effect was inhibited by carrageenan, a macrophage inhibitor, which indicates that macrophage stimulation by lentinan also plays a key role in inducing host defense activities. Actually, lentinan triggered the increased production of colony stimulating factor and interleukin-3 in correlation with the interleukin-l producing activity of macrophages. The stimulation of interleukin-1 production by lentinan augments the maturation of immature effector cells to mature cells, thus increasing the response sensitivity to lymphokines such as interleukin-2 or to other cytokines (39).

For instance, it was shown using BALB/C mice that a maturation of thymocytes to allokiller cells against EL-4 lymphoma is amplified by lentinan in the presence of interleukin-2 (39). Synergism with NK cell activator poly I:C on the augmentation of natural cytotoxicity was also observed in lentinan treated mice: non-adherent spleen cells from lentinan treated mice amplified cytotoxic activity with poly I:C *in vitro,* although, lentinan has no direct effect on NK cells *in vitro,* as it is known that lentinan does not exert cytotoxicity directly to tumor cells *in vitro* (40). Based upon these observations, it is most likely that lentinan stimulates host defense mechanisms to enhance the activities of host resistance against bacterial, viral, parasitic or fungal infection. Therefore, the application of lentinan and host defense potentiators to infectious diseases in humans is both rational and valuable.

REFERENCES

1. G. Chihara, Y. Maeda, J. Hamuro, T. Sasaki and F. Fukuoka, Inhibition of mouse sarcoma 180 by polysaccharides from *Lentinus edodes* (Berk.) Sing., Nature 222:687 (1969).
2. G. Chihara, J. Hamuro, Y. Y. Maeda, Y. Arai and F. Fukuoka, Fractionation and purification of the polysaccharides with marked antitumor activity, especially lentinan from *Lentinusedodes* (Berk.) sing., an edible mushroom. Cancer Res. 30:2776 (1970).
3. J. Hamuro, Y. Y. Maeda, Y. Arai, F. Fukuoka and G. Chihara, The significance of the higher structure of the polysaccharides, lentinan and pachymaran with regard to their antitumor activity. Chem. Biol. Interact. 3:69 (1971).
4. T. L. Bluhm and A. Sarco, The triple helical structure of lentinan, a linear β-(1→3)-D-glucan. Can. J. Chem. 55:293 (1977).
5. H. Saito, T. Ohki and T. Sasaki, A ^{13}C-nuclear magnetic resonance study of polysaccharide gels. Molecular architecture in the gels consisting of fungal, branched (1→3)-β-D-glucans (lentinan and schizophyllan) as manifested by conformational changes induced by sodium hydroxide, Carbohydrate Res. 74:227 (1979).
6. J. Zakany, G. Chihara and J. Fachet, Effect of lentinan on tumor growth in murine allogeneic and syngeneic hosts, Int J Cancer 25:371 (1980).
7. T. Suga, T. Shiio, Y.Y. Maeda and G. Chihara, Antitumor activity of lentinan in murine syngeneic and autochthonous hosts and its suppressive effect on 3-methyl-cholanthrene induced carcinogenesis, Cancer Res 44:5132 (1984).
8. W. C. Rose, F.C. Reed III, P. Siminoff and W.T. Bradner, Immunotherapy of Madison 109 lung carcinoma and other murine cancer using lentinan, Cancer Res 44:1368 (1984).
9. T. Suga, T. Yoshihama, Y. Tsuchiya, T. Shiio, Y. Y. Maeda and G. Chihara, Prevention of tumor metastasis and recurrence of DBA/2 MC.CS-T fibrosarcoma, MH-134 hepatoma and other murine tumors using lentinan. Int. J. Immunotherapy, 5:187 (1989).
10. T. Taguchi, H. Furue, T. Kimura, T. Kondo, T. Hattori, I. Ito and N. Ogawa, End-point result of a randomized controlled study on the treatment of gastro-intestinal cancer with a combination of lentinan and chemotherapeutic agents, in: "Rationale of Biological Response Modifiers in Cancer Treatment", T. Tsubura, T. Aoki and I. Urushizaki, eds., Excerpta Medica, Amsterdam (1985).
11. G. Chihara, Lentinan and its related polysaccharides as host defence potentiators: Their application to infectious diseases and cancer, in: "Immunotherapeutic Prospects of Infectious Diseases", K.N. Masihi and W. Lange, eds., p. 9, Springer-Verlag, Heidelberg (1990).
12. T. Kawamura and Y. Numasaki, Effect of lentinan to host resistance against Listeria infections, Proc Jpn Cancer Assoc 39:134 (1980).
13. K. Kanai, E. Kondo, P.J. Jacques and G. Chihara, Immunopotentiating effect of fungal glucans as revealed by frequency limitation of post-chemotherapy relapse in experimental mouse tuberculosis, Jpn J Med Sci Biol 33:283 (1980).
14. K. Matsuo, K. Nomoto, S. Shimotori and K. Takeya, Depression of protective mechanisms against microorganisms in tumor-bearing mice and its restoration by adjuvants, Jpn J Cancer Res (Gann) 68:456 (1977).

15. D. L. Williams, J.W. Browder and N.R. DiLuzio, Immunotherapeutic modification of *E. coli* induced experimental peritonitis and bacteremia by glucan, Surgery 93:448 (1983).

16. N. R. DiLuzio and D.L. Williams, Protective effect of glucan against systemic *Staphylococcus aureus* septicemia in normal and leukemic mice, Infect Immun 20:804 (1978).

17. J. A. Reynolds, M.D. Kastello, D.G. Hallington, C.L. Crabbs, C.L. Peters, J.L. Jemski, G.H. Scott and N.R. DiLuzio, Glucan induced enhancement of host resistance to selected infectious diseases, Infect Immun 30:51 (1980).

18. C. Hamada, Inhibition effect of lentinan on tumorigenesis of adenovirus type 12 in mice, in: "Manipulation of Host Defense Mechanisms", T. Aoki, I. Urushizaki and E. Tsubura, eds., p. 76, Excerpta Medica, Amsterdam (1981).

19. K. S.S. Chang, Lentinan-mediated resistance against VSV-encephalitis, Abelson virus-induced tumor, and trophoblastic tumor in mice, in: "Manipulation of Host Defence Mechanisms", T. Aoki, I. Urushizaki and E. Tsubura, eds., p. 88, Excerpta Medica, Amsterdam (1981).

20. D. L. Williams and N.R. DiLuzio, Glucan induced modification of murine viral hepatitis, Science 200:67 (1980).

21. T. S. Tochikura, H. Nakashima, Y. Kaneko, N. Kobayashi and N. Yamamoto, Suppression of human immunodeficiency virus replication by 3'-Azido-3'-deoxy-thymidine in various human hematopoietic cell lines in vitro: Augmentation of the effect by lentinan, Jpn J Cancer Res (Gann) 78:583 (1987).

22. M. Ito, M. Baba, A. Sato, R. Pauwels, E. DeClercq and S. Shigeta, Inhibitory effect of dextran sulfate and heparin on the replication of human immunodeficiency virus (HIV) *in vitro*, Antiviral Res 7:361 (1987).

23. M. Ito, M. Baba, K. Hirabayashi, T. Matsumoto, M. Suzuki, S. Suzuki, S. Shigeta and E. DeClercq, *In vitro* activity of mannan sulfate, a novel sulfated polysaccharide, against human immunodeficiency virus type 1 and other enveloped viruses, Eur J Clin Microbiol Infect Dis 8: 171 (1989)

24. L. Biesert, H. Suhartono, I. Winkler, C. Meichsner, M. Helsberg, G. Hewlett, V. Klimetzek, K. Molling, H.D. Schlunberger, E. Schrinner, H.D. Brede and H. Rubsamen-Waigmann, Inhibition of HIV and virus replication by polysulphated polyxylan: HOE/BAY 946, a new antiviral compound, AIDS 2:449 (1988).

25. O. Yoshida, H. Nakashima, T. Yoshida, Y. Kaneko, I. Yamamoto, K. Matsuzaki, T. Uryu and N. Yamamoto, Sulfation of the immunomodulating polysaccharide lentinan: a novel strategy for antivirals to human immunodeficiency virus (HIV), Biochem Pharmacol 37:2887 (1988).

26. Y. Kaneko, O. Yoshida, R. Nakagawa, T. Yoshida, M. Date, S. Ogihara, S. Shioya, Y. Matsuzawa, N. Nagashima, Y. Irie, T. Mimura, H. Shinkai, N. Yasuda, K. Matsuzaki, T. Uryu and N. Yamamoto, Inhibition of HIV-1 infectivity with curdlan sulfate *in vitro*, Biochem Pharmacol 39:793 (1990).

27. K. N. Brown, in: "Immunology of Parasitic Infections", Blackwell Scientific Publication, Oxford, p. 268, (1976).

28. D. W. Dresser and J. M. Phillips, The orientation of the adjuvant activities of *Salmonella typhosa* lipopolysaccharide and lentinan. Immunol. 27:895 (1974).

29. G. Dennert and D. Tucker, Antitumor polysaccharide lentinan, A T-cell adjuvant, J Natl Cancer Inst 51:1727 (1973).

30. J. E. Byram, A. Sher, J. DiPietro, F. von Lichtenberg, Potentiation of schistosome granuloma formation by lentinan - a T-cell adjuvant, Am J Pathol 94:201 (1979).

31. D. O. Adams, The granulomatous inflammatory response: A review. Am J Pathol 84:164 (1976).

32. T. R. White, R. C. A. Thompson, W. J. Penhale and G. Chihara, The effect of lentinan on the resistance of mice to *Mesocrestoides corti*. Parasitol Res 74:563 (1988).

33. M. Song and N. R. DiLuzio, Yeast glucan and immunotherapy of infectious diseases, in: "Lysosomes in Biology and Pathology" P. J. Jacques and I. B. Shaw, eds., North Holland Press, Amsterdam, (1979).

34. J. A. Cook, T. W. Holbrook and W. J. Dougherty, Prospective effect of glucan against visceral leishmaniasis in hamsters. Infect Immun 37:1261 (1982).

35. I. W. Browder, D. L. Williams, A. Kitahama and N. R. DiLuzio, Modification of post-operative *C. albicans* sepsis by glucan immunostimulation. Int J Immunopharmac 6:19 (1984).

36. J. Hamuro and G. Chihara, Lentinan, a T-cell oriented immunopotentiator: Its experimental and clinical applications and possible mechanism of immune modulation, in: "Immune Modulation Agents and Their Mechanisms", R. L. Fenichel and M. A. Chirigos, eds., Marcel Dekker, New York, (1985).

37. Y. Y. Maeda and G. Chihara, The effect of neonatal thymectomy on the antitumor activity of lentinan, carboxymethylpachymaran and zymosan, and their effect on various immune responses. Int J Cancer 11:153 (1973).

38. Y. Y. Maeda, J. Hamuro and G. Chihara, The mechanism of action of antitumor polysaccharides. The effect of antilymphocyte serum on the antitumor activity of lentinan. Int J Cancer 8:41 (1971).

39. Y. Akiyama, S. Kashima, T. Hayami, M. Izawa, K. Mitsugi and J. Hamuro, Immunological characteristics of antitumor polysaccharides, lentinan and its analogs, as immune adjuvants, in: "Manipulation of Host Defense Mechanisms", T. Aoki, I. Urushizaki and E. Tsubura, eds., Excerpta Medica, Amsterdam, (1981).

40. S. Kashima, T. Hayami, R. Yoshimoto, M. Izawa and J. Hamuro, Augentation of lymphokine-mediated natural killer cell activation by antitumor polysaccharide lentinan. Int J Immunopharmacol 4:269 (1982).

CYTOKINES AND ANTI-FUNGAL IMMUNITY

Julie Y. Djeu

Department of Medical Microbiology and Immunology
University of South Florida College of Medicine
Tampa, FL 33612

INTRODUCTION

Candida albicans is classified as an opportunistic pathogen because of its relatively low virulence in normal hosts without underlying disease (1). Indeed, it forms part of the normal flora of man. Superficial and oral infections by this organism, however, is commonplace in immunocompromised individuals and systemic infection oftens leads to an unfavorable prognosis, particularly in AIDS patients and other patients undergoing cytoreductive therapy (2-5). There is therefore a need to clarify both the infectious mechanisms of the organism and the host defense mechanism against it.

Innate resistance mechanisms provide an essential first line of defense against the spread of the fungus and are primarily aided by neutrophils and possibly by monocytes/ macrophages in eliminating the invading microbes during early exposure (1,6). Supporting data for the crucial role of neutrophils come from observations of high incidence of mucosal or systemic candidiasis in patients suffering from neutropenia due to prolonged chemo-radiotherapy (4,5) or from neutrophil dysfunction due to inherent disease, such as chronic granulomatous disease and leukocyte adhesion deficiency disease (7,8). Although neutrophils are essential end effector cells for defense against opportunistic fungi, other components of the immune system are also important in effective host resistance to *C. albicans* (9,10). In the course of infection, specific T cells and B cells are activated that provide cytokines and antibodies respectively for the activation and recruitment of the phagocytes and facilitated ingestion of the opsonized fungi. Neutrophil fungicidal action is primarily associated with generation of highly-reactive oxygen radicals from the respiratory burst, either directly or in conjunction with myeloperoxidase released from azurophilic granules (11,12). Non-oxidative killing or fungistasis can also be achieved by release of lysozyme, cationic peptides including defensins, and lactoferrin from the neutrophil granules (6,13,14). In addition to ingestion, extracellular killing of large hyphae can also occur via oxidants released from neutrophils at the point of contact (11).

However, in the face of high levels of anti-Candida antibodies and previous sensitization with *C. albicans*, certain types of patients still succumb to infection with this yeast. Other mechanisms other than T and B cell immunity may therefore play an important role in resistance to *C. albicans*. The purpose of this chapter is to review information generated in our laboratory on the key contribution of natural killer (NK) cells to control of Candida growth and to reveal a new role for neutrophils as an active component of the cytokine-generating machinery in the host.

Microbial Infections, Edited by H. Friedman *et al.*
Plenum Press, New York, 1992

Identification of Effector Cells with Anti-Candidal Activity

To achieve an in-depth evaluation of the NK/PMN axis that controls *C. albicans* growth, we first established a highly-sensitive radiolabel assay that makes use of the incorporation of ^{3}H-glucose into residual Candida after they have been preincubated with effector cells for 18 h at 37°C in triplicate wells of a 96 well microtiter plate (15). Percent growth inhibition of *C. albicans* was calculated by comparison of the ^{3}H-glucose incorporation into residual Candida in the absence and presence of effector cells. Components of human peripheral blood, i.e., phagocytes and lymphocytes, were individually tested for their ability to inhibit Candida growth *in vitro*. In addition, their capacity to respond to *C. albicans* to produce cytokines that could influence antifungal activity was investigated. The mononuclear cells, consisting of lymphocytes and monocytes, were isolated as a single band after Ficoll-Hypaque gradient centrifugation of peripheral blood of normal donors. The neutrophils were recovered from the top of the red blood cell pellet and lysed free of red blood cells by hypotonic shock with sterile, endotoxin-free distilled water for 30 sec. The mononuclear cells were further separated by plastic adherence to obtain monocytes. The remaining nonadherent lymphocytes were passed over a nylon wool column before they were subjected to discontinuous Percoll gradient centrifugation to purify LGL, that exhibit NK function against K562 tumor cells, from small, mature T cells.

Analysis of these cell subsets indicated that only the phagocytic cells, i.e., neutrophils and monocytes, had direct antifungal activity against *C. albicans* (16). Neutrophils, added to live *C. albicans* for 18 hr and then pulsed with ^{3}H glucose for 3 additional hours indicated that they were highly efficient in inhibiting Candida growth *in vitro*. The usual findings were that neutrophils could inhibit 90-95% of Candida growth at the effector/target ratio of 50:1 to 100:1. Frequently, complete elimination of Candida was seen at the 30:1 ratio, and significant antifungal activity was still evident at the 10:1 ratio. Lymphocytes, on the other hand, had no capacity to control Candida growth. Percoll-fractionated lymphocytes, highly enriched in large granular lymphocytes (LGL) that are responsible for NK activity, as well as purified T cells, were tested in the same manner as phagocytes for activity against *C. albicans in vitro*. Neither LGL nor T cells could inhibit fungal growth even at an extremely high effector/target ratio of 300:1, while the phagocytes could effectively do so at 3:1 or less.

Release of Neutrophil-Activating Factors by Large Granular Lymphocytes

Although LGL could not inhibit fungal growth and could not be made fungicidal by interferon-alpha (IFNa), IFNg, or interleukin 2 (IL2), which are cytokines that would normally activate NK function against tumor cells, they did respond directly to *C. albicans* in another potent way. The LGL were directly triggered by *C. albicans* to release cytokines that had the ability to activate PMN (16-18). Supernatants from LGL cultured for 24-72 hr with *C. albicans* were capable of enhancing PMN to inhibit fungal growth. These LGL supernatants were found to contain high levels of tumor necrosis factor (TNF), IFNg, and granulocyte-macrophage colony stimulating factor (GMCSF). Characterization of these cytokines was accomplished by the use of cytokine-specific biological assays and by neutralization with specific antibodies against each cytokine. TNF was detected by its ability to lyse WEHI-164-JD tumor cells in an 18 hr ^{51}Cr release assay. IFNg was detected by its ability to protect human WISH cells from infection and cytopathic effects of vesicular stomatitis virus. GMCSF was measured by its ability to support the CSF-dependent Mo7e cell line.

The 3 cytokines did not have an identical pattern of release from LGL. The peak of TNF protein release was under 24 hr while that of IFNg was at 48 hr and GMCSF continued to be released at peak levels even at day 7 after initial contact with the yeast. The phenotype of the cytokine-producing LGL was next assessed by serological depletion of various subpopulations in the LGL by incubation with specific monoclonal antibodies plus complement. Alternatively, negative selection was performed by panning with the various antibodies. First, these experiments indicated that removal of putative contaminating monocytes with anti-CD15 or contaminating T cells with anti-CD4 and CD8 had no effect on the ability of the LGL to produce all 3 cytokines. Second, the responsive LGL were further typed to be CD2+CD16+HLADR+. It should be noted that the DR+ LGL subset that released the cytokines were devoid of NK activity against K562 tumor cells. This suggests the existence of at least 2 subsets of LGL, the CD2+CD16+DR-LGL with spontaneous tumoricidal activity and the CD2+CD16+DR-+LGL with the capacity to release cytokines upon antigen stimulation.

The ability of monocytes and small T cells to produce cytokines was also evaluated. Monocytes readily responded to *C. albicans* by TNF and IFNa production but were unable to release GMCSF. The lack of GMCSF production by monocytes is not an intrinsic property because monocytes do release GMCSF upon stimulation by *Mycobacterium avium intracellulare* (19). In contrast, Percoll-purified T cells, in the absence of monocytes or LGL, could not produce any of the 3 cytokines when they were exposed to *C. albicans*.

Activation of PMN Function with TNF, IFNg and GMCSF

With the availability of recombinant cytokines, the role of TNF and IFNg in activation of PMN function was investigated in greater detail (15, 18). Recombinant TNF was able to rapidly activate PMN within 2 hr of incubation, with significant effects evident at 1-10 units/ml. Its continued presence was not required following a short pulse of PMN with TNF. Similar results were obtained with 1-10 units/ml of recombinant IFNg. Addition of both TNF and IFNg had a synergistic effect in that TNF or IFNg, at a concentration that did not activate PMN by itself, together caused potent stimulation of PMN function. Recombinant GMCSF was also an effective PMN stimulator, and its action was as rapid as that of the other cytokines.

Activtion of Monocytes by Recombinant IFNg, GMCSF, Interleukin 3 (IL3) and Monocyte-CSF (MCSF)

Studies of antifungal activity in monocytes were also initiated because these phagocytes constitute an important defense mechanism against *C. albicans*. Plastic-adherent monocytes were found to be as potent as PMN in inhibiting Candida growth (20). Antifungal activity of fresh monocytes was in the range of 75-85% at the effector/target ratio of 10:1 and significant activity was still visible at the 3:1 ratio. However, monocytes required at least 18 hr of incubation with cytokines to be activated, unlike PMN, whose responses took place within 2 hr. Of the 3 CSF that were studied, GMCSF and IL3 showed equal potency in activating monocytes. MCSF was able to stimulate monocytes, but to a lesser degree. Further analysis indicated that the CSF could reactivate aged monocytes, which progressively lost their antifungal activity with prolonged culture. In addition, GMCSF or IL3, added at the initiation of the monocyte culture, prevented the loss of function in monocytes, and the initial high antifungal activity was maintained up to 5 days, which was the last timepoint tested. In contrast, IFNg was only able to upregulate function in freshly-isolated monocytes but were unable to preserve the initial high level of function of monocytes. Other investigators have made similar findings on the ability of GMCSF and IFNg to activate fresh monocytes (21, 22) but we have new evidence that these two cytokines may not affect monocytes, especially during culture, in exactly the same manner. It appears that monocytes may have different pathways of activation, depending on the cytokine they are exposed to, and IFNg and CSF may provide the tools to dissect out these differences.

Role of IL-8 in Neutrophil Activation

Thus far, we had concentrated on functional activators of neutrophils that might be produced by LGL/NK cells. We chose to study the role of a chemotactic factor, IL8, because of its known specificity for PMN mobilization and not for monocyte movement (23, 24). We postulated that IL8 should activate PMN function besides acting as a chemoattractant. With the recent cloning of IL8 from LPS-activated monocytes, we acquired the recombinant IL8 and found it to be a potent activator of neutrophil function against *C. albicans* (25). IL8, at an optimal dose of 10 ng/ml, was highly efficient in activating PMN and this dose was similar to that reported for optimal chemotactic activity. Comparison of IL8 to the standard chemoattractant, F-met-leu-phe (FMLP), indicated that, although both were potent activators of PMN against Candida, the pathways by which antimicrobial activity may be triggered differed for the 2 PMN chemoattractants. While FMLP was a superior agent for inducing superoxide anion release and degranulation, IL8 did not itself induce superoxide anion release and barely primed PMN for FMLP-induced metabolic burst. The effect of IL8 appeared to be centered mainly on degranulation of lytic enzymes, as measured by b-glucuronidase release. Therefore, IL8 is not only a chemotactic factor but also a functional activator for PMN.

Neutrophil Production of TNF

In addition to NK regulation of PMN, we have also begun to question whether neutrophils can produce cytokines in response to *C. albicans*. The possibility that neutrophils are not bystanders responding to lymphokines and monokines but are actual active participants in the early immune response needs to be considered. For this series of experiments, it was important to check the purity of PMN prior to antigen stimulation and we found that our method of purification of PMN from the Ficoll/Hypaque gradient yielded a pure population of more than 99% PMN with no detectable monocyte or lymphocyte contamination (26). PMN was found to be a excellent source of TNF, albeit at a lower level than that produced by LGL or monocytes. The range of TNF produced by PMN exposed to *C. albicans* was 8-36 units/ml as compared to 100-200 units/ml in mononuclear cells. The kinetics of TNF induction from PMN, however, was the same as that from LGL or monocytes, with a steady rise from 8 h to 18 h. The release of TNF was by de novo synthesis because Actinomycin D which inhibits RNA synthesis and blockers of protein synthesis, i.e., cycloheximide or emetine, could interfere with TNF production from PMN stimulated with *C. albicans*. The cytokine was typed to be TNFa by specific neutralizing antibodies. It is to be noted that TNF has been reported to have chemotactic properties for PMN (27). Therefore, the finding of TNF production by PMN indicates that PMN may respond to stimuli, such as *C. albicans*, by rapid TNF release at a level that may not be toxic which can not only activate PMN in an autocrine loop but also recruit neighboring PMN to the site of infection.

Analysis of Activation Pathways for IL-8, GMCSF, and LPS in Neutrophils

We have focussed on the cytokines, IL-8 and GMCSF, for analysis of the antifungal mechanisms activated in PMN and not on TNF and IFNg because much information is already available on the last 2 cytokines. The effects of IL8 and GMCSF were compared to those of LPS, which is a known neutrophil activating agent. Of the activating agents tested, none could by themselves induce superoxide release. Only GMCSF and LPS could prime PMN for FMLP induction of superoxide anion release. With regards to degranulation, only IL8 could by itself induce degranulation, as measured by b-glucuronidase release. Lactoferrin release, on the other hand, could be readily induced by all the agents tested. Therefore, PMN appears to respond differently to various stimulating agents and may utilize one or more of the 3 mechanisms, i.e. oxidative radicals, degranulation of enzymes, and lactoferrin, to control Candida growth, depending on the activating agent.

SUMMARY

In summary, we have defined a unique host resistance circuit that has not previously been investigated. For the host control of an opportunistic fungal pathogen such as *C. albicans*, we have definitive evidence to indicate that LGL can respond to *C. albicans* by producing key cytokines, i.e. TNF, IFNg, and GMCSF, to activate neutrophil function against *C. albicans*. The cytokine-producing LGL differs from the spontaneous tumoricidal LGL by being DR+; otherwise other markers are identical, i.e., CD2+-CD16+CD4-CD8-CD15-. From the point of view of the neutrophils, they can respond to these cytokines readily within 2 hr of activation and may utilize any of the 3 antifungal pathways, i.e., oxidative radical production, enzyme degranulation, and lactoferrin release, to control Candida. It is of importance to note that TNF and GMCSF have also been shown to have chemotactic properties on neutrophils (27,28). Thus, the cytokines produced by LGL may have bifunctional roles for PMN, in not only activating them but in mobilizing them to the site of fungal invasion.

In addition, we have defined that *C. albicans* as well as the bacterial polysaccharide, LPS, can activate PMN to produce TNF. Since TNF is a neutrophil activating factor, this implies that neutrophils may self-regulate function in an autocrine manner or utilize released TNF to recruit neighboring PMN. The possibility exists that other cytokines may also be produced by neutrophils when activated with *C. albicans*. Future studies should indicate the true role of neutrophils in host resistance to infection and may lead to a new identity for neutrophils as an active participant in the afferent phase of the immmune response rather than an end effector cell population, waiting for outside signals to mobilize and activate them.

REFERENCES

1. F. C. Odds, in: "Candida and candidiasis. A review and bibliography," 2nd edition. W.A. Saunders, Philadelphia (1988).
2. R. S. Klein, C.A. Harris, C.B. Small, B. Moll, M. Lesser, and G.H.Friedman, Oral candidiasis in high risk patients as the initial manifestation of the Acquired Immunodeficiency Syndrome. N. Eng. J. Med. 33:354 (1984).
3. J. L. Rhoads, D.C. Wright, R.R. Redfield, and D.S. Burke, Chronic vaginal candidiasis in women with Human Immunodeficiency Virus Infection, J. Am. Med. Assoc. 257:3105 (1987).

4. Y. S. Cho, and H. Y. Choi, Opportunistic fungal infection among cancer patients, A ten year autopsy study. Amer. J. Clin. Path. 72:617 (1979).

5. M. C. Bach, A. Sahyoun, J.L. Adler, R.M. Schlesinger, J. Breman, P. Madras, F. P'eng, and A.P. Monaco, High incidence of fungus infections in renal transplantation patients treated with antilymphocyte and conventional immunosuppression, Transplant. Proc. 5:549 (1973).

6. S. J.Klebanoff, and R.A.Clark, in: "The Neutrophil:Function and Clinical Disorders. Amsterdam:Elsevier (1978).

7. M. H. Kim, G.E. Rodey, R.A. Good, R.A. Chilgren, and P.G. Quie, Defective candidacidal capacity of polymorphonuclear leukocytes in chronic granulomatous disease of childhood, J. Pediatrics. 75:300 (1969).

8. D. C.Anderson, and T.A. Springer, Leukocyte adhesion deficiency:an inherited defect in the MAC-1, LFA-1, and p150,95 glycoproteins. Ann. Rev. Med. 38:175 (1987).

9. J. W.Murphy, Natural Host Resistance Mechanisms against Systemic Mycotic Agents. in: Functions of the Natural Immune System. C.W. Reynolds and R.H. Wiltrout eds., Plenum Press, N.Y., p.149. (1990).

10. R. D.Diamond, Immune response to fungal infection. Rev. Infect. Dis. 11:S1600 (1989).

11. R. D.Diamond, R.A. Clark, and C.C. Haudenschild, Damage to *Candida albicans* hyphae and pseudohyphae by the myeloperoxidase system and oxidative products of neutrophil metabolism *in vitro*, J. Clin. Invest. 66:908 (1980).

12. R. I.Lehrer, and M.J. Cline, Leukocyte myeloperoxidase deficiency and disseminated candidiasis: The role of myeloperoxidase in resistance to candida infection, J. Clin. Invest. 48:1978 (1969).

13. R. I. Lehrer, K. M. Ladra, and R.B. Hake, Nonoxidative fungicidal mechanisms of mammalian granulocytes:demonstration of components with candidacidal activity in human, rabbit and guinea pig leukocytes, Infect. Immun. 11:1226 (1975).

14. T. Ganz, M.E. Selsted, D. Szklarek, S.S.L. Harwig, K. Daher, D.F. Bainton, and R.I.Lehrer, Defensins:natural peptide antibiotics of human neutrophils, Clin. Invest. 76:1427 (1985).

15. J. Y. Djeu, D.K. Blanchard, D. Halkias, and H. Friedman, Growth inhibition of *Candida albicans* by human polymorphonuclear neutrophils: activation by interferon-γ and tumor necrosis factor, J. Immunol. 137:2980 (1986).

16. J. Y. Djeu, and D.K.Blanchard, Regulations of human polymorphonuclear neutrophils (PMN) activity against *Candida albicans* by large granular lymphocytes via release of a PMN-activating factor, J. Immunol. 139:2761 (1987).

17. J. Y.Djeu, D.K. Blanchard, A.L. Richards, and H. Friedman, Tumor necrosis factor induction by *Candida albicans* from human natural killer cells and monocytes, J. Immunol. 141:4047 (1988).

18. D. K. Blanchard, M.B. Michelini-Norris, and J.Y. Djeu, Production of granulocyte-macrophage colony stimulating factor by large granular lymphocytes:role in activation of human neutrophil function, Blood 77:2259 (1991).

19. D. K. Blanchard, M.B. Michelini-Norris, C.A. Pearson, C.S. Freitag, and J.Y. Djeu, *Mycobacterium avium intracellulare* induces interleukin 6 from human monocytes and large granular lymphocytes, Blood 77:2218 (1991).

20. M. Wang, H. Friedman, and J.Y.Djeu, 1989. Enhancement of human monocyte function against *Candida albicans* by the colony-stimulating factors (CSF):IL3, granulocyte-macrophage-CSF, and macrophage-CSF, J. Immunol. 143:671.

21. P. D. Smith, C.L. Lamerson, S.M. Banks, S.S. Saini, L.M. Wahl, R.A. Calderone, and S.M. Wahl, Granulocyte macrophage colony stimulating factor augments human monocyte fungicidal activity for *Candida albicans,*. J. Infect. Dis. 161:999 (1990).

22. E. Brummer, and D.A.Stevens, Candidacidal mechanisms of peritoneal macrophages activated with lymphokines or gamma interferon, J. Med. Microbiol. 28:173 (1989).

23. K. Matsushima, K. Morishita, T. Yoshimura, S. Lavu, Y. Kobayashi, W. Lew, E. Appela, S.F. Kung, E.J. Leonard, and J.J. Oppenheim, Molecular cloning of cDNA for a human monocyte-derived neutrophil chemotactic factor (MDNCF) and the induction of MDNCF mRNA by interleukin 1 and tumor necrosis factor, J. Exp. Med. 167:1883 (1988).

24. M. Baggiolini, A. Walz, and S.L. Kunkel, NAP/IL8, a novel cytokine that activates neutrophils, J. Clin. Invest. 84:1045 (1989).

25. J. Y. Djeu, K. Matsushima, J.J. Oppenheim, K.Shiotsuki, and D.K. Blanchard, Functional activation of human neutrophils by recombinant monocyte-derived neutrophil chemotactic factor/IL-8, J. Immunol. 144:2205 (1990).

26. J. Y. Djeu, Serbousek, D. and Blanchard, D.K. Release of tumor necrosis factor by human polymorphonuclear leukocytes, Blood, 76:140 (1990).

27. W. Ji-Ming, L. Bersani, and A. Mantovani, Tumor necrosis factor is chemotactic for monocytes and polymorphonuclear leukocytes, J. Immunol. 138:1469 (1987).

28. J. M. Wang, S. Colella, P. Allavena, and A. Mantovani, Chemotactic activity of human recombinant granulocyte macrophage colony stimulating factor, Immunol., 60:439 (1987).

CRYPTOCOCCAL IMMUNITY AND IMMUNOSTIMULATION

Juneann W. Murphy

Department of Microbiology and Immunology
University of Oklahoma Health Sciences Center
Oklahoma City, OK 73190

Cell-mediated immunity (CMI) constitutes one of the major mechanisms by which certain infectious agents are cleared from the body. One disease that generally does not occur in hosts with adequate functional cell-mediated immune mechanisms is cryptococcosis. Recently, however, the numbers of immunocompromised and immunodeficient individuals has been rising due to the expanding use of immunosuppressive therapies and the spread of the human immunodeficiency virus (HIV). Concomitant with this a significant increase in the incidence of cryptococcosis has occurred (1-3). For example, prior to 1980 there were about 300 cases of cryptococcosis per year in the United States, and now cryptococcosis is diagnosed in 7% of all AIDS patients in the U.S.A. and in 20% of the black AIDS patients (3). Cryptococcosis ranks fourth in the infectious diseases and first in fungal diseases that are responsible for deaths in AIDS patients (1). Thus, there is more impetus now than ever before to gain a sufficient understanding of the protective anticryptococcal cell-mediated immune response and its regulation so that therapeutic protocols can be devised to supplement or replace the protective immune components in immunocompromised and immunodeficient patients.

Cryptococcosis is caused by the encapsulated, yeast-like organism, *Cryptococcus neoformans* which is found world-wide and is extremely prevalent in debris around pigeon roosts (4). When dehydrated, the organism becomes air borne and can be inhaled by humans (4). The smaller yeast cells gain entrance into the alveolar spaces of the lungs where they cause an infection that can be asymptomatic or that can be accompanied by mild pneumonia-like symptoms (4). In most individuals with normal resistance, the cryptococci remain localized in the lungs and are eventually eliminated resulting in a spontaneous resolution of the disease. On the other hand, in individuals who are immunocompromised or are otherwise susceptible to *C. neoformans*, the infection is not limited to the pulmonary compartment, and the organism spreads via the blood stream to extrapulmonary sites. *C. neoformans* has a predilection for the meninges; therefore, a high percentage of patients with disseminated cryptococcosis have meningitis which is fatal unless aggressively treated with antifungal drugs (4-6). Patients with disseminated cryptococcosis typically have measurable levels of cryptococcal antigens in their body fluids, and for many years cryptococcal antigen titers in serum and spinal fluid have been of prognostic value (7,8). For instance, increasing cryptococcal antigen titers were indicative of a poor prognosis; whereas, decreasing titers were a signal of recovery (7,8). However, these rules do not apply to AIDS patients with cryptococcosis. Although cryptococcal antigen titers in serum and spinal fluid of AIDS patients rise as the disease progresses, the titers usually climb to exceedingly high levels and do not readily decline even after apparently successful antifungal therapy (9). Therefore, in AIDS patients, cryptococcal antigen titers cannot be used as a prognostic tool (9).

The means by which the host defends itself against a *C. neoformans* infection have been studied by a number of investigators, and an understanding of the protective mechanisms is beginning to emerge (10-12). Based on human and animal studies, it is generally agreed that cell-mediated immunity is an essential factor in protective host defense in cryptococcosis (10-12). The mouse model has been extremely useful in defining the anticryptococcal resistance mechanisms and in establishing the means by which those defenses are regulated. With the CBA/J mouse model, the effects of circulating cryptococcal antigen on the protective anticryptococcal cell-mediated immune response have been elucidated (11). Cryptococcal antigen, when injected intravenously into CBA/J mice to simulate the levels of antigenemia in patients with systemic cryptococcosis, induces a cascade of suppressor cells that specifically down-regulate the protective anticryptococcal cell-mediated immune response (10-12). Most likely a similar induction of suppressor cells occurs in humans who have high serum levels of cryptococcal antigen. When the anticryptococcal cell-mediated immune response is depressed, whether by suppressor cells or other factors, then the organism has the advantage and progressive disease results (12). In AIDS patients whose CMI is significantly diminished due to their lack of CD4 T cells (13), cryptococci multiply uninhibited and produce large amounts of antigen (1-3,9). Since cryptococcal antigen has a relatively long half-life in the body (14), antigen accumulates in the body fluids and the levels often remain high even after antifungal drugs have limited the numbers of organisms in tissues (9). Continuously elevated cryptococcal antigen titers would favor induction of the T suppressor (Ts) cell cascade. The Ts cells then abrogate any residual anticryptococcal immunity in the AIDS or immunocompromised patients.

In studying anticryptococcal CMI and its regulation by suppressor cells, another regulatory cell population was discovered; however, this cell population augments the anticryptococcal cell-mediated immune response as measured by delayed-type hypersensitivity (DTH) rather than suppresses it (15). Since enhancement of the anticryptococcal DTH response directly correlates with increased protection (16-18, J. W. Murphy, unpublished data), elucidating the mechanisms by which this cell population modifies the cell-mediated immune response should provide the insights needed for developing protocols to enhance protection against *C. neoformans*. The regulatory cell that augments the anticryptococcal cell-mediated immune response is referred to as the Tamp cell and is the focus of this review (15,19).

Tamp cells are induced along with two other cryptococcal-antigen-specific cell populations by immunizing CBA/J mice with a 1:1 emulsion of cryptococcal antigen (CneF) and complete Freund's adjuvant (CFA) at two sites at the base of the tail (11,15). Six days after the immunization, the spleens of the mice contain three functionally distinct T cell populations (Figure 1) which are (i) the Tamp cells, (ii) the cells responsible for the DTH response (T_{DH} cells), and (iii) the third-order suppressor T (Ts3) cells that work in conjunction with another suppressor cell population (Ts2 cells) to inhibit the DTH response (11,15,19,21).

The presence of Tamp cells in spleens of immunized mice can be demonstrated by transferring the spleen cells from immunized mice to syngeneic recipient mice that are immunized with CneF-CFA at the time of spleen-cell transfer (15). Six days later

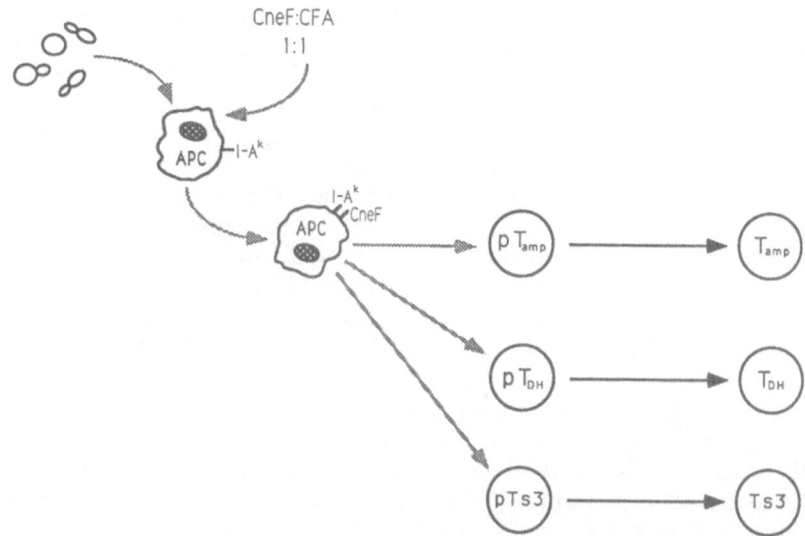

Fig. 1. Schematic drawing showing the cell populations induced by immunizing mice with cryptococcal antigen (CneF). Tamp cells amplify the
DTH response when transferred to mice at the time of immunization
of the mice with CneF in complete Freund's adjuvant. T_{DH} cells are
the cells responsible for the anticryptococcal DTH response. Ts3
cells are third-order suppressor cells. The solid black arrows indicate
aspects of the pathways that have been confirmed experimentally.
The gray arrows designate hypothetical aspects of the pathways.

the recipient mice are footpad challenged with CneF or saline, and 24 hr after footpad
challenge, the amount of swelling in the pads is measured. Mice given Tamp cells and
immunized with CneF:CFA display significantly higher DTH reactions to CneF
challenge, than do mice given naive spleen cells along with a similar immunization
(15,19). Serum from immunized mice given to recipient mice at the time of immunization does not amplify the anticryptococcal cell-mediated immune response (15), so
amplification of the DTH response is a cell-mediated reaction.

The Tamp cells have many characteristics in common with the T_{DH} cells
(15,19,20,22). Both the cryptococcal Tamp and T_{DH} cells are CD4$^+$ T cells. Also, like
the anticryptococcal T_{DH} cells, the Tamp cells are specific for cryptococcal antigen (19,
J. W. Murphy, unpublished data). The antigen specificity of the Tamp cells was
demonstrated by transferring cryptococcal Tamp cells to mice at the time of immunization with cryptococcal antigen, *Listeria monocytogenes*, or dinitrofluorobenzene (DNFB)
and footpad testing the animals at a later time with the antigen homologous to the
immunizing antigen (19). The anticryptococcal DTH response was significantly
augmented by the transfer of Tamp cells as compared to the immune control; however,
the anti-*Listeria* and anti-DNFB DTH responses in mice given the Tamp cells were
equivalent to responses in the respective immune control animals (19).

The kinetics of induction of the Tamp and T_{DH} cells are slightly different. Tamp
cells are not detectable in the immunized-donor mice until 5 days after immunization
(19); whereas, T_{DH} cells are present as early as 4 days after immunization in this model

(23). Tamp and T_{DH} cells have been distinguished from one another on the basis of their sensitivity to cyclosporin A (CsA) (15). T_{DH} cells are induced in the spleens of CsA-treated, immunized mice as demonstrated by the ability of the spleen cells from those mice to transfer anticryptococcal DTH to naive recipient mice (15). In contrast, Tamp cells are not induced in CsA-treated, immunized mice, since spleen cells from CsA-treated, immunized mice when transferred to recipient mice at the time of immunization of the recipient will not amplify the DTH response (15). The blocking of the induction of the Tamp cell was not due to the solvent in which the CsA was dissolved because spleen cells from solvent-treated, immunized mice do amplify the anticryptococcal DTH response of recipient mice (15).

Having demonstrated that solvent-treated, immunized mice had T_{DH} and Tamp cells in their spleens and CsA-treated, immunized mice had only T_{DH} cells present in their spleens, we were interested the abilities of the spleen cells from these two groups of mice to produce cytokines that could modulate the immune response. To assess the cytokines produced by spleen cells containing both T_{DH} and Tamp cells as compared to spleen cells containing T_{DH} cells alone, mice were treated with solvent or CsA from day -1 through day 5 and were immunized on day 0. Spleen cells were collected from each group of mice on days 1, 2, 4, and 6 after immunization and were cultured with medium alone, cryptococcal antigen (CneF), Concanavalin A (Con A), or phorbol myristate acetate (PMA) and Ca^{2+} ionophore for 48 hr at $37°$ C in 7% CO_2. At the end of the incubation, the supernatants were collected and assayed for interferon gamma (IFNγ) and tumor necrosis factor (TNF). On days 1 and 2 after either solvent or CsA treatment and immunization of mice, their spleen cells did not constitutively produce IFNγ nor could they be stimulated with CneF to produce IFNγ. However, by 4 days after immunization, spleen cells from solvent- or CsA-treated, immunized mice constitutively produced a small amount of IFNγ (700-800 pg/ml). When stimulated with cryptococcal antigen, the spleen cells from solvent-treated, immunized mice produced approximately 1700 pg/ml IFNγ; whereas, the spleen cells from the CsA-treated, immunized mice did not produce any more IFNγ when stimulated with CneF than they produced in medium alone. By 6 days following immunization, spleen cells from both groups of mice could be stimulated with CneF to produce IFNγ; however, the amount produced by the spleen cells from solvent-treated, immunized mice was significantly higher than the amount produced by the spleen cells from CsA-treated, immunized mice (2400 pg/ml as compared to 1700 pg/ml). The data indicate that spleen cell populations containing both T_{DH} and Tamp cells have a greater capacity to produce IFNγ after stimulation with cryptococcal antigen than do spleen cell populations containing only T_{DH} cells. Thus, it is possible that IFNγ produced by the transferred T_{DH} and Tamp cells after immunization of the recipient with cryptococcal antigen is responsible, at least in part, for amplifying the anticryptococcal DTH response in recipient mouse.

The abilities of spleen cells from solvent- or CsA-treated, immunized mice to produce IFNγ upon stimulation with the T cell mitogen, Con A, were also determined. At every time period that spleen cells were collected after immunization, the spleen cells from CsA-treated, immunized mice produced less IFNγ (20-25 ng IFNγ/ml) in response to Con A than did the spleen cells from the solvent-treated, immunized mice (45-50 ng IFNγ/ml). These data also support the idea that IFNγ may be a cytokine that plays a role in inducing the amplified anticryptococcal DTH response. It should also be noted that the profiles of IFNγ production after stimulation of *C. neoformans*-immune spleen cells with CneF are very different from the profiles of IFNγ production after stimulation of spleen cells from immunized mice with the T cell mitogen.

The concentrations of TNF in the supernatants from the various spleen cell preparations were also determined. There was no constitutive production of TNF by spleen cells taken from either solvent- or CsA-treated, immunized mice, and CneF did not stimulate any of the cell populations to produce TNF. TNF was produced, however, by spleen cells removed on days 1 and 2 from the solvent-treated, immunized group of mice after the cells were stimulated with PMA and Ca^{2+} ionophore (approximately 250 pg TNF/ml). By 4 and 6 days after immunization, spleen cells from solvent-treated, immunized mice responded to PMA-Ca^{2+} ionophore with only a slight increase in TNF production (approximately 140 pg/ml) over background levels (50 pg/ml). As one might expect based on data from other CsA-treatment models (24,25), spleen cells from CsA-treated, immunized mice did not produce measurable amounts of TNF after being stimulated *in vitro* for 48 hr with PMA-Ca^{2+}. Based on the fact that TNF was not produced *in vitro* after stimulation of immune spleen cells with CneF, TNF is not expected to play a role in inducing the amplified anticryptococcal DTH response. Furthermore, it appears that immunization with CneF actually diminishes the ability of spleen cells to produce TNF in response to potent stimulators such as PMA-Ca^{2+} ionophore.

Studies are currently in progress to determine the levels of interleukin-2 (IL-2) and interleukin-4 (IL-4) produced by spleen cells from solvent- or CsA-treated, CneF-immunized mice. Additional information on the cytokines produced by the cell populations containing the Tamp cells is required before the critical experiments can be done to demonstrate that the anticryptococcal cell-mediated immune response can be manipulated to enhance protection against *C. neoformans*.

REFERENCES

1. J. A. Kovacs, A. A. Kovacs, M. Polis, W. C. Wright, V. J. Gill, C. U. Tuazon, E. P. Gelmann, H. C. Lane, R. Longfield, G. Overturf, A. M. Macher, A. S. Fauci, J. E. Parrillo, J. E. Bennett, and H. Masur, Cryptococcosis in the acquired immunodeficiency syndrome, Ann. Int. Med. 103:533 (1985).

2. S. L. Chuck, and M. A. Sande, Infections with *Cryptococcus neoformans* in the acquired immunodeficiency syndrome, N. Engl. J. Med. 321:794 (1989).

3. J. Muller, Immunological aspects of cryptococcosis in AIDS patients, in: "Mycoses in AIDS Patients," H. van den Bossche, D. W. R. Mackenzie, G. Cauwenbergh, J. van Cutsem, E. Drouhet, and B. Dupont, eds., pp. 123-132, Plenum Press, New York (1990).

4. J. W. Rippon, *Medical Mycology.* Philadelphia: W. B. Saunders. pp. 582-609 3rd ed. (1988).

5. B. Dupont, I. Hilmarsdottir, A. Datry, M. Gentilini, P. Dellamonica, E. Bernard, S. Lefort, J. Frottier, P. Choutet, J. L. Vilde, and the French study group on fluconazole in cryptococcal meningitis in AIDS patients, Cryptococcal meningitis in AIDS patients, A pilot study of fluconazole therapy in 52 patients, in: "Mycoses in AIDS Patients," H. van den Bossche, D. W. R. Mackenzie, G. Cauwenbergh, J. van Cutsem, E. Drouhet, and B. Dupont, eds., pp. 287-303, Plenum Press, New York (1990).

6. S. A. Bozzette, R. A. Larsen, J. Chiu, M. A. E. Leal, J. Jacobsen, P. Rothman, P. Robinson, G. Gilbert, J. A. McCutchan, J. Tilles, J. M. Leedom, D. D. Richman, and the California Collaborative Treatment Group, A Placebo-controlled trial of maintenance therapy with fluconazole after treatment of cryptococcal meningitis in the acquired immunodeficiency syndrome, N. Engl. J. Med. 324:580 (1991).

7. M. A. Gordon, and D. K. Vedder, Serologic tests in diagnosis and prognosis of cryptococcosis, J. Am. Med. Assoc. 197:131 (1966).

8. R. D. Diamond, and J. E. Bennett, Prognostic factors in cryptococcal meningitis. A study of 111 cases, Ann. Intern. Med. 80:176 (1974).

9. D. W. R. Mackenzie, Cryptococcosis in the AIDS era, Epidem. Inf. 102:361 (1989).

10. J. W. Murphy, Cryptococcosis. in "Immunology of the Fungal Diseases," R. A. Cox, ed., pp. 93-138, CRC Press, Boca Raton, FL (1989).

11. J. W. Murphy, Immunoregulation in cryptococcosis, in: "Immunology of Fungal Diseases," E. Kurstak, ed., pp. 319-345, Marcel Dekker, Inc., New York (1989).

12. J. W. Murphy, Clearance of *Cryptococcus neoformans* from immunologically suppressed mice, Infect. Immun. 57:1946 (1989).

13. H. Masur, F. P. Ognibene, R. Yarchoan, J. H. Shelhamer, B. F. Baird, W. Travis, A. F. Suffredini, L. Deyton, J. A. Kovacs, J. Falloon, R. Davey, M. Polis, J. Metcalf, M. Baseler, R. Wesley, V. J. Gill, A. S. Fauci, and H. C. Lane, CD4 counts as predictors of opportunistic pneumonias in human immunodeficiency virus (HIV) infection, Ann. Intern. Med. 111:223 (1989).

14. J. E. Bennett, and H. F. Hasenclever, *Cryptococcus neoformans* polysaccharide: studies of serologic properties and role in infection, J. Immunol. 94:916, (1965).

15. P. L. Fidel, Jr., and J. W. Murphy, Effects of cyclosporin A on the cells responsible for the anticryptococcal cell-mediated immune response and its regulation, Infect. Immun. 57:1158 (1989).

16. L. K. Cauley, and J. W. Murphy, Response of congenitally athymic (nude) and phenotypically normal mice to *Cryptococcus neoformans* infection, Infect. Immun. 23:644 (1979).

17. T. S. Lim, and J. W. Murphy, Transfer of immunity to cryptococcosis by T-enriched splenic lymphocytes from *Cryptococcus neoformans* sensitized mice, Infect. Immun. 30:5 (1980).

18. C. H. Mody, M. G. Lipscomb, N. E. Street, and G. B. Toews, Depletion of CD4$^+$ (L3T4$^+$) lymphocytes in vivo impairs murine host defense to *Cryptococcus neoformans*, J. Immunol. 144:1472 (1990).

19. P. L. Fidel, Jr., and J. W. Murphy, Characterization of a cell population which amplifies the anticryptococcal delayed-type hypersensitivity response, Infect. Immun. 58:393 (1990).

20. J. W. Murphy, and J. W. Moorhead, Regulation of cell-mediated immunity in cryptococcosis. I. Induction of specific afferent T suppressor cells by cryptococcal antigen, J. Immunol. 128:276 (1982).

21. F. R. Khakpour, and J. W. Murphy, Characterization of third-order T suppressor cells (Ts3) induced by cryptococcal antigen(s), Infect. Immun. 55:1657 (1987).

22. K. L. Buchanan, P. L. Fidel, Jr., and J. W. Murphy, Effects of *Cryptococcus neoformans*-specific suppressor T cells on the amplified anticryptococcal delayed-type hypersensitivity response, Infect. Immun. 59:29 (1991)

23. J. W. Murphy, Effects of first-order *Cryptococcus*-specific T suppressor cells on induction of cells responsible for delayed-type hypersensitivity, Infect. Immun. 51:844 (1985).

24. A. Granelli-Piperno, M. Keane, and R. M. Steinman, Evidence that cyclosporine inhibits cell-mediated immunity primarily at the level of the T lymphocyte rather than the accessory cells, Transplantation 46:535 (1988).

25. M. Kinkhabwala, P. Sehajpal, E. Skolnik, D. Smith, V. K. Sharma, H. Vlassara, A. Cerami, and M. Suthanthiran, A novel addition to the T cell repertory. Cell surface expression of tumor necrosis factor/cachectin by activated normal human T cells, J. Exp. Med. 171:941 (1990).

IMMUNOMODULATORS AND FUNGAL INFECTIONS: USE OF

ANTIFUNGAL DRUGS IN COMBINATION WITH G-CSF

Yoshimasa Yamamoto, Katsuhisa Uchida*, Thomas W. Klein,
Herman Friedman, and Hideyo Yamaguchi*

Department of Medical Microbiology and Immunology
University of South Florida College of Medicine
Tampa, Florida, USA, and
Research Center for Medical Mycology*
Teikyo University School of Medicine, Tokyo, Japan

INTRODUCTION

It is widely accepted that one of the most important host-defense mechanisms against invading microorganisms is phagocytosis, followed by intracellular killing and digestion of the microorganism by phagocytic cells. Recently it was found that survival, growth and differentiation of progenitor phagocytic cells are controlled by hemopoietic colony-stimulating factors such as granulocyte-macrophage colony-stimulating factor (GM-CSF), granulocyte colony-stimulating factor (G-CSF), macrophage colony-stimulating factor (M-CSF) or multi colony-stimulating factor (IL-3) (1). G-CSF induces hemopoietic precursor cells to proliferate and differentiate into neutrophils, which are one of the most important first defense lines to microbial infections (2). Recently our studies have shown that G-CSF activates the antimicrobial activity of peripheral neutrophils against pathogenic yeast such as *Candida albicans in vitro* (3). Furthermore, it has also been demonstrated that G-CSF treatment of neutropenic mice can protect following infection with fungi (3). These previous results suggest that G-CSF might be a useful cytokine for treatment of neutropenic patients to protect from secondary fungal infections. To treat fungal infections of neutropenic patients, antifungal drugs must often be used as a therapeutic agent. However, there is little information about the usefulness of G-CSF in combination with antifungal drugs against fungal infections. The present studies were designed to investigate the therapeutic efficacy of four antifungal drugs, i.e., amphotericin B (AMPH), flucytosine (5-FC), fluconazole (FCZ) and itraconazole (ICZ) in combination with recombinant human G-CSF against *C. albicans* infection in neutropenic mice induced with cyclophosphamide. The administration of G-CSF in combination with either FCZ or ICZ markedly enhanced the therapeutic effect of antifungal drugs against Candida infection. However, G-CSF administered to mice in combination with either 5-FC or AMPH resulted in no significant enhancement. These results indicate that effectiveness is dependent upon the kind of antifungal drugs used.

Microbial Infections, Edited by H. Friedman *et al.*
Plenum Press, New York, 1992

MATERIALS AND METHODS

Animals

Female outbred mice, ICR, were obtained from Charles River Japan, Kanagawa, Japan. The mice were 5 to 7 weeks of age at the time of the experiments.

Neutrophils

Human neutrophils were isolated from leukocyte buffy coats obtained from healthy volunteers. The white cell layer on the surface of the red blood cell pellet after Ficoll-Hypaque (Pharmacia Fine Chemicals, Piscataway, NJ) centrifugation was collected and treated with 10 volumes of ACK solution (0.83% NH_4CO, 0.1% $KHCO_3$, 0.1mM Na_2EDTA) in order to lyse the red blood cells. The leukocytes were washed twice in Hanks' balanced salt solution (HBSS) and suspended in RPMI 1640 medium (Gibco Laboratories, Madison, WI) containing 1% heat inactivated fetal calf serum (FCS, Hyclone Laboratories, Logan, UT). Centrifuged preparations of neutrophil showed greater than 98% purity and 99% viability by Giemsa staining and trypan blue dye exclusion, respectively.

Cytokine and Antifungal Agents

Recombinant human G-CSF was kindly provided by Kirin Brewery Co., Tokyo, and was produced in an *E. coli* expression system and purified to homogeneity using HPLC as described previously (4). The CSF activity of the protein was 2.1×10^8 u/mg. Endotoxin was not detected in the CSF preparation by the *Limulus* amoebocyte lysate assay (E-toxate, Sigma Chemical Co., St. Louis, MO). Itraconazole, fluconazole, and amphotericin B were also kindly provided by Janssen-Kyowa Co., Ltd., Pfizer Pharmaceutical Inc. and Squibb Japan, respectively. Flucytosine was purchased from Sigma Chemical Co. FCZ and ICZ were suspended in polyethylene glycol 200 (Sigma). AMPH and 5-FC were dissolved in 5% glucose solution and distilled water, respectively.

Growth Inhibition

Inhibition of *C. albicans* growth by the neutrophils was accomplished by the following procedures: briefly, 50 µl of the neutrophil suspension (1×10^6/ml) in 1% FCS-RPMI 1640 medium was added to triplicate wells of a 96-well flat bottomed microplate. Then 20 µl of the appropriate concentrations of G-CSF and 50 µl *C. albicans* at 1×10^4/ml were added to wells containing effector cells, yielding effector:target (E:T) ratio of 100:1. After the cell mixtures were incubated for 15 hr at 37°C in 5% CO_2, 80 µl 0.25% saponin (Sigma) was added to lyse effector cells. The number of viable yeast in the wells was determined by the plate count method. The cell lysates were diluted with HBSS and plated on Sabouraud's dextrose agar plates. The plates were incubated at 37°C for 48 hr and then counted for colony forming units (CFU). Saponin and G-CSF had no effect on Candida viability.

Animal Study

Mice were injected intraperitoneally (i.p.) with or without 200 mg cyclophospha-mide (CY, Shionogi Pharmaceuticals, Osaka, Japan) per kg of body weight. One day after CY injection, the mice were injected subcutaneously (s.c.) once a day with 15 µg/kg G-CSF or vehicle phosphate buffered saline, pH 7.2 for three days. *C. albicans* were injected intravenously (i.v.) one day after the final injection of G-CSF or

Table 1. Effect of G-CSF on anti-Candida activity of cultured human neutrophils[a]

	Concentration (ng/ml)	Number of viable yeast in culture: CFU x 10^4
Candida only		146 ± 21^b
Neutrophils		20 ± 3
Neutrophils + G-CSF	1000	2.8 ± 1.0 $p<0.001^c$
	10	4.4 ± 1.1 $p<0.001$
	0.1	6.2 ± 1.6 $p<0.03$

[a]Neutrophils (5×10^4) were treated with graded doses (0.1 to 1,000 ng/ml) of G-CSF and simultaneously infected with 5×10^2 *C. albicans* cells. After 15 hr incubation, the neutrophil cultures were lysed with saponin and the number of viable yeast (CFU) in the cell lysates determined by the plate count method.
[b]Mean $\pm$ SD for 10 donors.
[c]Significantly different from non-treated neutrophil cultures.

vehicle. Antifungal drugs were administered to mice once a day for 7 days, beginning on the day of infection. AMPH and 5-FC were administered subcutaneously. ICZ and FCZ were administered orally. In the case of some experiments, mice were treated with G-CSF and antifungal drugs simultaneously, beginning the day of infection for 7 days. Death was determined up to three weeks after infection. The number of viable yeasts in the kidneys of infected mice was measured by the following procedures: mice were injected i.v. with 2×10^4 *C. albicans* 4 days after CY injection and treated with or without G-CSF and/or FCZ. After sacrifice one to five days later, the number of viable yeasts in homogenates of kidneys of individual mice were prepared in sterile saline and plated on Sabouraud's dextrose agar plates and the number of CFU counted 48 hr after incubation at 37°C.

Statistical Analysis

The ED50 of antifungal drugs in *C. albicans* infected mice was calculated by the Probit analysis method (5). The statistical significance of differences between treated and control groups was examined by Student's *t* test.

RESULTS AND DISCUSSION

Neutrophil Activation

Recently it has been reported that G-CSF activates mature neutrophils to promote reactive oxygen production in response to f-MLP (6-8) and up-regulate certain surface receptors of neutrophils (9, 10). From these reports, it was expected that G-CSF can activate neutrophils, even the mature ones, to induce anti-microbial activity. Our recent studies demonstrated that recombinant G-CSF activated isolated peripheral neutrophils and induced anti-*C. albicans* activity (3). Table 1 shows the summary of *in vitro* anti-*C. albicans* activity of isolated human peripheral neutrophils incubated with G-CSF. The isolated human peripheral neutrophils were cultured with various concentrations of G-CSF (1,000 -0.1 ng/ml) and also infected with *C. albicans* simultaneously. These neutrophil cultures were incubated for 15 hr at 37°C in 5% CO_2. At the end of incubation, the number of viable Candida in the neutrophil cultures was measured after lysis of the neutrophils with saponin. Neutrophils themselves could restrict the growth of Candida and the percentage of inhibition was almost 87%. G-CSF treated neutrophils restricted the Candida growth more strongly than non-treated control neutrophils, i.e., almost 10 times more. Even a low concentration of G-CSF, 0.1 ng/ml, induced a significant restriction of the growth of Candida in neutrophil cultures. However, G-CSF itself did not have any anti-fungal activity. Taken together with these results, this data and our previous published data suggest that G-CSF can activate peripheral neutrophils directly, including reactive oxygen metabolism, surface receptor modulation, phagocytosis and anti-microbial activity.

Protective Activity of G-CSF against Candida Infection

The efficacy of G-CSF treatment on the level of neutrophils in neutropenic animals has been demonstrated by several groups (11, 12). From these reports, it was expected that G-CSF treatment would protect against infections in immunocompromised hosts. Matsumoto et al. have recently reported the protective effect of G-CSF in microbial infections in neutropenic mice (13, 19). Cohen et al. have also demonstrated that treatment of neutropenic hamsters with recombinant G-CSF induced resistance to

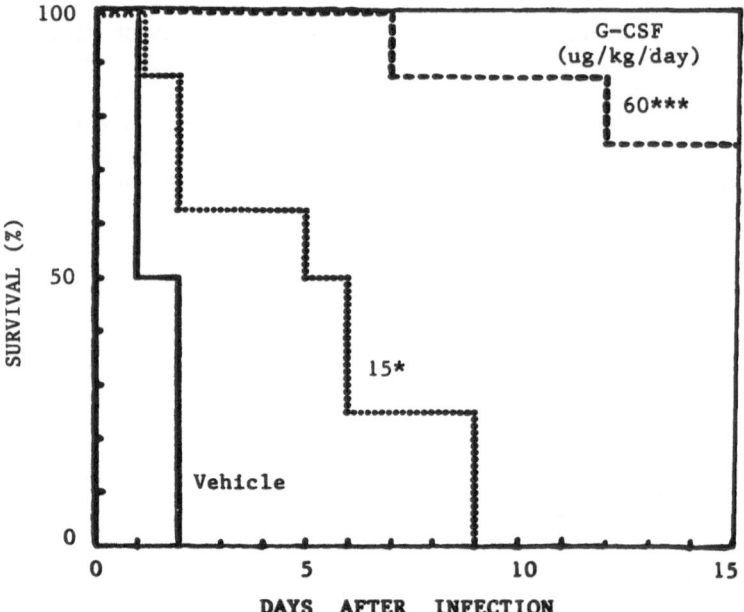

Fig. 1. Protective effect of G-CSF against *C. albicans* infection in CY-pretreated mice. Mice were inoculated intravenously with *C. albicans* (1 x 10⁵ cells/ mouse) 4 days after CY injection. G-CSF (dashed lines) or vehicle (solid line) was given intravenously to mice once a day for 3 days starting one day after CY injection. Each group consisted of 8 mice. *;p < 0.05, ***;p < 0.001: significantly different from the group of mice injected with vehicle only.

lethal infection with *S. aureus* (14). Our recent studies also demonstrated that G-CSF treatment significantly protected not only against Candida but also Aspergillus and Cryptococcus infections in neutropenic mice (3). The representative data of the protective effect of G-CSF against Candida infection in neutropenic mice are shown in Figure 1. The results showed G-CSF treatment with even a low concentration of 15 µg/kg/day still significantly protected against death from Candida infection and this protection occurred in a dose-dependent manner.

Combination with Antifungal Agents

Antifungal drugs are necessary therapy for fungal infections in immunocompromised patients. Immunotherapeutic agents are also necessary for immunologic improvement of immunocompromised patients. For example, G-CSF has been shown to restore the number of peripheral neutrophils in neutropenic animals (11, 12). Many antifungal drugs, such as antifungal antibiotics and chemical antifungal agents, have been developed and widely used against fungal infections. ICZ (15) and FCZ (16) are triazoles which are chemically synthesized antifungal agents. AMPH is used as a therapeutic agent for deep seated mycoses and this drug still remains the drug of choice for treatment of most invasive mycoses (17). 5-FC is an antimetabolite from the fluoropyridine series and also shows marked activity against pathogenic yeasts (18). These antifungal drugs have powerful therapeutic activity against fungal infections *in vivo*; however, these antifungal drugs have many side effects.

Combination therapy of antifungal drugs with immunotherapeutic agents may reduce the dose of antifungal drugs and reduce many side effects. However, up to now, combination therapy with immunotherapeutic agents such as G-CSF and antifungal drugs has not been investigated well. In the study reported here, designed mainly to investigate the therapeutic efficacy of the combination therapy with G-CSF and antifungal drugs in Candida infected mice, mice were injected with cyclophosphamide (CY) to induce neutropenia. The number of blood neutrophils was reduced to approximately one fifth of normal levels by 3 to 4 days after injection of CY (data not shown). Infection with 2×10^4 cells/mouse of *C. albicans* was performed 4 days after the CY injection. Various concentrations (0.1 - 10 mg/kg/day) of FCZ were given orally to mice once a day for 7 days starting one hour after infection. Panel A of Figure 2 shows the results of this experiment. FCZ itself has powerful therapeutic activity against Candida infection in neutropenic mice; however, it required a relatively high concentration, i.e., 1 mg/kg, for protection of mice against the Candida infection.

The results of G-CSF treatment in combination with FCZ are shown in panel B of Figure 2. The administration of G-CSF (15 µg/kg/day for 3 days) was begun one day after CY-treatment. G-CSF control (no-treatment with FCZ) showed significant prolongation of survival compared with infection only (no-G-CSF and no-FCZ, vehicle in panel A). When G-CSF treatment was combined with FCZ therapy in CY-treated mice, remarkable protection against Candida infection was evident, especially when combined with 0.1 mg/kg FCZ.

The number of viable Candida in the kidneys of treated infected mice, with or without G-CSF and/or FCZ, was measured. As shown in Figure 3, there were restricted numbers of viable yeasts in the kidneys of mice treated with both FCZ and G-CSF, as compared to mice treated with FCZ or G-CSF alone. This tendency of combined therapy to restrict CFU in the kidneys was stronger than that in mice given monotherapy. These data correlated well with results of the mortality course of mice receiving combination therapy.

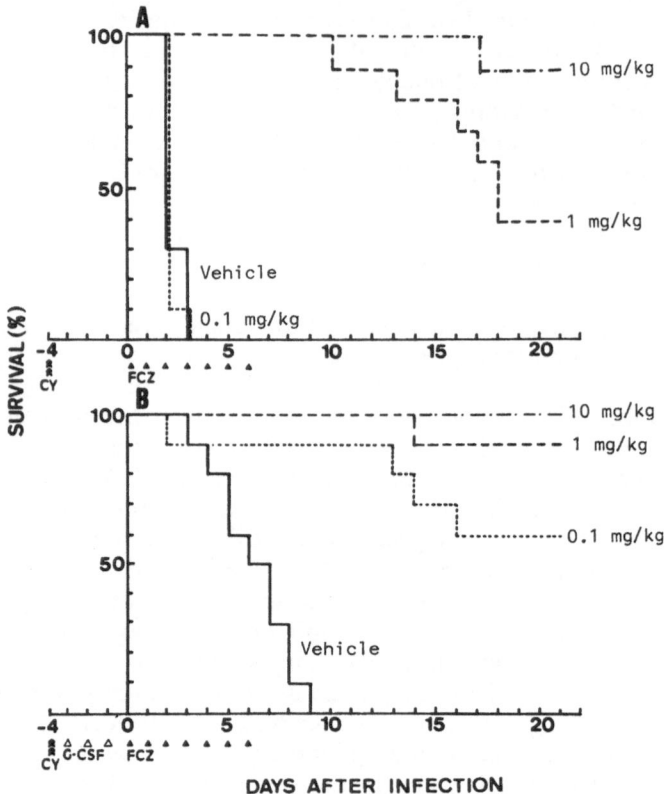

Fig. 2. Combination effect of G-CSF and FCZ against
C. albicans infection in CY-pretreated mice.
Mice were inoculated intravenously with *C.
albicans* (2×10^4 cells/mouse) 4 days after CY
injection. G-CSF or vehicle was given subcutaneously to mice once a day for 3 days starting
one day after CY injection. FCZ was given
orally to mice once a day for 7 days starting on
the day of infection. Each group consisted of
10 mice.

The combination therapy with G-CSF and FCZ in competent mice was also
studied. After infection with *C. albicans* (2.5×10^6 cells/mouse), mice were given
various concentrations of FCZ and G-CSF once a day for 7 days, starting on the day of
infection. Panel A of Figure 4 shows the results of mortality curves of Candida infected
mice treated with FCZ alone. The concentration of *C. albicans* needed to cause
infection in competent mice was 100 times more than that needed by CY-treated mice
(2.5×10^6 cells/mouse vs. 2×10^4 cells/mouse). The concentration of FCZ required for
protection against Candida infection in competent mice was 10 mg/kg. However, when
FCZ treatment was combined with G-CSF, markedly enhanced survival of mice was
evident, even at a low concentration of 1 mg/kg FCZ, as shown in panel B of Figure
4. These results suggest that combination therapy is useful even when treatment is
started after infection.

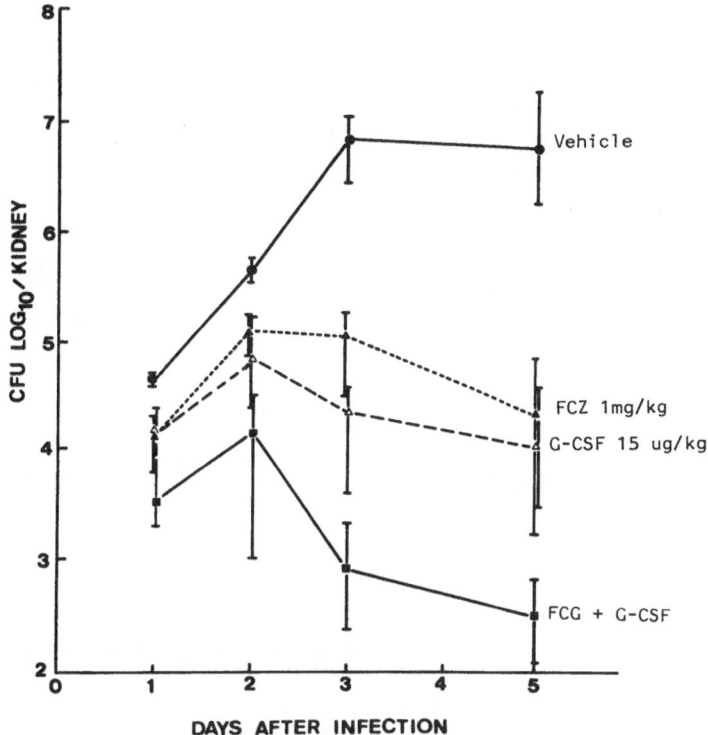

Fig. 3. Influence of the combination therapy with G-CSF
and FCZ on the number of viable *C. albicans* in
the kidney of CY-pretreated mice. Mice were
inoculated intravenously with *C. albicans* (2×10^4
cells/mouse) 4 days after CY injection. G-CSF or
vehicle was given subcutaneously to mice once a
day for 3 days starting one day after CY injection.
FCZ was given orally to mice once a day for 7
days starting on the day of infection. The results
are expressed as means of 3 mice ± SE.

The effect of ICZ in combination with G-CSF was studied in competent mice
infected with Candida. The results are shown in Figure 5. In this experiment it was
also demonstrated that combination therapy was more effective than monotherapy in
Candida infected mice when ICZ was used as the antifungal drug. For example,
treatment with 10 mg ICZ/kg in combination with G-CSF significantly prolonged
survival time and increased the percentage of survival from 0% to 60% in comparison
with that of monotherapy. On the other hand, combination therapy with G-CSF and
either AMPH or 5-FC showed either minimum or no enhancement in therapy after
Candida infection. Table 2 shows a summary of the effect of both G-CSF and these
antifungal drugs in Candida infected mice. The ED50 for both FCZ and ICZ was
markedly enhanced by combination with G-CSF in either competent or CY-treated
mice. For example, the ED50 for FCZ was much less than that of FCZ alone, i.e.,
almost 100 times lower (2.4 mg/kg vs. 0.03 mg/kg) when combined with G-CSF in CY-
treated mice. In contrast, the ED50 for both AMPH and 5-FC in combination with G-
CSF was not much different from that of these antifungal drugs alone in either

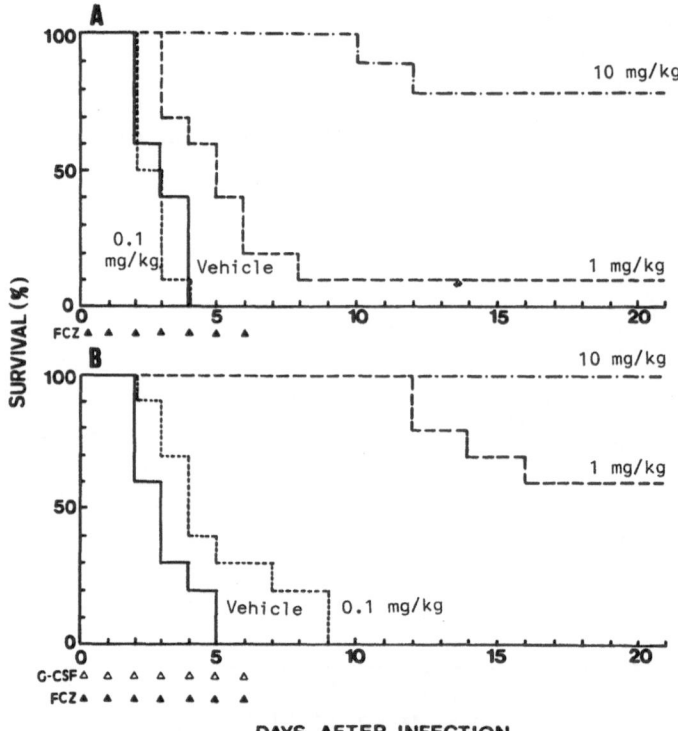

Fig. 4. Combination effect of G-CSF and FCZ against
C. albicans infection in competent mice. Mice
were inoculated intravenously with *C. albicans*
(2.5×10^6 cells/mouse). FCZ and vehicle given
orally to mice once a day for 7 days starting on
the day of infection. G-CSF was given subcutaneously to mice once a day for 7 days. Each
group consisted of 10 mice.

competent or neutropenic mice infected with *C. albicans*. These results indicate that
the effectiveness of the combination of an immunotherapeutic agent and antifungal
drugs is dependent upon the type of antifungal drug used. The effectiveness of G-CSF
treatment in combination with the antifungal drugs was demonstrated when combined
with triazoles such as ICZ and FCZ, but not with AMPH and 5-FC.

SUMMARY

The therapeutic efficacy of four antifungal drugs, i.e., amphotericin B, flucytosine,
fluconazole and itraconazole, in combination with recombinant human G-CSF against
Candida albicans infection in competent and neutropenic mice was studied. G-CSF
treatment enhanced the anti-*C. albicans* activity of isolated peripheral neutrophils *in
vitro*. The administration of G-CSF to mice augmented resistance against Candida
infection in neutropenic mice. Furthermore, administration of G-CSF in combination
with either FCZ or ICZ markedly enhanced the therapeutic effect of these antifungal

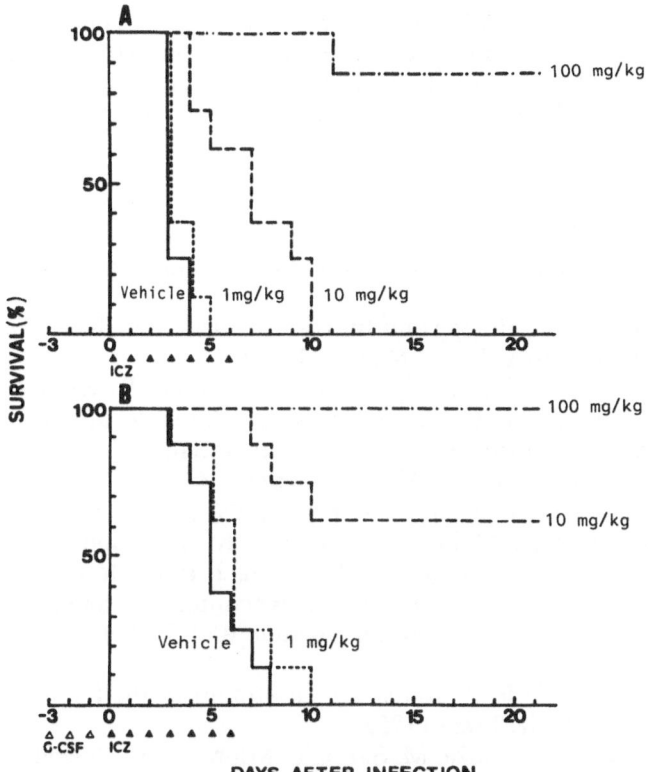

Fig. 5. Combination effect of G-CSF and ICZ against
C. albicans infection in competent mice. Mice
were inoculated intravenously with C. albicans
(2.5 × 10⁶ cells/mouse). G-CSF was given
subcutaneously to mice once a day for 3 days
starting one day after CY injection. ITZ and
vehicle given orally to mice once a day for 7
days starting on the day of infection.

Table 2. Summary of the combination effect of G-CSF
and antifungal drugs against C. albicans infection in competent and CY-pretreated mice[a]

| Mice | G-CSF | ED50 (mg/kg) | | | |
		Fluconazole (FCZ)	Itraconazole (ICZ)	Amphotericin B (AMPH)	Flucytosin (5-FC)
No treated	-	1.1	38.0	3.2	24.0
	+ 60ug/kg	0.18	5.2	1.4	24.0
CY-treated	-	2.4	32.2	4.2	16.0
	+ 15ug/kg	0.032	1.0	0.7	27.0

[a]ED50 for antifungal drugs against C. albicans infected
mice was calculated by the Probit analysis method (5).

drugs against Candida infection in both competent and neutropenic mice. However, G-CSF administered to mice in combination with either 5-FC or AMPH resulted in no significant enhancement. These results indicate that the effectiveness of combination therapy with G-CSF and antifungal drugs depends upon the antifungal drug used.

REFERENCES

1. N. A. Nicola and M. Vadas, Hemopoietic colony-stimulating factors, Immunol. Today 5:76 (1984).
2. D. Metcalf, The molecular biology and functions of the granulocyte-macrophage colony-stimulating factors, Blood 67:257 (1986).
3. Y. Yamamoto, K. Uchida, T. Hasegawa, H. Friedman, T. W. Klein, and H. Yamaguchi, Recombinant G-CSF induces anti-*Candida albicans* activity in neutrophil cultures and protection in fungal infected mice, in: "Recent Progress in Antifungal Chemotherapy," H. Yamaguchi, G. S. Kobayashi, and H. Takahashi, eds., Marcel Dekker, New York, (1991).
4. L. M. Souza, T. C. Boone, J. Gabrilove, P. H. Lai, K. M. Zesbo, D. C. Murdock, V. R. Chazin, J. Bruszewsk, H. Lu, K. K. Chen, J. Barendt, E. Platzer, M. A. S. Moore, R. Mertelsman, and K. Welte, Recombinant human granulocyte colony-stimulating factor: Effects on normal and leukemic myeloid cells, Science 232:61 (1986).
5. D. J. Finney, "Probit Analysis. A Statistical Treatment of the Sigmoid Response Curve," Cambridge Univ. Press (1952).
6. S. Kitagawa, A. Yuo, L. M. Souza, M. Saito, Y. Miura, and F. Takaku, Recombinant human granulocyte colony-stimulating factor enhances superoxide release in human granulocyte stimulated by chemotactic peptide, Biochem. Biophys. Res. Commun. 144:1143 (1987).
7. A. Yuo, S. Kitagawa, A. Okabe, Y. Komatsu, S. Itoh, and F. Takaku, Recombinant human granulocyte colony-stimulating factor repairs the abnormalities of neutrophil in patients with myelodysplastic syndromes and chronic myelogenous leukemia, Blood 70:404 (1987).
8. R. Sullivan, J. D. Griffin, E. R. Simons, A. I. Schafer, T. Meshulan, J. P. Fredette, A. K. Mass, A.-S. Gadenne, J. L. Leavitt, and D. A. Melnick, Effects of recombinant human granulocyte and macrophage colony-stimulating factors on signal transduction pathways in human granulocytes, J. Immunol. 139:3422 (1987).
9. A. N. Buckle, and N. Hogg, The effect of IFN-gamma and colony-stimulating factors on the expression of neutrophil cell membrane receptors, J. Immunol. 143:2295 (1989).
10. A. Yuo, S. Kitagawa, A. Ohsaka, M. Ohta, K. Miyazono, T. Okabe, A. Urabe, M. Saito, and F. Takaku, Recombinant human granulocyte colony-stimulating factor as an activator of human granulocytes: Potentiation of responses triggered by receptor-mediated agonists and stimulation of C3bi receptor expression and adherence, Blood 74:2144 (1989).
11. K. Welte, M. A. Bonilla, A. P. Gillion, T. C. Boone, G. K. Potter, J. L. Gabrilove, M. A. S. Moore, R. J. O'Reiley, and L. M. Souza, Recombinant human granulocyte colony-stimulating factor: Effects on hematopoiesis in normal and cyclophosphamide-treated primates, J. Exp. Med. 165:941 (1987).
12. M. Tamura, K. Hattori, H. Nomura, M. Oheda, N. Kubota, I. Imazeki, M. Ono, Y. Ueyama, S. Nagata, N. Shirafuji, and S. Asano, Induction of neutrophilic granulocytosis in mice by administration of purified human native granulocyte colony-stimulating factor (G-CSF), Biochem. Biophys. Res. Commun. 142:454 (1987).

13. M. Matsumoto, S. Matsubara, T. Matsuno, M. Tamura, K. Hattori, H. Nomura, M. Ono, and T. Yokota, Protective effect of human granulocyte colony-stimulating factor on microbial infection in neutropenic mice, Infect. Immun. 55:2715 (1987).

14. A. M.Cohen, D. K. Hines, E. S. Korach, and B. J. Ratzkin, *In vivo* activation of neutrophil function in hamsters by recombinant human granulocyte colony-stimulating factor, Infect. Immun. 56:2861 (1988).

15. J. VanCutsem, F. VanGerven, M.-A. Van de Ven, M. Borgers, and P. A. J. Janssen, Itraconazole, a new triazole that is orally active in aspergillosis, Antimicrob. Agents Chemother. 26:527 (1984).

16. J. F.Ryley and R. G. Wilson, ICI 153,066, a new orally active antifungal agent, in: "Program and abstracts of the 22nd Interscience Conference of Antimicrobial Agents and Chemotherapy," ASM, Washington D. C. (1982).

17. E. Drouhet and B. Dupont, Evolution of antifungal agents: Past, present, and future, Rev. Infec. Dis. 9(Sup 1):s4 (1987).

18. E. Grunberg, E. Titsworth, and M. Bennett M, Chemotherapeutic activity of 5-fluorocytosine, Antimicrob. Agents Chemother. 3:566 (1963).

19. M. Matsumoto, S. Matsubara, T. Matsuno, M. Ono, and T. Yokota T, Protective effect of recombinant human granulocyte colony-stimulating factor (rG-CSF) against various microbial infections in neutropenic mice, Microb. Immunol. 34:765 (1990).

BIOLOGIC RESPONSE MODIFIERS AS ANTIVIRALS IN

IMMUNOSUPPRESSED HOSTS

Page S. Morahan and Angelo J. Pinto

Department of Microbiology and Immunology,
The Medical College of Pennsylvania, Philadelphia, PA 19129

ABSTRACT

A wide variety of immunomodulators/biologic response modifiers (BRM) have been demonstrated to provide broad spectrum antiviral activity against both RNA and DNA viruses in several animal species. Dramatic decreases in mortality, reduced virus titers in tissues, and reduced histopathology can be produced. The antivirally effective agents include microbially derived materials, polyanions, cytokines, and chemically diverse small molecular weight chemicals. Antiviral efficacy with BRM treatment has been shown in numerous kinds of immunosuppression, emphasizing the potential for BRM treatment in immunocompromised patients. The greatest protective effects are observed with prophylactic or early therapeutic treatment. BRMs act indirectly, most likely by activating cells and/or inducing antiviral mediators early in the course of viral pathogenesis. In general, viral specific immune responses in BRM-treated and infected mice are absent or similar to those in untreated mice. Because BRMs are pleiotropic in their immunomodulatory effects, it has been difficult to establish whether one cell type or mediator is critical for the broad spectrum antiviral activity. Interferon appears to be critical for some small molecular weight synthetic compounds, but does not appear to explain all the antiviral activity of certain large molecular weight polyanions. Whether there is a unified antiviral mechanism among different BRMs remains to be determined.

INTRODUCTION

Since the mid 1960's, a variety of microbially derived substances, synthetic small molecular weight molecules or polymers, and proteins produced by recombinant DNA technology have been documented to produce antiviral effects in experimental infections (Table 1) (1-6). It is not clear whether there is any common structure activity relationship among the diverse BRMs that is responsible for antiviral activity (2), and continued investigation is needed to define the exact chemical structures required for antiviral activity of BRMs.

Despite impressive antiviral effects, BRMs have yet to achieve their clinical potential (Table 2). This is due in part to the need to treat early, toxicity of some

Microbial Infections, Edited by H. Friedman *et al*.
Plenum Press, New York, 1992

BRMs, and finding both suppression and enhancement of disease in certain experimental systems (7,8). Several new developments, however, are likely to lead to wider clinical use of BRMs (Table 2). It is now possible to treat infections earlier than previously possible. There is continued development of immunomodulators and cytokines with more narrow immunomodulatory effects (5,9). Finally, immunomodulation, in combination with chemotherapeutic antimicrobial drugs, is being proposed for use in severe viral and other microbial infections, especially in immunocompromised hosts (35,37-39). Synergy has been demonstrated between the antiviral drugs, ribavirin and azidothymidine, and several immunomodulators (e.g. poly ICLC, CL246,738, Ampligen, ABPP, 7-thia-8-oxoguanosine) (10,11,35), and between several nucleoside analogs and alpha interferon (12,39) in experimental models of viral infections. Combining antivirals with different mechanisms of action may prove particularly useful in AIDS patients, or patients being treated with immunosuppressive regimens.

Table 1. Examples of Immunomodulators with Antiviral Efficacy

NAME	COMPOSITION	SOURCE
Microbials		
C. parvum	Killed bacterial vaccine	Burroughs Wellcome
TDM	Trehalose dimycolate	Ribi Chem
MPL	Detoxified monophosphoryl lipid A	Ribi Chem
OK432	Killed Streptococcal vaccine	Chugai
Synthetics		
7-Th-8-OxG	7-Thia 8-oxyguanosine	Nucleic Acids Res.
CL246, 738	3,6-Bis (2-piperidinoethoxy) acridine	Lederle
S26308/R837	Quinilonamine	3M/Riker
MVE-2	Maleic anhydride divinyl ether copolymer	Hercules
ABPP	2-NH2-5-Br-6-phenyl-4 (3H) pyrimidinone	Upjohn
ABMP	2-NH2-5-Br-6-methyl-4 (3H) pyrimidinone	Upjohn
AIPP	2-NH2-5-I-6-phenyl-4 (3H) pyrimidinone	Upjohn
Ampligen	Mismatched poly rI:rC	Johns Hopkins/HEM
Poly ICLC	Polyribonucleotide-lysine carboxymethylcellulose	NCI
Recombinant DNA Derived		
rMuIFN-G	Murine gamma interferon	Genentech
rHuIFN-A A/D	Human alpha A/D interferon	Hoffman LaRoche
rMuIFN-B	Murine beta interferon	Toray Industries
rHuTNF-A	Human tumor necrosis factor alpha	Genentech
rHuCSF-M	Human Mφ colony stimulating factor	Cetus

Features of Antiviral Action Mediated by BRMs

The antiviral action of BRMs exhibits similar features (2,3) (Table 3). Activity is nonspecific against both RNA and DNA viruses, and is expressed in many animal species (1-5). The antiviral activity appears to be mediated indirectly, through effects on the host, because: (i) BRMs do not inhibit viral replication *in vitro* at concentrations that are effective *in vivo*; (ii) there is much greater efficacy of antiviral action with

Table 2. Clinical Potential for BRMs as Antiviral

```
1. Reasons why BRMs have not achieved clinical potential
   - Need for very early treatment
   - Lack of early viral diagnostic tests
   - Drug toxicity
   - Pleiotropic effects and side effects
   - Pharmacokinetics, drug delivery of polymers

2. Potential for antiviral efficacy in the next decade
   - More selective BRMs, less side effects
   - Mechanism of action better understood
   - Combination treatment
   - BRMs + chemotherapeutic antivirals
   - BRMs with different modes of action
   - Use in immunocompromised patients
   - Pharmacokinetics, drug delivery of polymers better understood
   - Earlier diagnosis and therefore earlier treatment possible
```

prophylactic treatment than with therapeutic treatment; and (iii) BRM treatment can provide systemic, long term (weeks) protection as well as local protection against viral infection (7). Antiviral activity of BRMs is generally expressed very early in the course of viral pathogenesis (3). Prophylaxis is usually most effective, but early therapeutic administration can prove equally effective (4-6,13), and this may be accompanied by induction of normal virus-specific immune responses.

All of these features are consistent with the concept that BRM antiviral activity is mediated through induction of nonspecific host responses that inhibit viral growth or spread early in viral pathogenesis. The question that remains is whether there is a common nonspecific host resistance mechanism that is induced by the chemically distinct BRMs, a mechanism that is effective against viruses with diverse replication strategies and pathogenesis.

Mechanisms of Antiviral Action of Biologic Response Modifiers

Most BRMs that exhibit antiviral activity are immunomodulators that exert widespread effects on nonspecific and specific immune responses, inflammation, coagulation, and the central nervous system (2,4,14-16). This multitude of physiologic effects on the host has made it difficult to establish whether there is one essential, unified mechanism of antiviral activity. A common antiviral mechanism may exist, but at sites or cells other than those commonly examined (e.g. peripheral blood, spleen). Measurements are needed to assess changes in the cells in the tissue relevant to viral infection (e.g. immunofluorescence, image analysis, in situ hybridization, polymerase chain reaction). Also, antiviral activity of host responses such as NK cells or macrophages may not be reflected by the assays commonly used (antitumor assays), and methods need to be developed for the direct measurement of extrinsic and intrinsic cellular antiviral activity (15).

Two experimental approaches useful in delineating specific immune responses to microbial infections have received some attention with BRMs -the transfer and selective depletion of effector cells or mediators. Studies on transfer of immunomodulator-activated cells have focused on macrophages, and have shown that transfer of about a mouse equivalent of BRM-activated, but not inflammatory, peritoneal cells from adult mice protected recipient mice against Friend leukemia virus (FLV) or herpes simplex virus (HSV) (17-19). With our increasing knowledge about the heterogeneity of mononuclear phagocytes (20), it would be helpful to transfer more highly purified and stably labelled cell populations (21,22), and follow effects on viral pathogenesis using molecular virologic techniques.

Antiviral Efficacy of BRMs in Immunocompromised Hosts

There have been numerous studies documenting antiviral efficacy of immunomodulators in mice suppressed of various cell populations (Table 4). In some cases, even though natural resistance was severely depressed, immunomodulator treatment gave complete protection. For example, while local silica treatment destroyed peritoneal cells and decreased natural antiviral resistance to HSV over 100 fold, treatment with two immunomodulators increased resistance even more in the immunocompromised mice than in normal mice (23). In other situations, immunomodulator treatment was effective when measured against an infection burden in the immunosuppressed animal that was equivalent to that in the immunocompetent controls (24,25). Efficacy of BRMs against HSV, cytomegalovirus, or foot and mouth disease

Table 3. Features of Antiviral Activity of BRMs

1. Nonspecific against RNA and DNA viruses (e.g. polyanions vs RNA picorna-, alpha-, flavi-, retro-, rhabdo-, bunyaviruses and DNA herpes-, adeno-, papova-, poxviruses)

2. Mediated indirectly by effects on host
  - no direct antiviral action in vitro
  - systemic as well as local protection
  - prophylactic efficacy versus therapeutic

3. Expressed early in viral pathogenesis
  - early inhibition of viral replication
  - low specific humoral, cellular immune responses
  - lack of resistance of survivors to reinfection
  - action in immunosuppressed mice

virus infections has been demonstrated in naturally immunodeficient states, such as that observed in neonatal mice (23,26,27), or beige mice congenitally deficient in NK cell function (28). The effectiveness of the streptococcal OK-432 vaccine against cytomegalovirus was partially lost when mice were treated with anti-asialo GM serum to deplete NK cells (27,36). However, efficacy of several other BRMs has been demonstrated to be maintained in the face of NK cell depletion, such as in animals treated with 89Sr to cause deficiency in circulating monocytes and granulocytes and NK cells (28,31), or treated with antibody to NK cells (32,33) (Morahan et al., unpublished observations). Antiviral activity has also been reported in mice rendered deficient in T cell responses (25,29,36), treated with immunosuppressive agents such as cyclophosphamide (30), or treated with silica or other agents known to affect tissue macrophages (23,34). Of particular interest are the observations that the retroviral infection itself, or HSV superinfection, can be suppressed in mice that are profoundly immunocompromised by infection with FLV or murine AIDS (29,37-39).

Antiviral efficacy in profoundly immunosuppressed hosts gives additional support to the usefulness of BRMs in viral infections in immunocompromised patients. The data also suggest that no single cell population may be essential for BRM-induced antiviral activity. This interpretation must be qualified, however, by the experimental inability to completely eliminate the cells. Only small numbers of mediator-producing

Table 4. BRM Action in Immunosuppressed Hosts

```
1.  Antiviral activity maintained in:
    - neonatal mice >8 days old
    - beige mice

2.  Antiviral activity maintained in mice treated with:
    - silica
    - 89Sr
    - Adult thymectomy, lethal irradiation, bone marrow reconstitution
    - Anti-lymphocyte serum
    - Dichloromethylene diphosphonate encapsulated in liposomes
    - Cyclophosphamide
    - Anti asialo GM1 antibody
    - Anti NK 1.1 monoclonal antibody
    - Estrone

3.  Antiviral activity maintained in mice infected with:
    - Friend leukemia virus
    - Mouse AIDS
    - AKR lymphoma virus
```

cells at critical sites may be necessary, if antiviral activity is mediated by a cascade of soluble substances.

Role of Interferon in BRM-mediated Antiviral Efficacy

The soluble antiviral mediators that have received the most attention for providing a possible unified mechanism of BRM-induced antiviral activity are the interferons. An interferon-mediated mechanism would be consistent with the common features of BRM-induced antiviral activity: viral nonspecificity, greater activity with prophylaxis than with therapeutic administration, and early effects in viral pathogenesis. Early observations suggested that interferons were not the common antiviral mechanism of all BRMs (2,3). There is a lack of correlation between BRM-induced circulating interferon and antiviral activity (3,40) or between kinetics of interferon induction and duration of protection[7]. There is antiviral efficacy with doses of BRMs that do not induce circulating interferon (40), lack of abrogation of protection with treatment of animals with anti-interferon serum (2,3), and lack of antiviral efficacy with transfer of serum (17).

More recent investigations, using higher titered anti-interferon sera, have suggested that interferon may be essential for the antiviral efficacy of some immunomodulators (Table 5). Sarzotti et al. (41) reported that treatment of mice with anti-alpha or with anti-beta interferon antibodies abrogated the antiviral efficacy produced by prophylactic CL246,738 treatment against Semliki forest virus (SFV) infection. They concluded that alpha and beta interferon were independently required for antiviral activity of this BRM. We have recently confirmed the importance of alpha/beta interferon for CL246,738 (40). Alpha/beta interferon induction also appeared to be essential for the prophylactic antiviral activity of the ABMP pyrimidinone (40) or 7-thia-8-oxyguanosine (32) against SFV infection, or quinilonamine against Rift Valley fever virus (Kende et al., personal communication). Antiviral efficacy of the polyribonucleotide poly IC against Banzi virus infection appeared to require induction of alpha interferon; treatment of mice with antiserum against alpha/beta interferon, but not against beta interferon, abrogated protection (Barnhardt, Gangemi, Mayer and Ghaffar, personal communication).

Thus, antiviral efficacy of at least five synthetic immunomodulators of diverse chemical structure (e.g. ABMP pyrimidinone, CL246,738, 7-thia-8-oxyguanosine, quinilonamine, poly IC) appears to require either or both alpha and beta interferons, when measured against arboviruses that are sensitive to interferon in vitro. Questions that remain are whether alpha or beta interferon is the critical antiviral mediator, what cells produce the interferon, and whether interferon induction by these immunomodulators is their common mechanism of antiviral action against other viruses.

Alpha/beta interferon also appears to be at least partially involve in the antiviral efficacy against SFV infection of two polyanionic immunomodulators, MVE-2 and the mismatched polyribonucleotide Ampligen (40). The data with MVE-2 are interesting since the early observations with the more polydisperse parent material, pyran, indicated that interferon was probably not important (2,3). Whether the results reflect differences in chemistry or the potency and specificity of the anti-interferon antibody remain to be determined.

The antiviral efficacy against SFV for at least one immunomodulator, poly ICLC, has not been able to be abrogated by treatment of mice with very high titers of anti-interferon antibody (40). Antiviral protection remained although induction of

Table 5. Role of IFN in BRM-mediated Antiviral Action

BRM	Virus	IFN AB	Action	Reference
7-Th-8-OxG	SFV	alpha/beta	Lost	32
	SFV	beta	OK	32
	SFV	gamma	OK	32
CL246,738	SFV	alpha	Lost	41
	SFV	beta	Lost	41
	SFV	alpha/beta	Lost	40
ABMP	SFV	alpha/beta	Lost	40
S26308/R837	RVFV	alpha/beta	Lost	Kende et al. Personal commun.
Poly IC	Banzi	alpha/beta	Lost	Barnhart et al.
	Banzi	beta	OK	Personal commun.
Ampligen	SFV	alpha/beta	Partially Lost	40
MVE-2	SFV	alpha/beta	Partially Lost	40
Poly ICLC	SFV	alpha/beta	OK	40

circulating interferon by poly ICLC was almost completely eliminated, and natural host resistance to the virus was reduced.

Considering the selective cell depletion and mediator depletion data, there are several possibilities for the antiviral activities of the polyanions, poly ICLC, Ampligen and MVE-2. Cells not affected by the selective depletion method may function as direct or indirect (through induction of interferon or other cytokines) antiviral effectors. Small numbers of cells remaining after depletion may be all that are required to start an amplifying cascade of cytokines that exert direct antiviral action. Interferon may be induced at critical sites and consumed rapidly by local cells, and therefore not be affected by neutralizing antibody. Furthermore, double stranded RNA molecules, such as Ampligen and poly ICLC, can directly induce 2'-5' oligoadenylate synthetase (which is normally induced by interferon) which then induces RNAse L which degrades viral mRNAs. Investigation of combinations of selective cell and mediator depletion methods may distinguish among these possible BRM-induced antiviral mechanisms.

ACKNOWLEDGEMENTS

This work was partially supported by a grant from the National Institute of Allergy and Infectious Diseases (AI 25751), a contract from the Army (DMAD 17-86-6-6117), and a contract from the Office of Naval Research (ONR N00014-82-K-0069). Some of the information in this review has been published in the review listed in Reference 6.

REFERENCES

1. P. S. Morahan, E.R. Leake, D.J. Tenney, and M. Sit, Comparative analysis of modulators of nonspecific resistance against microbial infections, in: "Immunopharmacology of Infectious Diseases: Vaccine Adjuvants and Modulators of Nonspecific Resistance," J. Majde, ed., Alan R. Liss, New York, 313-324 (1987).

2. P. S. Morahan, Anionic polymers and polysaccharides: overview of interferon inducing ability, antitumor activity and mechanisms of action, in: "Augmenting Agents in Cancer Therapy" E.M. Hersh, et al., eds., Raven Press, New York, 185-192 (1985).

3. M. C. Breinig, and P.S. Morahan, Interferon inducers: polyanions and others, in: "Interferon and Interferon Inducers", D.A. Stringfellow, ed., Marcel Dekker, New York, 239-261 (1980).

4. A. J. Pinto, P.S. Morahan, M.A. Brinton, Comparative study of various immunomodulators for macrophage and natural killer cell activation and antiviral efficacy against exotic RNA viruses, Int. J. Immunopharm. 10:197-209 (1988).

5. A. J. Pinto, P.S. Morahan, M.A. Brinton, D. Stewart, and E. Gavin, Comparative therapeutic efficacy of recombinant interferons alpha, beta and gamma against alphatogavirus, bunyavirus, flavivirus and herpesvirus infections, J. Interf. Res. 10:293-298 (1990).

6. P. S. Morahan, and A.J. Pinto, An historic overview of biologic response modifiers as antivirals, Canad. J. Infec. Dis., in press (1991).

7. G. B. Schuller, P.S. Morahan, and M. Snodgrass, Inhibition and enhancement of Friend leukemia virus by pyran copolymer, Cancer Res. 35:1915-1920 (1975).

8. R. P. Warren, J.D. Morrey, R.A. Burger, K.M. Okleberry, and R.W. Sidwell, Murine retroviral disease enhancing effects of a pyrimidinone immunomodulator, The Pharmacol. 31:166-169 (1989).

9. H. E. Broxmeyer, and R. Vadham, Preclinical and clinical studies with the hematopoietic colony stimulating factors and related interleukins, Immunol. Res. 8:185-201 (1989).

10. R. W. Sidwell, J.H. Huffman, D.F. Smee, J. Gilbert, M. Kende, and J.H. Huggins, Utilizing BRMs in combination with antivirals against experimentally induced virus infections, Canad. J. Infec. Dis. In press (1991).

11. M. Kende, H.W. Lupton, W.L. Rill, H.B. Levy, and P.G. Canonico, Enhanced therapeutic efficacy of poly ICLC and ribavirin combinations against Rift valley fever virus infection of mice, Antimicrob. Ag. Chemoth. 31:986-990 (1987).

12. L. R. Crane, and J.C. Sunstrom, Enhanced efficacy of nucleoside analogs and recombinant alpha interferon in weanling mice lethally infected with herpes simplex virus type 2, Antiviral. Res. 9:1-10 (1988).

13. D. F. Smee, J.H. Huffman, J. Coombs, J.W. Huggins, and R.W. Sidwell, Prophylactic and therapeutic activities of 7-thia-8-oxyguanosine against Punta Toro virus infections in mice, J. Biol. Resp. Mod. in press (1991).

14. P. S. Morahan, and D.M. Murasko, Viral Infections, in: "Natural Immunity in Disease Processes," D.S. Nelson, ed., Academic Press, Sydney 557-586 (1988).

15. P. S. Morahan, J.R. Connor, and K.R. Leary, Viruses and the versatile macrophage, Brit. Med. Bull. 41:15-21 (1985).

16. H. O. Besedovsky, and A. Del Ray, Physiologic implications of the immune-neuroendocrine network, in: "Psychoneuroimmunology," R. Ader, D.L. Felton, and N. Cohen, eds., Academic Press, New York (1991).

17. G. B. Schuller, and P.S. Morahan, Cellular and serum involvement in protection against Friend leukemia virus, Cancer Res. 37:4064-4069 (1977).

18. M. C. Breinig, L.L. Wright, M.B. McGeorge, and P.S. Morahan, Resistance to vaginal or systemic infection with herpes simplex virus type 2, Arch. Virol. 57:25-34 (1978).

19. L. A. Glasgow, J. Firschbach, S.M. Bryant, and E.R. Kern, Immunomodulation of host resistance to experimental viral infections in mice: effects of *Corynebacterium acnes*, *Corynebacterium parvum*, and Bacille Calmette-Guerin, J. Infec. Dis. 135:763-770 (1977).

20. P. S. Morahan, A. Volkman, M. Melnicoff, and W.L. Dempsey, Macrophage heterogeneity, in: "Macrophages and Cancer," G. Heppner, and A. Fulton, eds., CRC Press, Boca Raton, FL 1:25 (1989).

21. M. J. Melnicoff, P.S. Morahan, B.D. Jensen, E.W. Breslin, and P.K. Horan, *In vivo* labelling of resident peritoneal macrophages. J. Leuk. Biol. 43:387-397 (1988).

22. H. Rosen, and S. Gordon, Adoptive transfer of fluorescence-labelled cells shows that resident peritoneal macrophages are able to migrate into specialized lymphoid organs and inflammatory sites in the mouse, Eur. J. Immunol. 20:1251-1258 (1990).

23. P. S. Morahan, E.R. Kern, and L.A. Glasgow, Immunomodulator induced resistance against herpes simplex virus, Proc. Soc. Exp. Biol. Med. 154:615-620 (1977).

24. D. J. Giron, P.T. Allen, F.F. Pindak, and J.P. Schmidt, Inhibition by estrone of the antiviral protection and interferon elicited by interferon inducers in mice. Infec. Immun. 3:318-322 (1971).

25. P. S. Morahan, and R.S. McCord, Resistance to herpes simplex type 2 virus induced by an immunopotentiator (pyran) in immunosuppressed mice. J. Immunol. 115:311-313 (1975).

26. J. Y. Richmond, and C.H. Campbell, Influence of divinyl ether maleic anhydride (pyran) on foot and mouth disease virus infection: effect on adsorption and multiplication in mouse tissues, Arch. Ges. Virus forsch. 36:232-239 (1972).

27. K. Ebihara, and Y. Minamishima, Protective effect of biological response modifiers on murine cytomegalovirus infection. J. Virol. 51:117-122 (1984).

28. P. S. Morahan, P.H. Coleman, S.S. Morse, and A. Volkman, Resistance to infections in mice with defects in the activities of mononuclear phagocytes and natural killer cells: effects of immunomodulators in beige mice and 89Sr-treated mice. Infec. Immun. 37:1079-1085 (1982).

29. M. S. Hirsch, P.H. Black, M.L. Wood, and A.P. Monaco, Effects of pyran copolymer on leukemogenesis in immunosuppressed AKR mice. J. Immunol. 111:91-95 (1973).

30. C. Ishihara, J. Iida, N. Mizukoshi, N. Yamamoto, K. Yamamoto, K. Yato, and I Azuma, Effects of N alpha-acetylmuramyl-L-alanyl-D-isoglutamine-N epsilon-stearoyl-L-lysine on resistance to herpes simplex virus type 1 infection on cyclophosphamide treated mice, Vaccine 7:309-313 (1989).

31. P. S. Morahan, W.L. Dempsey, A. Volkman, and J. Connor, Antimicrobial activity of various immunomodulators: independence from normal levels of circulating monocytes and natural killer cells. Infec. Immun. 51:87-93 (1986).

32. D. F. Smee, H.A. Alaghammandam, A. Jin, B.S. Sharma, and W.B. Jolley, Roles of interferon and natural killer cells in the antiviral activity of 7-thia-8-oxyguanosine against Semliki Forest virus infections in mice, Antiviral. Res. 13:19-28 (1990).

33. J. F. Bukowski, K.W. McIntyre, H. Yang, and R.M. Welsh, Natural killer cells are not required for interferon-mediated prophylaxis against vaccinia or murine cytomegalovirus infections, J. Gen. Virol. 2219-2222 (1987).

34. A. J. Pinto, D. Stewart, N. van Rooijen, and P.S. Morahan, Selective depletion of liver and splenic macrophages using liposomes encapsulating the drug dichloromethylene diphosphonate: effects on antimicrobial resistance, J. Leuk. Biol. In press (1991).

35. O. Launay, M. Sinet, P. Varlet, and J.J. Pocidalo, Mouse model of retroviral infection: early combination therapy of azidothymidine with synthetic double-stranded RNA (poly I).(poly C), in: "Animal Models of AIDS," H. Schellekens and M.C. Horzinek, eds., Elsevier Press, NY, 311-321 (1990).

36. M. Okada, and Y. Minamishima, The efficacy of biologic response modifiers against murine cytomegalovirus infection in normal and immunodeficient mice. Microb. Immunol. 31:45-57 (1987).

37. J. D. Morrey, R.P. Warren, K.M. Okleberry, R.A. Burger, M.A. Chirigos, and R.W. Sidwell, Effect of imexon treatment on Friend virus complex infection using genetically defined mice as a model for HIV-1 infection. Antiviral. Res. 15:51-66 (1991).

38. R. M. Cozens, and H.K. Hochkeppel, Murine AIDS: effect of the muramyl-tripeptide derivative, MTP-PE, and interferon alpha against herpes simplex virus superinfection, in: "Animal Models of AIDS,: H. Schellekens and M.C. Horzinek, eds., Elsevier Press, NY 323-330 (1990).

39. J. D. Gangemi, R.M. Cozens, E. DeClerq, J. Balzarini, and H.K. Hochkeppel, 9-(2-phosphonylmethoxyethyl) adenine in the treatment of murine acquired immunodeficiency disease and opportunistic herpes simplex virus infections. Antimicrob. Ag. Chemother. 33:1864-1868 (1989).

40. P. S. Morahan, A.J. Pinto, D. Stewart, D.M. Murasko, and M.A. Brinton, Varying role of alpha/beta interferon in the antiviral efficacy of synthetic immunomodulators against Semliki Forest virus infection, Antiviral. Res. In Press (1991).

41. M. Sarzotti, D.H. Coppenhaver, I.P. Singh, J. Poast, and S. Baron, The in vivo antiviral effect of CL246,738 is mediated by the independent induction of IFN-A and IFN-B, J. Interfer. Res. 9:265-274 (1989).

STIMULATION OF HOST-DEFENSE MECHANISM WITH SYNTHETIC ADJUVANTS AND RECOMBINANT CYTOKINES AGAINST VIRAL INFECTION IN MICE

Ichiro Azuma, Chiaki Ishihara*, Joji Iida,
Yung Choon Yoo, Kumiko Yoshimatsu, and Jiro Arikawa

Institute of Immunological Science, Hokkaido University,
Sapporo, Japan, *School of Veterinary Medicine, Rakuno-
Gakuen University, Ebetsu, Japan

SUMMARY

The efficacy of synthetic immunoadjuvants and recombinant cytokines for the potentiation of host-resistance against virus infection was investigated using mouse models infected with Sendai virus and herpes simplex type 1 virus (HSV). The synthetic MDP derivative, MDP-Lys(L18), and recombinant cytokines, IL-1β, IFN-γ, G-CSF and GM-CSF were shown to be effective for the stimulation of nonspecific protection against Sendai virus infection in mice. Both MDP-Lys(L18) and GM-CSF were effective for the protection against HSV infection in cyclophosphamide (CY)-treated mice. B30-MDP was suggested to be useful as an immunoadjuvant for the potentiation of antigenicity of recombinant or component vaccines.

INTRODUCTION

It has been reported that immunostimulants (immunoadjuvants) stimulate not only immune responses but also host resistance against cancer and microbial infections. The authors have reported the biochemical and immunological properties of various kinds of bacterial fractions such as the cell-wall skeletons (CWSs) of *Mycobacterium bovis* BCG, *Nocardia rubra*, *Propionibacterium acnes* and *Listeria monocytogenes*. We have also shown that the peptidoglycan moiety of those bacterial CWSs plays the most important role for the expression of adjuvant effect. Especially BCG-CWS and *N. rubra*-CWS were shown to be effective for the prolongation of survival periods of cancer patients and for the protection against viral infections (1). In 1974, it was shown that the minimal structure requirement of the bacterial peptidoglycans for the adjuvant activity was *N*-acetylmuramyl-L-alanyl-D-isoglutamine (MDP) (2,3). In the last two decades, a variety of MDP analogs and derivatives have been chemically synthesized and their immunological properties have been examined. In addition to MDPs, we have shown the adjuvant activities of other synthetic compounds such as cord factor (trehalose dimycolate, TDM), lipid A, chitin and poly amino acids and their analogs and derivatives (1, 4, 5).

Microbial Infections, Edited by H. Friedman *et al.*
Plenum Press, New York, 1992

In this report we will summarize our results on the stimulation of host-resistance with synthetic adjuvants and several recombinant cytokines against viral infections in mice.

MATERIALS AND METHODS

Mice

Specific pathogen-free, male inbred BALB/c slc mice were obtained from Japan SLC, Inc., Hamamatsu, Japan, and maintained in the Laboratory of Animal Experiment, Institute of Immunological Science, Hokkaido University, under laminar-flow conditions. All mice were used at the age of 4-5 wks for the study on Sendai virus and HSV infections. Water and a pelleted diet (Nihon Nosan Kogyo Co., Ltd., Yokohama, Japan) were supplied *ad libitum*.

Adjuvants and Cytokines

MDP and its derivatives, 6-O-stearoyl-MDP (6) and N^α-acetylmuramyl-L-alanyl-D-isoglutaminyl-N^ϵ-stearoyl-L-lysine [MDP-Lys(L18)] (7) (Fig. 1A) were kindly donated by the Research Institute, Daiichi Pharmaceutical Co., Ltd., Tokyo, Japan. MDP-Lys(L-18) was dissolved in phosphate-buffered saline (PBS) at a concentration of 500 µg/ml. A dose of 100 µg of the adjuvant was administered either intranasally, subcutaneously, intraperitoneally or intravenously (through the venous plexus) under light Ketamine (Ketalar-50, Sankyo Co., Ltd., Tokyo, Japan) anaesthesia. B30-MDP (8) [6-O-(2-tetradecyl-hexadecanoyl)-muramyl-dipeptide, Fig. 1B] was obtained from Daiichi Pharmaceutical Co., Ltd., Tokyo, Japan. Recombinant mouse interferon-γ (IFN-γ) (0.7 mg/ml, specific activity 10^7 U/mg), prepared by Schering-Plough Corporation, was donated by the Suntory Co., Ltd. (Osaka, Japan). Recombinant human granulocyte colony-stimulating factor (G-CSF) (50 µg/ml, specific activity 3 x 10^7 U/mg) was supplied by Chugai Pharmaceutical Co., Ltd., Tokyo, Japan. Recombinant human interleukin-1β (IL-1β) (0.1 mg/ml, specific activity 2 x 10^6 U/ml) was supplied by Otsuka Pharmaceutical Co., Ltd., Tokushima, Japan. Recombinant granulocyte-macrophage colony-stimulating factor (GM-CSF) (1.17 mg/ml, specific activity 10^9 U/mg) was provided by Sumitomo Pharmaceutical Co, Ltd., Osaka, Japan. These recombinant cytokines were diluted with phosphate-buffered saline (PBS) containing 0.3% bovine serum albumin (BSA). In the preliminary experiment, we observed that the diluent did not have any effect on protection against Sendai virus or herpes simplex virus (HSV) infections.

Protection against Sendai Virus Infection

Details of the methods have been reported previously (9). The Sendai virus was purchased from Flow Laboratories Inc., Rockville, MD, USA. This virus was passaged for 11 generations in suckling C3H/He mice; after the 11th passage the lungs were homogenized in PBS and the supernatant fluid was dispersed in ampoules in 1 ml amounts, frozen and stored as the stock virus suspension at -70°C until use. For each experiment, an ampoule was thawed and 0.03 ml of the virus suspension was administered intranasally under light Ketamine anaesthesia. The survivors were monitored for up to 21 days after infection. Probability values were calculated by applying Mann-Whitney U probability test to the mean survival times of the treated group and that of the control group.

A

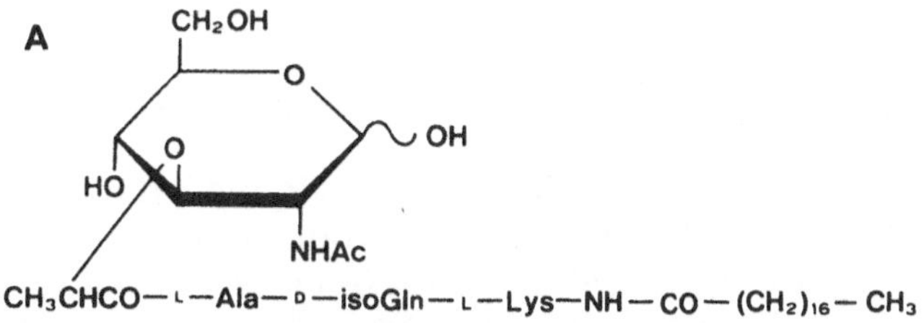

MDP-Lys (L18) : Abbreviation
Romurtide : Generic Name
Nopia : Brand Name
DJ-7041 : Code No. (Daiichi Pharmaceutical Co., Ltd.)

B

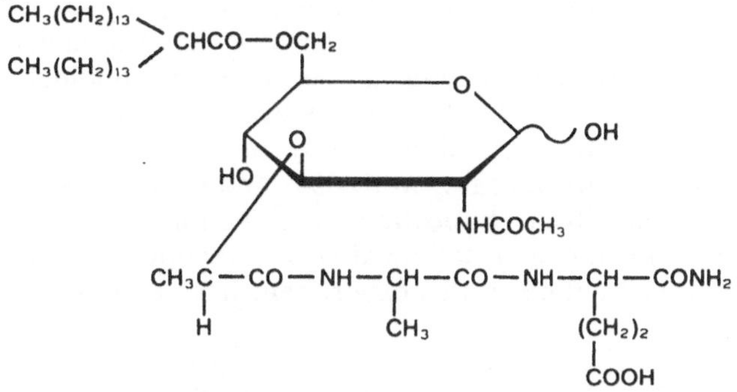

Fig. 1. Chemical structure of MDP-Lys(L18) (1A) and B30-MDP (1B).

Protection against (HSV) Infection

Details of the methods have been reported previously (10). HSV type 1 strain MacIntyre ATCC VR535 was provided by Dr. H. Sakaoka, School of Dentistry, Hokkaido University. A $10^{2.4}$ plaque-forming units (p.f.u.) of HSV was injected intravenously in the mice which had received intraperitoneally cyclophosphamide (CY, Sionogi Pharmaceutical Co., Ltd., Osaka, Japan) at a dose of 4 mg per mouse 1 day before infection. Statistical analysis was performed as described above.

Immunization of Recombinant Hantaan Virus Antigen with B30-MDP in Mice

The insect cells (Sf-9), infected with the recombinant baculovirus were used as immunogen. The infected Sf-9 cells (10^7; 1.2 mg total protein) in 0.5 ml of PBS containing 1.0 mg of B30-MDP and 200 μg of squalane oil were ground in a glass homogenizing vessel with a Teflon pestle. Saline (0.5 ml) containing 1% HCO-60, 5.6% d-Mannitol was added to the vessel, and the mixture was ground again for 3 min. Five-wk-old female BALB/c mice were injected i.m. with 0.2 ml of the mixture. Booster injection was given at 2 wks after the first injection with the same mixture. Mixtures without B30-MDP or without infected cells were prepared and inoculated as above. Mice were bled 1 and 2 wks after the first injection and 2 wks after the booster injection.

RESULTS

Effect of MDP-Lys(L18) on Host Resistance against Sendai Virus Infection in Mice (11)

Intranasal pretreatment with MDP, L18-MDP and MDP-Lys(L18) dissolved in PBS augmented the nonspecific resistance to intranasal infection with Sendai virus in mice. Intranasal administration of MDP-Lys(L18) and L18-MDP gave 100% and 88% survival rates, respectively. MDP also showed a 71% survival rate; the survival rate of the control group was 38%. Intranasal administration of MDP-Lys(L18) (10 μg) 3 days and 1 day before intranasal challenge with Sendai virus resulted in 100% survival rates; however, administration of MDP-Lys(L18) 1 day after the infection was 20% and control group was 20%, under this experimental condition. It was also shown that intranasal administration of 10 μg and 1 μg of MDP-Lys(L18) showed 86% and 83% survival rate, respectively, but the survival rate of the group treated with 0.1 μg of MDP-Lys(L18) was 29% and the control group was 29%. The effect of administration route of MDP-Lys(L18) was examined using the same experimental model. Intranasal administration of MDP-Lys(L18) was highly effective even at a dose of 10 μg per mouse (100% survival rate). When treated with the same dose by the intravenous or intraperitoneal route, the survival rates were 60% and 67%, respectively. The survival rate of control mice was 20%. Subcutaneous injection of MDP-Lys(L18) was not effective (33%) for protection against Sendai virus infection in comparison with that of control group (20%). The details of this experiment have been reported and discussed previously (11).

Protective Activity of Cytokines on Sendai Virus Infection (12)

IFN-γ. Mice infected with Sendai virus died with severe pneumonitis 7-15 days after infection. In the first experiment, we compared the protective activity of IFN-γ when administered intranasally or intravenously. All mice that had received 10, 10^2, and 10^3 U of IFN-γ intranasally 3 days before infection survived for 21 days after infection, whereas 10^3 U IFN-γ intravenously given was not effective (survival rate 13%). The survival rate of the control group was also 13%. We examined the effect of the timing of IFN-γ administration on its protective activity against Sendai virus infection. Although intranasal administration with 100 U IFN-γ either 5 days, 3 days or 1 day before infection showed potent protective activity (88%, 88% and 83%, respectively), post-infection administration (simultaneously with and 1 day after infection) had no effect (13%). The survival rate of the control group was 25%.

G-CSF and GM-CSF. When the mice were given intranasally 2.0 μg G-CSF 1 day before infection, the survival rate (75%) was significantly higher than that of the control group (0%). Treatment 3 days before infection showed a slight protective activity (43%) and all the mice (which were injected simultaneously or 1 day after infection) died within 21 days of infection. Subcutaneous or intravenous administration of 2.0 μg G-CSF 1 day before infection was not effective for protection against Sendai virus infection. Although the data are not shown, 2.0 μg G-CSF was the minimal effective dose by intranasal administration for affording resistance to infection under our experimental conditions, the survival rate of the mice that received 2.0 μg G-CSF subcutaneously or intravenously 1 day before infection was similar to that of the control group (14%, 14% and 0%, respectively). Intranasal administration of GM-CSF augmented host resistance to infection either 1 or 3 days before infection (100% survival rates).

IL-1β. The results presented above show that intranasal administration of IFN-γ and G-CSF before infection afforded a higher rate of protection against Sendai virus infection than intravenous or subcutaneous administration. Neither simultaneous administration nor post-administration of either cytokine was effective in increasing resistance to Sendai virus infection. As IL-1β has been shown to have some therapeutic effect in controlling microbial infection in mice, we examined its protective or therapeutic activity against Sendai virus infection by intranasal administration. Pretreatment with 0.2 μg IL-1β either 3 or 1 day before infection showed 100% protection against infection (control group 15% survival). Although the simultaneous administration of IL-1β (2 hrs after infection) was remarkably effective(86% survival), the post-administration of 0.2 μg IL-1β either 1 or 3 days after infection was not effective.

Protective Activity of MDP-Lys(L18) and Cytokines on HSV Infection in Cyclophosphamide (CY)-Treated Mice (10, 12)

MDP-Lys(L18). Administration of CY to mice 1 day before infection with sublethal doses of HSV type 1 lowered their resistance to infection, resulting in fatal HSV infection within 5-9 days after infection. CY treatment 3 or 8 days prior to HSV infection did not impair resistance. In contrast, in normal mice, no deaths occurred even if the mice were infected with the same dose of HSV. Treatment with MDP-Lys(L18) through subcutaneous, intraperitoneal or intravenous routes restored the impaired host resistance state of CY-treated mice to HSV. The survivors on day 21 after the infection of mice treated with MDP-Lys(L18) through intravenous, intraperitoneal and subcutaneous routes and that of controls with and without CY were 57, 51, 86, 29 and 100%, respectively.

Restorative activity of MDP-Lys (L18) to infection with HSV type 1 in CY-treated immunocompromised mice was induced when animals were treated intravenously twice with MDP-Lys(L18) at a 100 and a 50 μg dose, and subcutaneously at a 100 μg dose. Treatment twice intravenously with 10 and 1 μg of the adjuvant, or subcutaneously with 20 and 4 μg, showed no restorative effect on host resistance against HSV type 1 infection. The survivors on day 21 after the infection of mice treated intravenously twice with MDP-Lys(L18) at a 100, 50, 10 or 1 μg dose, and that of controls with and without CY, were 57, 29, 33, 14, 0 and 100%, respectively.

Treatment with MDP-Lys(L18) before infection of mice which were immunosuppressed by CY injection resulted an increase in survival time compared with the CY-treated control. However, treatment with MDP-Lys(L18) after challenge had no restorative effect on survivals or survival time. The survivors on day 21 of infection of mice treated with MDP-Lys(L18) on days 7, 3, 1 before and days 1 and 3 after infection were 33, 33, 43, 17 and 0%, respectively. Those of controls with and without CY were 0 and 83%, respectively.

To determine if the treatment of MDP-Lys(L18) had a significant depressed inhibitory effect on HSV growth in CY-treated mice, virus titer in the infected mice was assayed. A preliminary experiment indicated that intravenous infection with HSV into CY-treated mice showed peak virus titer in the liver on days 4 to 5, decreasing to an undetectable level on day 7. The virus titer in the brain was increased on day 7. Thus we examined HSV titer in the liver on day 4 of the infection to determine whether MDP-Lys(L18) restored the host resistance against HSV growth. Mean HSV titer in the liver was significantly higher in CY-treated mice than that in mice without CY. The administration of MDP-Lys(L18) in the CY-treated mice caused a restoration of host resistance, and the HSV growth was significantly suppressed not only in mice infected with $10^{2.4}$ p.f.u. but also in those with $10^{3.4}$ p.f.u.

Cytokines. Subcutaneous administration of MDP-Lys(L18) 1 and 3 days before infection with HSV almost completely protected the mice that had received CY intraperitoneally 1 day before infection. To evaluate the efficacy of IFN-γ, G-CSF, GM-CSF and IL-1β against HSV infection, various doses of the cytokines were administered subcutaneously four times a day before infection. It was shown that MDP-Lys(L18) has a protective action (87% survival rate), whereas IFN-γ (100 U), G-CSF (2 μg) and IL-1β (0.2 μg or 0.02 μg) were not effective. GM-CSF (0.2 μg) showed significant protective activity (75% survival rate, in comparison with 12% survival in CY control).

In addition to MDP-Lys(L18), we have reported that TDM (13), chitin derivative [especially 70% deacetylated chitin (DAC-70) (14)] and polyamino acids (5) [especially poly-arginine (poly-Arg), poly-lysine (poly-Lys) and poly-ornitinine (poly-Orn)] were effective in protecting against Sendai virus infection in mice.

Immunostimulation in Mice with B30-MDP to Recombinant Hantaan Virus Envelope Protein Expressed by Recombinant Baculovirus System

Hemorrhagic fever with renal syndrome (HFRS) is a rodent borne viral disease characterized by fever, renal disorder and hemorrhagic manifestations (15). The causative virus, Hantaan virus, was classified as *Hantavirus genus* of the *Bunyaviridae* family (16). Since outbreaks of HFRS are still being reported from countries throughout Asia and Europe, attempts to develop a vaccine have been made by several groups (17). Although the traditional killed vaccines for Hantaan virus have generally been effective in animal experiments (18), recombinant-expressed proteins have been investigated as a potential vaccine because of its advantages in avoiding a biohazard problem when a large amount of pathogenic viruses is prepared. The Hantaan virus envelope protein gene has been successfully expressed in baculovirus recombinant (19), however, immunogenicity of the recombinant protein was relatively low (20). Therefore, the efficacy of the protein as a vaccine is not fully characterized. In the present study, adjuvant activity of B30-MDP in the induction of humoral and cellular immunity to the Hantaan virus with the recombinant envelope (rENV) protein was characterized by a mouse model.

To examine the antigenicity of the rENV as immunogen, antibody titers of immunized BALB/c mice were measured by indirect immunofluorescent antibody (IFA) test with Vero cells infected with Hantaan virus as antigen. All the mice immunized with the rENV protein produced antibodies which reacted with authentic viral protein. The antibody titers determined by IFA test ranged from 1:20 to 1:320 at 4 wks after immunization. The results indicated that the recombinant protein expressed by the recombinant baculovirus system was able to elicit antibodies in mice. However, no statistical difference of antibody titers in both groups immunized with rENV together with or without B30-MDP was observed. Neutralizing or protective activities of the immune serum are under examination.

To examine the adjuvant effect of B30-MDP in the induction of cellular immune response, we compared the ability of spleen cells from mice immunized with or without B30-MDP to confer protection to syngeneic suckling mice. Spleen cells were prepared from mice immunized with rENV protein admixed with or without B30-MDP or immunized with B30-MDP alone. Spleen cells from untreated mice were also used as control. A passive protection study was carried out as follows: A group of seven BALB/c suckling mice (less than 24 hrs after birth) was inoculated s.c. with 10^4 focus forming units (40 LD_{50}) of Hantaan virus. Twenty-four hours after the inoculation, spleen cells (5×10^6 cells) were transferred i.p. into the mice. The animals were monitored for 42 days for survival rates. Spleen cells obtained from those immunized with rENV protein admixed with or without B30-MDP conferred protection of recipient suckling mice from fatal Hantaan virus challenge. Survival rate of mice receiving the immune spleen cells prepared by immunization with rENV protein admixed with B30-MDP was higher (70%) than those obtained by immune spleen cells from mice immunized with rENV protein alone (43%). In addition, immunization with B30-MDP alone was found effective to induce antiviral activity mediated by spleen cells. In contrast, all the mice receiving the spleen cells from unimmunized mice died within 20 days post infection.

These preliminary results demonstrated that B30-MDP is effective for stimulating the immunogenicity of mouse lymphoma cells (L5178-ML25) and recombinant protein, and especially for the induction of cellular immunity. Further study on basic conditions for preparing the immunogen mixture, such as purification and concentration of the expressed protein and the procedure for admixing with B30-MDP and antigen will be required.

DISCUSSION

It is well recognized that nonspecific host stimulation with immunostimulants is one of the most important factors in the treatment of cancer patients and a variety of immunostimulants have been developed (1, 21-22). Since the discovery of MDP by the groups of Ellouz (2) and Kotani (3), numerous biological activities have been reported (23). Chedid and co-workers (24) reported that MDP and its derivatives stimulated host resistance against *Klebsiella pneumoniae* infection in mice. Several MDP derivatives and related compounds such as murabutide, stearoyl-MDP derivatives, MTP-PE, FK-565, FK-156 and R.P. 40 639 have been found to have host stimulating activities against bacterial infections in experimental models (21, 22). Matsumoto and co-workers (25-28) also examined 64 acy-MDP derivatives as adjuvants for the stimulation of host resistance against nonspecific bacterial infection using a sepsis type infection model with *Escherichia coli* in mice, and selected 2 MDP derivatives, L18-MDP and MDP-Lys(L18). The stimulatory effects and their mechanisms of action of both stearoyl-MDP derivatives were examined in detail (28, 29). Results of the protective experiments against Sendai virus infection in mice suggest that MDP-Lys(L18) is effective not only for the prevention of the bacterial and fungal infections but also for the viral infection. It was also shown that MDP-Lys(L18) was a potent inducer of cytokines such as IL-1, IL-6, CSFs, TNF, IFN-γ and PGE2 in mice and human (30, 31) and was suggested that these cytokines might play important roles for the stimulation of host resistance against infections. Following these results, we have examined the protective effect of recombinant cytokines for the nonspecific resistance of host against the Sendai virus and HSV type 1 infections in mice.

Using the Sendai virus infection model in mice, we have shown that intranasal administration of IL-1β, G-CSF and GM-CSF as well as IFN-γ is effective for protection. The protection afforded by IFN-γ, G-CSF and IL-1β against this infection seemed to depend on the route or the timing of administration. Intranasal administration of these cytokines is likely to cause an inflammatory response, or to activate the immune system at the administration site (lungs) and consequently to stimulate host resistance against the viral infection. Intranasal administration of MDP-Lys(L18) was able to suppress the growth of Sendai virus during the early phase of infection. These findings also suggest that the protection against Sendai virus infection afforded by intranasal administration of IFN-γ, G-CSF, GM-CSF and IL-1β as well as MDP-Lys(L18) may be attributable to the activation of alveolar macrophages or neutrophils.

It was shown that the administration of MDP-Lys(L18) restored the protective activity against HSV type 1 infection in mice which were immunosuppressed by treatment with CY. When MDP-Lys(L18) was administered twice, at more than 50 μg per injection, intravenous, intraperitoneal or subcutaneous treatments were all effective. The treatment of MDP-Lys(L18) 1, 3 or 7 days prior to infection was effective, but treatment 1 or 3 days after infection produced no effect. This prophylactic, but not therapeutic, activity of MDP-Lys(L18) has also been shown in some bacterial and viral infections. HSV injected intravenously replicated first in the liver and the fulminant growth was inhibited by MDP-Lys(L18) treatment as early as 4 days after the infection in CY-treated mice. Although the exact protective mechanisms of MDP-Lys(L18) is not clarified yet, macrophages in the liver activated by MDP-Lys(L18) might be an important factor in the suppression of HSV growth, and it might result in less spreading of HSV into the brain. MDP-Lys(L18) has been shown to possess no anti-HSV activity by itself *in vitro*.

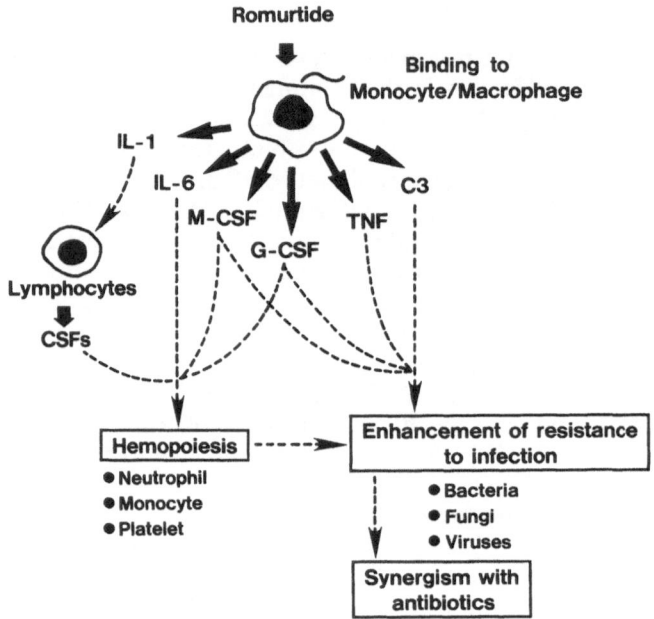

Fig. 2. Possible mode of action of MDP-Lys(L18)

Summing up these experimental studies, the mode of action of MDP-Lys(L18) was elucidated as shown in Fig. 2. It was suggested that the several kinds of cytokines produced by the action of MDP-Lys(L18) on macrophages/monocytes may be key elements in the hemopoiesis on neutrophils, monocytes and platelets which cause the stimulation of resistance to bacterial, fungal and viral infection. MDP-Lys(L18) was also shown to have synergistic action with antibiotics in bacterial and fungal infections in experimental models.

Recently, it was shown that MDP-Lys(L18) (Romurtide, generic name) was effective in cancer patients who were treated with radiation therapy and anti-cancer chemotherapeutics for the restoration of decreased neutrophils and platelets (32, 33).

During clinical trials, it was observed that the incidence of infectious diseases in the MDP-Lys(L18)-treated group was lower than that in the control group.

We previously have reported the chemical synthesis and adjuvant activity of B30-MDP and its function as adjuvant for the induction of allogeneic killer T cells (8). Kataoka et al. (34, 35) have reported that B30-MDP is a potent adjuvant for the induction of cellular immunity against line 10 hepatoma and B cell leukemia EN-L2C in strain 2 guinea pigs. Nerome et al. (36) found that B30-MDP is effective for the enhancement of humoral and cellular immune responses against influenza subunit vaccine in mice.

In conclusion, immunoadjuvants play very important roles in the prevention of viral infections via potentiation of specific and nonspecific immunities in experimental models and human.

ACKNOWLEDGMENT

This work was supported in part by Grants-in-Aid for Cancer Research from the Japanese Ministry of Education, Science and Culture; from the Japanese Ministry of Health and Welfare for Comprehensive 10-Year Strategy for Cancer Control; and for Scientific Research from the Japanese Ministry of Education, Science and Culture; by for the Osaka Foundation for Promotion of Clinical Immunology; and also by Grant-in-Aid for Special Project Research from Hokkaido University, Japan. The authors wish to thank Ms. M. Araki for her secretarial assistance.

REFERENCES

1. I. Azuma, Immunological and biochemical properties of bacterial fractions and related compounds with special reference to BCG cell wall skeleton and N. rubra cell wall skeleton, in: "Molecular and Cellular Networks for Cancer Therapy," Y. Yamamura and I. Azuma, eds., Excerpta Medica, Tokyo, pp. 83-104 (1989).

2. F. Ellouz, A. Adam, R. Ciorubaru, and E. Lederer, Minimal structural requirements for adjuvant activity of bacterial peptidoglycan subunits, Biochem. Biophys. Res. Commun. 59: 1317 (1974).

3. S. Kotani, Y. Watanabe, F. Kinoshita, T. Shimono, I. Morisaki, T. Shiba, S. Kusumoto, Y. Tarumi, and K. Ikenaka, Immunoadjuvant activities of synthetic N-acetylmuramyl-peptides or amino acids, Biken J. 18: 105 (1975).

4. I. Azuma, K. Nishimura, and S. Tokura, The immunological properties of chitin derivatives. - A Review, in: "Degradation and Biocompatibility of Synthetic Degradable Polymers", D.F. Williams, ed., CRC Press, London, in press.

5. J. Iida, N. Nishi, I. Saiki, N. Mizukoshi, C. Ishihara, S. Tokura, and I. Azuma, Macrophage activation and host augmentation against Sendai virus infection with synthetic polypeptides in mice, Int. J. Immunopharm. 11: 249 (1989).

6. S. Kusumoto, S. Okada, K. Yamamoto, and T. Shiba, Synthesis of 6-O-acyl derivatives of immunoadjuvant active N-acetylmuramyl-L-alanyl-D-isoglutamine, Bull. Chem. Soc. Jpn. 51: 2122 (1978).

7. R. Moroi, K. Yamazaki, T. Hirota, S. Watanabe, K. Kataoka, and M. Ichinose, Physicochemical properties of muroctasin, Arzneim.-Forsch./Drug Res. 38 (II): 953 (1988).

8. S. Kusumoto, M. Inage, T. Shiba, I. Azuma, and Y. Yamamura, Synthesis of long chain fatty acid esters of N-acetyl-L-alanyl-D-isoglutamine in relation to antitumor activity, Tetrahed. Lett. 49: 4899 (1978).

9. C. Ishihara, N. Hamada, K. Yamamoto, J. Iida, I. Azuma, and Y. Yamamura, Effect of muramyl dipeptide and its stearoyl derivatives on resistance to Sendai virus infection in mice, <u>Vaccine</u> 3: 370 (1985).

10. C. Ishihara, J. Iida, N. Mizukoshi, N. Yamamoto, K. Yamamoto, K. Kato, and I. Azuma, Effect of N^{α}-acetylmuramyl-L-alanyl-D-isoglutaminyl-N^{ϵ}-stearoyl-L-lysine on resistance to herpes simplex virus type-1 infection in cyclophospham-ide-treated mice, <u>Vaccine</u> 7: 309 (1989).

11. C. Ishihara, N. Mizukoshi, J. Iida, K. Kato, K. Yamamoto, and I. Azuma, Suppression of Sendai virus growth by treatment with N^{α}-acetylmuramyl-L-ala-nyl-D-isoglutaminyl-N^{ϵ}-stearoyl-L-lysine in mice, <u>Vaccine</u> 5: 295 (1987).

12. J. Iida, I. Saiki, C. Ishihara, and I. Azuma, Protective activity of recombinant cytokines against Sendai virus and herpes simplex virus (HSV) infections in mice, <u>Vaccine</u> 7: 229 (1989).

13. F. Numata, K. Nishimura, H. Ishida, S. Ukei, Y. Tone, C. Ishihara, I. Saiki, I. Sekikawa, and I.Azuma, Lethal and adjuvant activities of cord factor (trehalose-6,6'-dimycolate) and synthetic analogs in mice, <u>Chem. Pharm. Bull.</u> 33: 4544 (1985).

14. K. Nishimura, S. Nishimura, N. Nishi, I. Saiki, S. Tokura, and I. Azuma, Immunological activity of chitin and its derivatives, <u>Vaccine</u> 2: 93 (1984).

15. World Health Organization, Hemorrhagic fever with renal syndrome: memorandum from a WHO meeting, <u>Bull. World Health Organ.</u> 61: 269 (1983).

16. C. S. Schmaljohn and J.M. Dalrymple, Analysis of Hantaan virus RNA: evidence for a new genus of Bunyaviridae, <u>Virology</u> 131: 482 (1983).

17. J. W. LeDuc, Epidemiology of hemorrhagic fever virus, <u>J. Infect. Dis.</u> 11 Suppl. 4: s730 (1989).

18. K. Yamanishi, O. Tanishita, M. Tamura, H. Asada, K. Kondo, M. Takagi, I. Yoshida, T. Konobe, and K. Fukai, Development of inactivated vaccine against virus causing haemorrhagic fever with renal syndrome, <u>Vaccine</u> 6: 278 (1988).

19. C. S. Schmaljohn, J. Arikawa, J.M. Dalrymple, and A.L. Schmaljohn, Expression of the envelope glycoproteins of Hantaan virus with vaccinia and baculovirus recombinants, <u>in</u>: "Genetics and Pathogenicity of Negative Strand Viruses", D. Kolakofsky and B. Mahy, eds., Elsevier Biomedical Press, Amsterdam (1989).

20. C. S. Schmaljohn, Y.-K.C. Chu, A.L. Schmaljoh, and J.M. Dalrymple, Antigenic subunits of Hantaan virus expressed by baculovirus and vaccinia virus recombinants, <u>J. Virol.</u> 64: 3162 (1990).

21. I. Azuma, Development of immunostimulants in Japan, <u>in</u>: "Immunostimulants: Now and Tomorrow", I. Azuma and G. Jollès, eds., <u>Jpn. Sci. Soc. Press,</u> Tokyo/Springer-Verlag, Berlin, pp.41-56, (1987).

22. H. Werner, Immunostimulants: The western scene, <u>in</u>: "Immunostimulants: Now and Tomorrow", I. Azuma and G. Jollès, eds., <u>Jpn. Sci. Soc. Press,</u> Tokyo/-Springer-Verlag, Berlin, p.3 (1987).

23. A. Adam and E. Lederer, Muramyl peptides, immunomodulator, sleep factors, and vitamins, <u>Med. Res. Rev.</u> 4: 111 (1984).

24. L. Chedid, M. Parant, P. Lefrancier, J. Choay, and E. Lederer, Enhancement of nonspecific immunity to *Klebsiella pneumoniae* infection by a synthetic immunoadjuvant (*N*-acetylmuramyl-L-alanyl-D-isoglutamine) and several analogs, <u>Proc. Natl. Acad. Sci.</u> USA 74: 2089 (1977).

25. K. Matsumoto, H. Ogawa, T. Kusama, O. Nagase, N. Sawaki, M. Inage, S. Kusumoto, T. Shiba, and I. Azuma, Stimulation of nonspecific resistance to infection induced by 6-*O*-acyl muramyl dipeptide analogs in mice, <u>Infect. Immun.</u> 32: 748 (1981).

26. K. Matsumoto, H. Ogawa, O. Nagase, T. kusama, and I. Azuma, Stimulation of nonspecific resistance to infection induced by muramyl dipeptides, Microbiol. Immunol. 25: 1047 (1981).

27. K. Matsumoto, T. Otani, T. Une, Y. Osada, H. Ogawa, and I. Azuma, Stimulation of nonspecific resistance to infection induced by muramyl dipeptide analogs substituted in the γ-carboxy group and evaluation of N^α-muramyl dipeptide-N^ϵ-stearoyllysine, Infect. Immun. 39: 1029 (1983).

28. K. Matsumoto, Y. Osada, T. Une, T. Otani, H. Ogawa, and I. Azuma, Anti-infectious activity of the synthetic muramyl dipeptide analogue MDP-Lys (L18), in: Immunostimulants: Now and Tomorrow," I. Azuma and G. Jollès, eds., Jpn. Sci. Soc. Press, Tokyo/Springer-Verlag, Berlin, p.79-97 (1987).

29. T. Otani, T. Une, and Y. Osada, Stimulation of nonspecific resistance to infection by muroctasin, Arzneim.-Forsch./Drug Res. 38 (II): 969 (1988).

30. I. Saiki, S. Saito, C. Fujita, H. Ishida, J. Iida, J. Murata, A. Hasegawa, and I. Azuma, Induction of tumoricidal macrophages and production of cytokines by synthetic muramyl dipeptide analogs, Vaccine 6: 238 (1988).

31. F. Yamaguchi, M. Akasaki, and W. Tsukada, Induction of colony-stimulating factor and stimulation of stem cell proliferation by injection of muroctasin, Arzneim.-Forsch./Drug Res. 38 (II): 980 (1988).

32. E. Tsubura, T. Nomura, H. Niitani, S. Osamura, T. Okawa, M. Tanaka, K. Ota, H. Nishikawa, T. Masaoka, M. Fukuoka, A. Horiuchi, K. Furuse, M. Ito, K. Nagai, T. Ogura, M. Kozuru, N. Hara, K. Hara, M. Ichimaru, and K. Takatsuki, Restorative activity of muroctasin on leukopenia associated with anticancer treatment, Arzneim.-Forsch./Drug Res. 38 (II): 1070 (1988).

33. S. Sakamoto, T. Okawa, and N. Ogawa, Therapeutic effect of muroctasin on cancer patients with leukopenia during radiation therapy, Shin-yaku to Rinsho 38: 1407 (1989) (in Japanese).

34. T. Kataoka and T. Tokunaga, A synthetic adjuvant effective in inducing antitumor immunity, Jpn. J. Cancer Res. (Gann) 79: 817 (1988).

35. T. Kataoka, M. Kinomoto, M. Takegawa, and T. Tokunaga, Effect of a synthetic adjuvant for inducing anti-tumor immunity, Vaccine in press.

36. K. Nerome, Y. Yoshioka, M. Ishida, K. Ikuma, T. Oka, T. Kataoka, A. Inoue, and A. Oya, Development of a new type of influenza subunit vaccine made by muramyldipeptide-liposome: enhancement of humoral and cellular immune responses, Vaccine 8: 503 (1990).

ADF (ADULT T CELL LEUKEMIA-DERIVED FACTOR)/HUMAN THIOREDOXIN AND VIRAL INFECTION: POSSIBLE NEW THERAPEUTIC APPROACH

H. Masutani[1], H. Nakamura[1], Y. Ueda[1], Y. Kitaoka[1], T. Kawabe[1], S. Iwata[1], A. Mitsui[2], and J. Yodoi[1]

[1]Department of Biological Responses, Institute for Virus Research, Kyoto University, Kyoto, 606; [2]Central Research Laboratory, Aji-no-moto Co. Ltd., Kawasaki, Kanagawa, Japan

SUMMARY

ADF (adult T-cell leukemia-derived factor), originally defined as an inducer of interleukin 2 receptor/α (IL-2R/α), is a homologue of thioredoxin. ADF is constitutively produced by human lymphoid cell lines transformed by human T-lymphotropic virus type I (HTLV-I) or Epstein-Barr virus (EBV). ADF augments the proliferation of HTLV-I and EBV transformed cells as an autocrine growth factor. These data are indicative of the possible involvement of ADF in virus-related transformation of cells and their autocrine growth. On the other hand, thioredoxin contains a redox active disulfide and has a reducing activity in the presence of thioredoxin reductase and NADPH. To clarify the role of ADF/thioredoxin system in the viral transformation, we tested the effect of 13-cis-retinoic acid (RA), which is a competitive inhibitor of thioredoxin reductase, on the growth of ADF high producing cells. The expression of IL-2R/α on HTLV-I (+) cells was suppressed by RA. RA dose-dependently reduced the cell number and viability of ADF high producing lymphoid cells. Moreover, it had a suppressive effect on the proliferation of ADF high producing cells. It is suggested that RA has an inhibitory effect on the activation and the growth of cells producing ADF and that inhibition of the ADF/thioredoxin system may be a new therapeutic approach for retrovirus-related disorders.

INTRODUCTION

Adult T cell leukemia derived factor (ADF) was originally reported as an IL-2 receptor α chain inducer constitutively produced by HTLV-I transformed T cells (1-3). cDNA cloning of ADF has shown a remarkable homology between ADF and *Escherichia coli* thioredoxin (4). Subsequent studies on mammalian thioredoxin system have indicated that ADF is a human homologue of thioredoxin (5, 6, 7).

While thioredoxin was described in *E. coli* as an important coenzyme required for the conversion of deoxyribonucleotides from ribonucleotides, thioredoxin in the eukaryotic system has been shown to be involved in many cellular processes as a potent endogenous thiol-related reducing agent (8-10). An autocrine growth factor, 3B6/IL-1, produced by an EBV transformed B lymphoblastoid cell line 3B6 was recently proved to be identical to ADF (11-13). The production of ADF was enhanced or induced in association with the transformation by lymphotropic viruses such as HTLV-I and EBV (13).

As a human counterpart of thioredoxin, ADF may facilitate cytokine mediated signal transduction through the activation of target proteins by dithiol-dependent reduction (10). Indeed, the redox regulating mechanism maintains the activity of certain cellular proteins which interact with RNA or DNA (14, 15).

It was of interest to study the effect of the inhibitor of thioredoxin/thioredoxin reductase system, because thioredoxin system is deeply involved in cellular and viral activation processes. Schallreuter and Wood reported that 13-cis-retinoic acid (RA) is a competitive inhibitor of thioredoxin reductase (16). RA has various physiological functions including morphogenesis, cell differentiation, and keratinization (17). Retinoic acid receptors (RAR, RXR), and cellular retinoic acid binding protein (CRABP), have been considered to be the intracellular mediator of RA (18) and recently other target molecules of RA have also been suggested by several groups (19).

We show here that 13-cis RA inhibited the NADPH dependent reducing activity of the ADF/thioredoxin-thioredoxin reductase system and the growth of ADF high producing HTLV-I transformed cells, indicating that the inhibition of the thioredoxin system may be a possible therapeutic approach against viral transformation.

MATERIAL AND METHODS

Reducing Activity of ADF

Thioredoxin reductase purified from rat liver and recombinant ADF was prepared as described previously (10, 20). Thioredoxin reductase (final concentration ; 0.4U/ml), and human recombinant ADF were added to an assay mixture consisting of 0.14mM insulin (Sigma, St Louis, MO) and 0.27mM NADPH in 0.1M Tris-HCl (PH 7.5) and 2mM EDTA. After 10-70 min, the degraded insulin was quantitated by the decrease of NADPH content in absorbance at 340nm.

Western Blotting

ATL-2 and MT-2 cell lines were established from HTLV-I infected peripheral T lymphocytes of ATL patients. Jurkat is an HTLV-I (-) T cell line. 3B6 is an EBV-transformed B-lymphoblastoid cell line, while Jijoye is an EBV (+) Burkitt-derived B-lymphoblastoid B cell line without transformation. U937 is a human monoblastic cell line. These cells were maintained in RPMI 1640 medium containing 10% heat-inactivated fetal calf serum (FCS) and antibiotics (100U/ml penicillin and 100μg/ml streptomycin) in 95% humidity in 5% CO_2 at 37°C. Polyclonal antibodies from rabbit were raised by immunization with the synthetic peptides of the C-terminal 28 mer of ADF protein. The antibody was purified using an affinity column conjugated with recombinant ADF. Precise procedure of the antibody purification and Western blot analysis was previously described (3, 21). 10^7 cells were solubilized in 100μl of the lysing buffer (pH7.2, 10mM Tris-HCl, 150mM NaCl) containing 0.5% Nonidet P-40, 1mM PMSF (phenylmethysulfonyl fluoride) and 0.1TIU/ml of aprotinin. Two μg protein of each lysate was electrophoresed on 15% SDS polyacrylamide gels under reduced condition. After electro-transfer to an Immobilon PVDF membrane (Millipore, Bedford, MA), it was immunostained with 3.5μg/ml anti-ADF antibody by the avidin-biotin-alkaline phosphatase complex method (Vector Laboratories, Burlingame,

ADF		Val	Lys	Gln	Ile	Glu	Ser	Lys	Thr	Ala	-	-	Phe	Gln	Glu	Ala	Leu	Asp
E. Coli thioredoxin	Ser	Asp	Lys	Ile	Ile	His	Leu	-	Thr	Asp	Asp	Ser	Phe	Thr	Asp	Leu	Val	Lys
		Ala	Ala	Gly	Asp	Lys	Leu	Val	Val	Val	Asp	Phe	Ser	Ala	Thr	<u>Trp</u>	<u>Cys</u>	<u>Gly</u>
		Ala	Asp	Gly	Ala	Ile	Leu	Val	-	-	Asp	Phe	Trp	Ala	Glu	<u>Trp</u>	<u>Cys</u>	<u>Gly</u>
		<u>Pro</u>	<u>Cys</u>	<u>Lys</u>	<u>Met</u>	<u>Ile</u>	Lys	Pro	Phe	Phe	His	Ser	Leu	Ser	-	Glu	Lys	Tyr
		<u>Pro</u>	<u>Cys</u>	<u>Lys</u>	<u>Met</u>	<u>Ile</u>	Ala	Pro	Ile	Leu	Asp	Gln	Ile	Ala	Asp	Glu	-	Tyr
		Ser	Asn	Val	Ile	Phe	Leu	Glu	Val	Asp	Val	Asp	Asp	Cys	Gln	Asp	Val	Ala
		Gln	Gly	Lys	Leu	Thr	Val	Ala	Lys	Asp	-	Gln	Asn	Pro	Gly	Thr	Ala	Pro
		Ser	Glu	Cys	Glu	Val	Lys	Cys	Met	Pro	Thr	Phe	Gln	Phe	Phe	Lys	Lys	Gly
		Lys	Tyr	Ile	Gly	Arg	Gly	Ile	-	Pro	Thr	Leu	Leu	Leu	Phe	Lys	Asn	Gly
		Gln	Lys	Val	Gly	Glu	Phe	Ser	Gly	Ala	Asn	-	Lys	Glu	Lys	Leu	-	Glu
		Gln	Val	Ala	Ala	Thr	Lys	Val	Gly	Ala	Leu	Ser	Lys	Gly	Gln	Leu	Lys	Glu
		Ala	Thr	Ile	-	Asn	Glu	Leu	Val									
		Phe	Leu	Asp	Ala	Asn	-	Leu	Ala									

Fig. 1. Structural homology between ADF and thioredoxin. Upper: the amino acid sequence of ADF; Lower: the amino acid sequence of *E. coli* thioredoxin. The underlines indicate the active site of the reducing activity.

CA). As a control the same concentration of normal rabbit IgG was used for the primary antibody.

Immunohistochemical Staining

Specimens were fixed in 10% neutral formalin and embedded in paraffin. Sections were deparaffinized in toluene and treated with 3% hydrogen peroxide in methanol to block endogenous peroxidase activity, then incubated with normal goat serum for blocking of non-specific binding. The staining procedure was performed by the avidin-biotin-peroxidase complex method using a Biomeda Histoscan Kit (Biomeda Co., Foster City, CA). Preparations were incubated with the primary antibody (rabbit anti-ADF peptide IgG (2.0 mg/ml) or normal rabbit IgG). Counter staining was performed with hematoxylin.

Thymidine Incorporation Assay

Cells were cultured in 96-microwell plates (Nunc, Napervill, IL) with samples at the concentration of 1×10^6/ml with or without 13-cis RA (Sigma Co., St. Louis, Mo). ^{3}H-thymidine (^{3}H-TdR; Amersham, Tokyo) was added (final concentration 37KBq/ml) 6 hr prior to the harvest. After 72 hr culture, the incorporation of ^{3}H-thymidine was measured by a liquid scintillation counter (Aloka, Tokyo).

RESULTS AND DISCUSSION

ADF and Thioredoxin System

There is a structural homology between ADF and *E. coli* thioredoxin, especially on the active site (Try- Cys- Gly- Pro-Cys) of reducing activity (Figure 1) (4). As shown in Figure 2, recombinant ADF had an insulin-reducing activity as determined by the

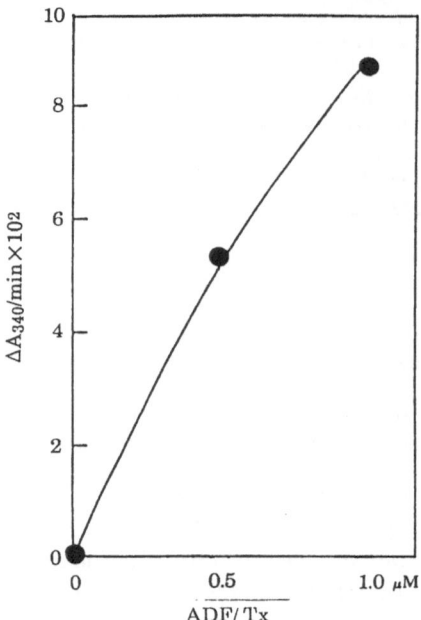

$$\text{NADPH} \diagdown \diagup \text{TxR} \diagdown \diagup \text{ADF}\diagup\text{Tx} \stackrel{\text{SH}}{\diagdown_{\text{SH}}} \diagdown \diagup \text{Insulin}$$

$$\text{NADP}^{\cdot} \diagup \diagdown \text{TxR} \diagdown \text{ADF}\diagup\text{Tx} \stackrel{S}{\underset{S}{|}} \diagup \diagdown \text{Insulin}\downarrow$$

Δ A 340/min

Fig. 2. Reducing activity of ADF/thioredoxin. Thioredoxin reductase (final concentration; 0.4U/ml),
and serially diluted human recombinant ADF
were added to an assay mixture consisting of
0.14mM insulin and 0.27mM NADPH in 0.1M
Tris-HCl (PH 7.5) and 2mM EDTA. After 10-70
min, the decrease of the absorbance at 340nm
was measured.

decrease of NADPH content. Recombinant ADF dose-dependently cleaved disulfide
bonds of insulin molecules in the presence of human or rat thioredoxin reductase and
NADPH (Figure 2) (10). In addition, ADF and human placental thioredoxin is
identical in antigenicity (Holmgren et al., personal communication.). Therefore, ADF
was considered to be a human counterpart of thioredoxin.

Previous studies have shown that the ADF/thioredoxin system is one of the major
endogenous reducing systems, modulating protein-protein and protein-nucleic acid
interactions through the reducto/oxidation of protein cysteine residues (3, 8, 10, 13).
In addition to its disulfide reducing activity, recombinant ADF was recently proven to
have a radical scavenging activity, as well as protein disulfide isomerase (PDI) activity
(Mitsui et al., manuscript in preparation). Furthermore, ADF has recently been shown
to be closely related or identical to T-hybridoma-derived growth factor MP6-BSF (22),
and eosinophil cytotoxic enhancing factor (ECEF) (23).

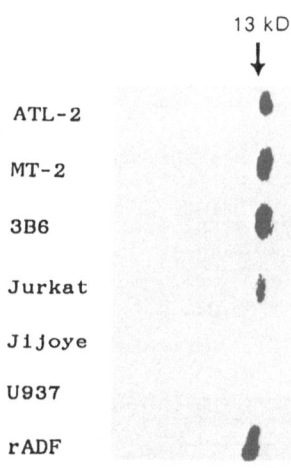

Fig. 3. Western blot analysis of ADF with
affinity purified anti-ADF antibody.
2μg protein/lane of the cell lysates of
ATL-2, MT-2, 3B6, Jurkat, jijoye, and
U937 cells were subjected to 15%
SDS-polyacrylamide gel electrophoresis. 0.75μg of recombinant ADF was
used as a control. The gel was transferred to PVDF membrane, which was
then incubated with anti-ADF antibodies, and developed as described in
Materials and Methods.

ADF and Viral Transformation

The production of ADF was markedly enhanced in association with the
transformation by lymphotrophic viruses such as HTLV-I and EBV (Figure 3).
Lymphocyte transformation by EBV or HTLV-I is associated with ADF gene activation
(4, 13). An autocrine growth factor 3B6-IL-1 produced by EBV transformed B cell lines
is identical to ADF (11-13). There is also a remarkable correlation between ADF
production and DNA replication of HPV in human cervical carcinoma tissues examined
by the *in situ* hybridization (24). These data collectively indicated that ADF is involved
in the viral transformation by HTLV-I, EBV and HPV.

Recently, we found that the growth of these ADF high producer cell lines was
highly dependent on thiol compounds including cysteine, 2-ME and reduced glutathione
(GSH), in addition to ADF (21). Furthermore, there is accumulating evidence for the
importance of ADF/thioredoxin in gene activation. Our recent data also show specific
and reversible regulation of the activity of the inducible enhancer binding protein,
NF-kB, by ADF/thioredoxin, through the reducto-oxidation of the dithiol-disulfide bond
(submitted, Kawabe et al.). The interaction of NF-kB with the NF-kB binding sequence
in the long terminal repeat (LTR) of HIV was found to be highly dependent on redox
regulation by ADF/thioredoxin *in vitro* (submitted, Okamoto et al.). Therefore, these
data strongly indicated the involvement of ADF/thioredoxin in the viral transformation
process and the constitutive IL-2R/α expression in HTLV-I transformed cells (Figure
4).

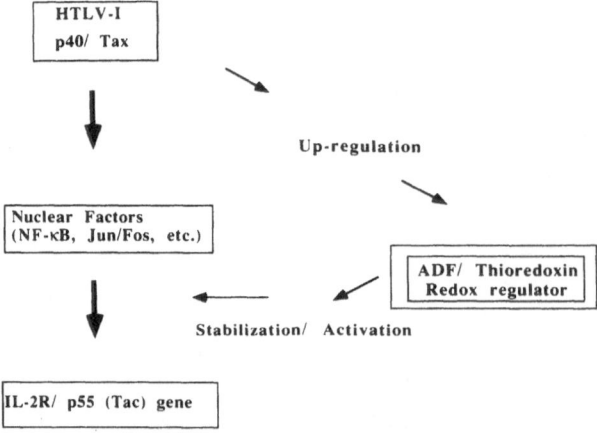

Fig. 4. Model for the involvement of ADF/thio-
redoxin in viral transformation. Possible
mechanism of IL-2R/p55(Tac) gene
activation in virus-induced lymphocyte
transformation.

ADF High Producing Cells in Lymph Nodes

ADF high producing cells (ADFh cells) in the paracortical area of lymph nodes
shared morphology and distribution pattern with the interdigitating dendritic cells.
Macrophages were poorly stained with anti-ADF antibody (Figure 5). Indeed,
constitutive ADF expression was seen on S-100 positive dendritic-like cells in both fetal
and adult thymic medulla (Go et al., unpublished observation).

In the atrophic and destroyed lymph nodes of AIDS patients, ADFh dendritic cells
were depleted nearly completely and their distribution was altered markedly (submitted,
Masutani et al.). Because dendritic cells, interdigitating dendritic cells and macrophages
are regarded as the reservoir of the persisting HIV replicating in lymph nodes (25), the
progressive immune dysfunction in AIDS seems to be associated with the depletion of
ADFh dendritic cells. If the constitutive production of ADF/thioredoxin is required for
the effective replication of HIV in dendritic cells, it is a tempting speculation that HIV
replication is dependent on the host thiol reducing system. The depletion of
paracortical ADFh cells may be a consequence of the breakdown of the cellular
reducing environment, possibly due to explosively rapid HIV replication. Elucidation
of the role of the ADF/thioredoxin system in the replication of HIV and HTLV-I may
provide important clues to establishing our strategy for the control of these viral
diseases parasitizing the host immune system.

Inhibitor of Thioredoxin/Thioredoxin Reductase System

To clarify the role of the ADF/thioredoxin system in the transformation of
HTLV-I (+) T cells, we have tested the effect of 13-cis RA, which is a competitive
inhibitor of thioredoxin reductase. As shown in Figure 3, ATL-2 cells were high
producers of ADF with constitutive IL-2R/α receptor expression on the cell surface.

In the insulin degradation assay for reducing activity, 13-cis RA dose-dependently
blocked the decrease of NADPH as measured by the absorbance at 340nm, while all
trans-RA did not (Figure 6). These data directly demonstrated that 13-cis RA inhibited
the reducing activity of the thioredoxin/thioredoxin reductase system.

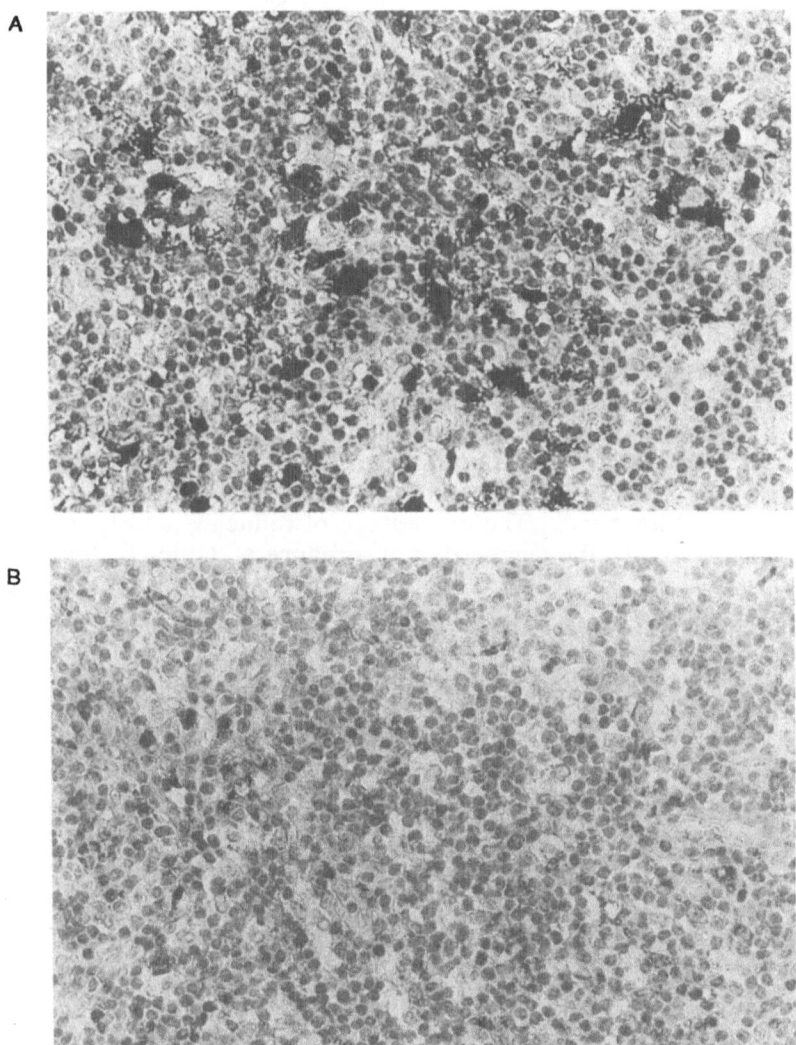

Fig. 5. Immunohistochemical staining of normal lymph nodes with anti-ADF antibody. Specimens were deparaffinized and immunostained with: **A.** rabbit anti-ADF peptide IgG, (x100); **B.** normal rabbit IgG, (x100) as described in Materials and Methods. Counter staining was performed with hematoxylin.

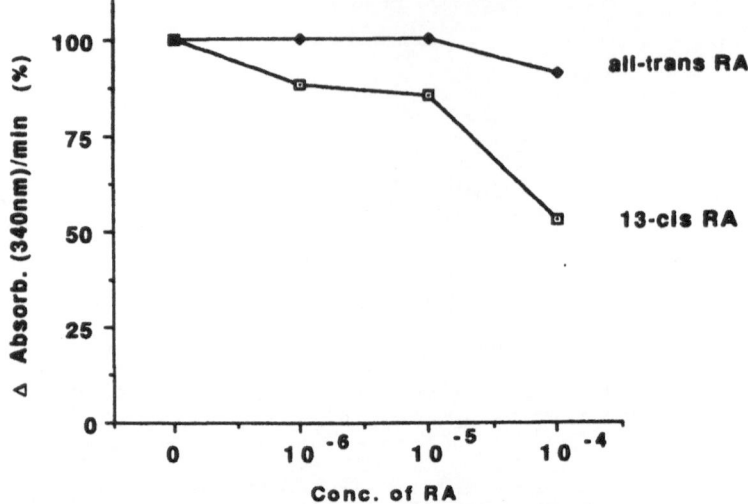

Fig. 6. Inhibition of NADPH dependent reducing activity by 13-cis RA. 0.4U/ml thioredoxin reductase, and human recombinant ADF were added to an assay mixture consisting of 0.14mM insulin and 0.27mM NADPH in 0.1M Tris-HCl (PH 7.5) and 2mM EDTA with 13-cis RA or all-trans RA. The percentage of reducing activity measured by the decrease in absorbance at 340nm is shown.

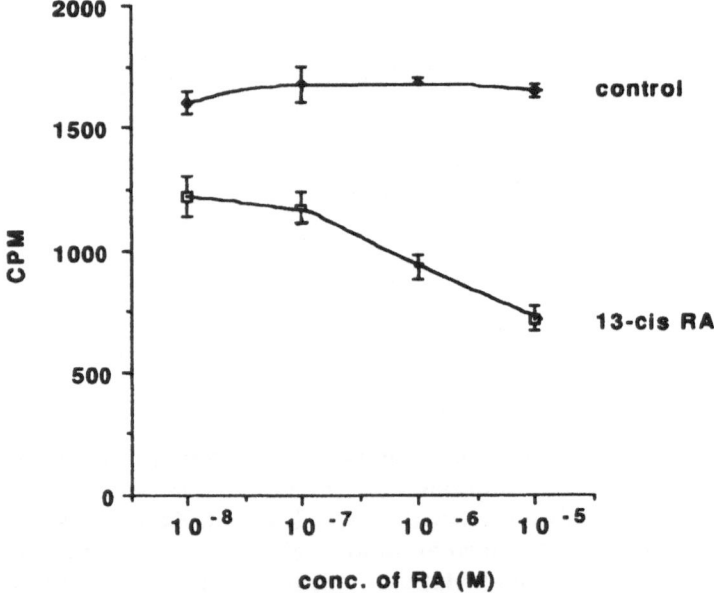

Fig. 7. Inhibition of ^{3}H-thymidine uptake of HTLV-I (+) ATL-2 cells by the administration of 13-cis RA. ATL-2 cells were cultured in 96-microwell plates with samples at the concentration of 1×10^6/ml with 13-cis RA or ethanol. After 72 hr culture, the incorporation of ^{3}H-thymidine was measured as described in Materials and Methods.

In addition, 13-cis RA had suppressive effect on the proliferation of ATL-2 cells (Figure 7), dose-dependently reducing the cell number and viability of these cells (data not shown). The level of IL-2R/α expression on ADFh cells was also reduced by 13-cis RA as determined by flowcytometric analysis (data not shown). These data suggested that RA had an inhibitory effect on the activation and growth of the cells dependent on ADF. The question whether the inhibitory effect of RA is due to the inhibition of thioredoxin reductase activity should be clarified by studies with other inhibitors of the thioredoxin system and synthetic compounds of retinoids.

Concluding Remarks

These data collectively indicate that inhibition of the ADF/thioredoxin system might be a new therapeutic approach toward retrovirus-related disorders.

ACKNOWLEDGEMENTS

The authors are grateful for the excellent scientific help by Dr. Y. Nanbu, scientific discussion with Dr. A. Yamauchi, and technical help by Ms. R. Kasahara, T. Suemoto and K. Kamata. The authors are also grateful to Dr. M. Maeda for kindly providing ATL cell lines. This work was supported by a Grant-in Aid for Scientific Research and Special Project Research Cancer Bioscience from Ministry of Education, Science, Culture of Japan, the Life Science Research Project of Institute of Physical and Chemical Research (RIKEN) and Ministry of Health and Welfare.

REFERENCES

1. K. Teshigawara, M. Maeda, K. Nishino, T. Nikaido, T. Uchiyama, M. Tsudo, M. Wano, and J. Yodoi, Adult T leukemia cells produce a lymphokine that augments interleukin 2 receptor expression, J. Mol. Cell. Immunol. 2:75 (1985).
2. M. Okada, M. Maeda, Y. Tagaya, Y. Taniguchi, K. Teshigawara, T. Yoshiki, T. Diamanstein, K. A. Smith, T. Uchiyama, T. Honjo, and J. Yodoi, TCGF(IL-2)--receptor inducing factor(s). II. Possible role of ATL-derived factor (ADF) on constitutive IL-2 receptor expression of HTLV-I(+) T cell lines, J. Immunol. 135:3995 (1985).
3. Y. Tagaya, M. Okada, K. Sugie, T. Kasahara, N. Kondo, J. Hamuro, K. Matsushima, C. A. Dinarello, and J. Yodoi, IL-2 receptor/Tac(p55) inducing factor:Purification and characterization of ATL-derived factor (ADF), J. Immunol. 140:2613 (1988).
4. Y. Tagaya, Y. Maeda, A. Mitsui, N. Kondo, H. Matsui, J. Hamuro, N. Brown, K. Arai, K. Yokota, H. Wakasugi, and J. Yodoi, ATL-derived factor (ADF), an IL-2 receptor/Tac inducer homologous to thioredoxin; Possible involvement of dithiol-reduction in the IL-2 receptor induction, EMBO J. 8:757 (1989).
5. S. W. Jones and K-C. Luk, Isolation of a chicken thioredoxin cDNA clone, J. Biol. Chem. 263:9607 (1988).
6. R. S. Johnson, W. R. Mathews, K. Biemann, and S. Hopperl, Amino acid sequence of thioredoxin isolated from rabbit bone marrow determined by tandem mass spectrometry, J. Biol. Chem. 263:9589 (1988).
7. E. E. Wollman, L. Auriol, L. Rimsky, A. Shaw, J. P. Jaquot, P. Wingfield, P. Graber, F. Dessaros, P. Robin, F. Gilbert, J. Bertoglio, and D. Fradelizi, Cloning and expression of cDNA for human thioredoxin, J. Biol. Chem. 263:15506 (1988).

8. A. Holmgren, Thioredoxin, Ann. Rev. Biochem. 54:237 (1985).

9. T. C. Laurent, E. C. Moore, and P. Reichard, Enzymatic synthesis of deoxyribo-nucleotides. IV. Isolation and characterization of thioredoxin, the hydrogen donor from *Escherchia Coli* B, J. Biol. Chem. 258:13658 (1983).

10. Y. Tagaya, H. Wakasugi, H. Masutani, H. Nakamura, S. Iwata, A. Mitsui, S. Fujii, N. Wakasugi, T. Tursz, and T. Yodoi, Role of ATL-derived factor (ADF) in the normal and abnormal cellular activation:Involvement of dithiol related reduction, Molecular Immunol. 27:1279 (1990).

11. H. Wakasugi, R. Rimsky, Y. Mahe, A. M. Kamel, D. Fradelizi, T. Tursz, and J. Bertoglio, Epstein-Barr virus-containing B-cell line produced an interleukin 1 that it uses as a growth factor, Proc. Natl. Acad. Sci. USA 84:804 (1987).

12. N. Wakasugi, Y. Tagaya, H. Wakasugi, A. Mitsui, M. Maeda, J. Yodoi, and T. Tursz, ADF/Thioredoxin produced by both HTLV-I and EBV transformed lymphocytes acts as an autocrine growth factor and synergizes with IL-1 and IL-2, Proc. Natl. Acad. Sci. USA. 87:8282 (1990).

13. J. Yodoi and T. Tursz, ADF:An endogenous reducing protein homologous to thioredoxin:Involvement in lymphocyte immortalization by HTLV-1 and EBV, Adv. Cancer Res. (in press).

14. M. W. Hentze, T. A. Rouault, J. B. Jarford, and R. D. Klausner, Oxidation-reduction and the molecular mechanism of a regulatory RNA-protein interaction, Science 244:357 (1989).

15. C. Abate, L. Patel, F. J. Rausher III, and T. Curran, Redox regulation of Fos and Jun DNA-binding activity *in vitro*, Science 249:1157 (1990).

16. K. U. Schallreuter and J. M. Wood, The stereospecific suicide inhibition of human melanoma thioredoxin reductase by 13-cis-retinoic acid, Biochem. Biophys. Res. Commun. 160:573 (1989).

17. J. Brockes, Reading the retinoid signals, Nature 345:766 (1990).

18. R. Blomhoff, M. H. Green, T. Berg, and K. R. Norum, Transport and storage of Vitamin A, Science 250:399 (1990).

19. J. L. Goldstein and M. S. Brown, Regulation of the mevalonate pathway, Nature 343:425 (1990).

20. M. Luthman and A. Holmgren, Rat liver thioredoxin and thioredoxin reductase: Purification and characterization, Biochemistry 21:6628 (1982).

21. A. Yamauchi, H. Masutani, Y. Tagaya, H. Nakamura, T. Inamoto, K. Ozawa, and J. Yodoi, Lymphocyte transformation and dithiol compounds; The role of ADF/thioredoxin as an endogenous reducing agent, Mol. Immunol. (in press).

22. M. Carlsson, C. Sunderstrom, M. Bengtsson, T. H. Totterman, A. Rozen, and K. Nilsson, Interleukin 4 strongly augments or inhibits DNA synthesis and differentiation of B-type chronic lymphocyte leukemia cells depending on the co-stimulatory activation and progression signals, Eur. J. Immunol. 19:913 (1989).

23. D. S. Silberstein, M. H. Ali, S. L. Baker, J. F. David, Human eosinophil cytotoxicity-enhancing factor. Purification, physical characteristics, and partial amino acid sequence of an active polypeptide, Eur. J. Immunol. 19:913 (1989).

24. S. Fujii, Y. Nanbu, H. Nonogaki, I. Konishi, T. Mori, H. Masutani, J. Yodoi, Co-expression of ATL-derived factor (ADF), a human thioredoxin homologue, and human papilloma virus (HPV) DNA in neoplastic cervical squamous epithelium, Cancer (in press).

25. P. Biberfeld, A. Ost, A. Porwit, B. Sandstedt, G. Pallesen, B. Bottiger, L. Morfelt-Mansson, and G. Biberfeld, Histopathology and immunohistology of HTLV-III/LAV related lymphadenopathy and AIDS, Acta. Pathol. Microbiol. Scand. Ser. C. 95:47 (1987).

ANTIVIRAL AND ADJUVANT ACTIVITY OF IMMUNOMODULATOR

ADAMANTYLAMIDE DIPEPTIDE

K. Noel Masihi[1], Beate Rohde-Schulz[1], Karel Masek[2], and Bram Palache[3]

Robert Koch Institute[1], Federal Health Office, Berlin, FRG and Institute of Pharmacology[2], Czechoslovak Academy of Sciences, Prague, CSFR and Duphar[3], Weesp, The Netherlands

INTRODUCTION

The advent of acquired immunodeficiency syndrome has ushered in a renaissance in the field of antiviral research. Many compounds possessing inhibitory activity against different viruses have been described. New drugs such as azidothymidine have been added to the limited arsenal of licensed antiviral agents but serious side effects associated with many preparations have restricted their widespread use. In an effort to reduce adverse reactions, combination regimens consisting of antiviral drugs and synergistically acting immunomodulators are being increasingly investigated.

In an innovative approach, a group of compounds were synthesized where 1 amino-amantadane moiety was linked to the essential L-alanine-D-isoglutamine portion of immunomodulator muramyl dipeptide (MDP) (Figure 1).

Amantadine is a primary symmetric amine with an interesting tricyclic structure that has been extensively employed in humans since 1966 for the prophylaxis and chemotherapy of influenza (1) and Parkinson's disease (2). Chemoprophylaxis and therapy with amantadine is important in high risk immunodeficient persons unable to mount a satisfactory antibody response to influenza vaccine. Amantadine has also been used for the treatment of herpes simplex sciatica (3) and postherpetic neuralgia (4). Antiviral activity of amantadine against dengue virus (5), Semliki Forest virus (6), vesicular stomatitis virus (7) and a moderate effect against hepatitis A virus has been reported in cell cultures (8). Protean manifestations of biological activities of MDP and analogs include the stimulation of nonspecific resistance against influenza viruses (9, 10), murine hepatitis virus (11), herpes simplex virus (12), vaccinia virus (12), Sendai virus (13) and Semliki Forest virus (14). Adamantylamide dipeptide (AdDP) thus is a novel hybrid entity combining pertinent components of both an antiviral and an immunomodulator in a single synthetic compound (15). The profile of the immunobiological activities of AdDP was investigated using *in vitro* and *in vivo* models of influenza virus.

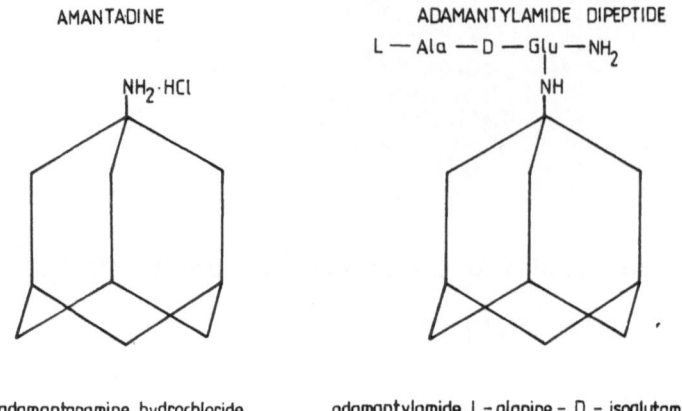

Fig. 1. Structure of amantadine and adamantylamide dipeptide.

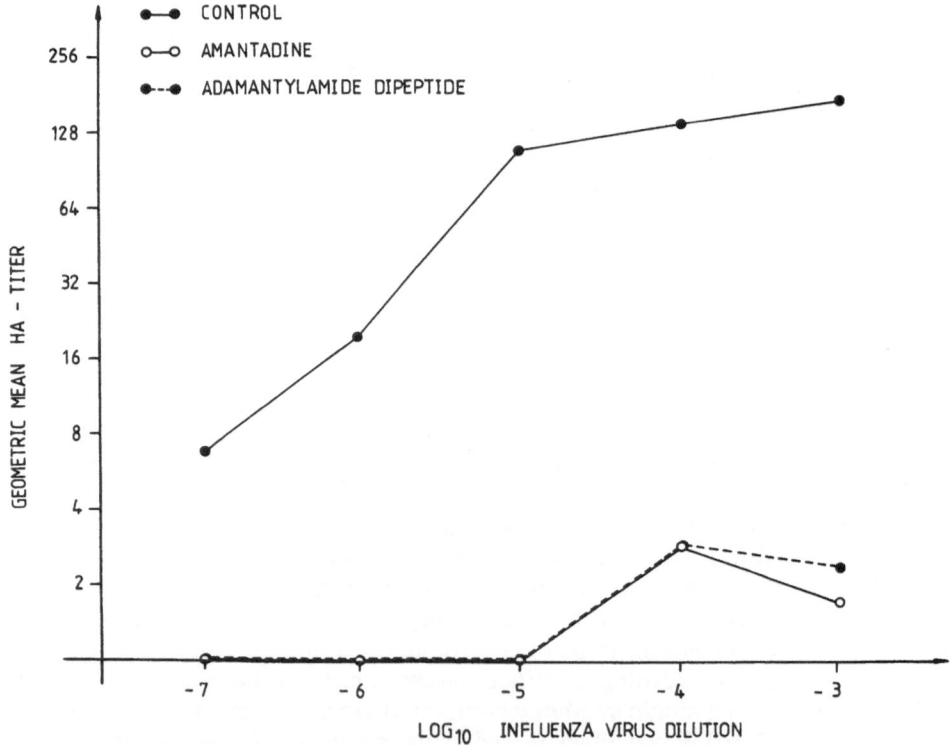

Fig. 2. Effect of 50 µg/ml of adamantylamide dipeptide or amantadine on the replication of influenza virus in MDCK cells.

Table 1. Effect of adamantylamide dipeptide on influenza virus replication in MDCK cells.

Compound	Dose (μg/ml)	TCID$_{50}$
AdDP	50	$10^{-4,500}$
	100	$10^{-3,500}$
	150	$10^{-3,600}$
Amantadine	50	$10^{-3,500}$
None	---	$10^{-6,249}$

ANTIVIRAL EFFECTS OF ADAMANTYLAMIDE DIPEPTIDE AND AMANTADINE ON THE INFECTIVITY OF INFLUENZA VIRUS

The effect of AdDP and amantadine on the infectivity of influenza virus was investigated using the sensitive Madin-Darby canine kidney (MDCK) cells (16). Serial ten-fold dilutions of influenza virus in serum-free, but trypsin-containing Eagle's Minimum Essential Medium (MEM), were added to confluent monolayers of MDCK cells established in tissue culture plates. MEM containing the needed amount of AdDP or amantadine was added to each well and incubated for 48 hr. Influenza virus infection and replication were detected by quantitating the viral hemagglutinin released into the culture medium. Chick red blood cell hemagglutination (HA) assay was used for this purpose. The geometric means of the HA titres detected in at least three parallel cultures were calculated. The tissue culture infectious dose 50% (TCID$_{50}$) was determined by the method of Reed and Muench. The results, presented in Figure 2, show that 50 μg/ml of either AdDP or amantadine completely inhibited the infection and replication of influenza virus inoculated at 10^{-5} to 10^{-7} dilutions. In the control cultures the same dilutions produced detectable viral hemagglutination activity as is shown in Figure 2. The effect of different concentrations of AdDP is shown in Table 1. Amantadine was used only at 50 μg/ml since amounts exceeding this concentration were found toxic for the sensitive Madin-Darby canine kidney cells. The results show that compared to controls, the TCID$_{50}$ in AdDP and amantadine treated cultures were significantly reduced. The efficacy of AdDP was comparable to that of amantadine and even the highest dose (150 μg/ml) that was tested did not produce toxic effects.

ANTIBODY RESPONSE TO PRIMARY AND SECONDARY IMMUNIZATION WITH INFLUENZA SUBUNIT VACCINES ALONE OR WITH IMMUNO-MODULATORS

Influenza A/Sichuan/2/87 (H3N2) and influenza B/Beijing/1/87 subunits were prepared under good manufacturing practice conditions at Duphar. The amount of hemagglutinin present in the subunit vaccines was quantitated using the single radial immunodiffusion test. Groups of 10 mice were injected subcutaneously with either influenza A/Sichuan/2/87 (H3N2) or influenza B/Beijing/1/87 subunits using a dose of 2.5 μg hemagglutinin, alone or mixed with 100 μg AdDP or 50 μg Al(OH)$_3$. On day 28, mice received a booster injection containing the same dose of the respective subunit

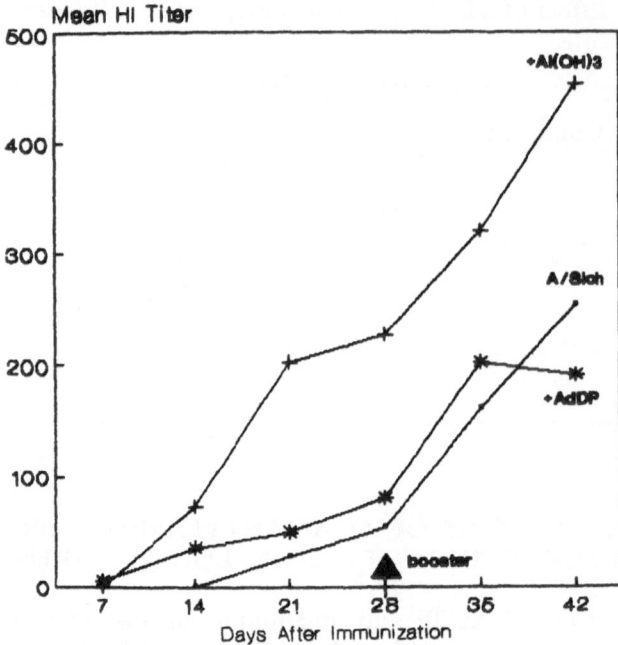

Fig. 3. Geometric mean hemagglutination-inhibition (HI) antibody in response to A/Sichuan/2/87 (H3N2) subunits alone or with immunomodulators.

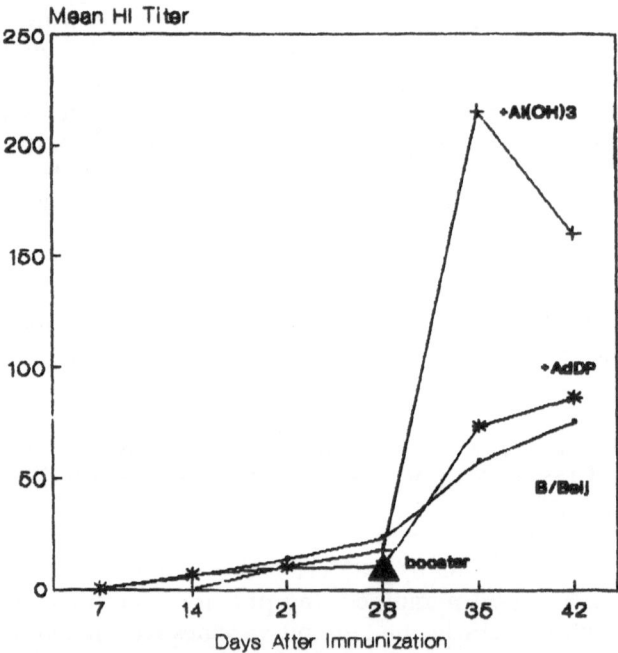

Fig. 4. Geometric mean hemagglutination-inhibition (HI) antibody in response to B/Beijing/1/87 subunits alone or with immunomodulators.

vaccine without the immunomodulators. Blood samples were collected on days 7, 14, 21, 28, 35 and 42 after the primary immunization. Infectious influenza A/Sichuan/2/87 and influenza B/Beijing/1/87 were grown in the allantoic cavity of 11-day embryonated eggs. Hemagglutination-inhibition (HI) tests were performed in microtiter plates using 0.5% chicken red blood cells and 4 hemagglutinating units. Influenza B/Beijing/1/87 antigen was ether-split for increased sensitivity. All sera were pretreated with receptor-destroying enzyme to remove nonspecific inhibitors.

The results presented in Figure 3 show that antibody titers remained comparable in animals receiving either A/Sichuan subunits alone or mixed with AdDP. In contrast, HI titers after the primary and the secondary immunizations were markedly elevated in animals administered A/Sichuan vaccine containing Al(OH)$_3$.

A low level of antibody was induced in all groups after primary immunization with B/Beijing subunits, even when mixed with immunomodulators, as is shown in Figure 4. The HI titers following the secondary immunization were significantly elevated by addition of Al(OH)$_3$ but not AdDP (17).

PROTECTION AGAINST AEROSOL INFLUENZA INFECTION AFTER IMMUNIZATION WITH INFLUENZA SUBUNIT VACCINES ALONE OR WITH IMMUNOMODULATORS

Groups of 20 mice were administered influenza A/Sichuan/2/87 (H3N2) or B/Beijing/1/87 influenza virus subunits alone or mixed with 100 µg AdDP or 50 µg Al(OH)$_3$. One month after the immunization, 10 animals were given a booster injection with the same dose of respective subunit vaccine alone. The remaining mice were given an aerosol of mouse-pathogenic influenza A/PR/8/34 (H1N1) virus. A small particle aerosol of lethal influenza virus was generated for the challenge infection in a Middlebrook Airborne Infection Apparatus (Tri-R Instruments, Rockville Center, USA) as described previously (9). Mortality was recorded daily. Animals receiving the booster injection were administered A/PR/8/34 aerosol infection 2 weeks after the

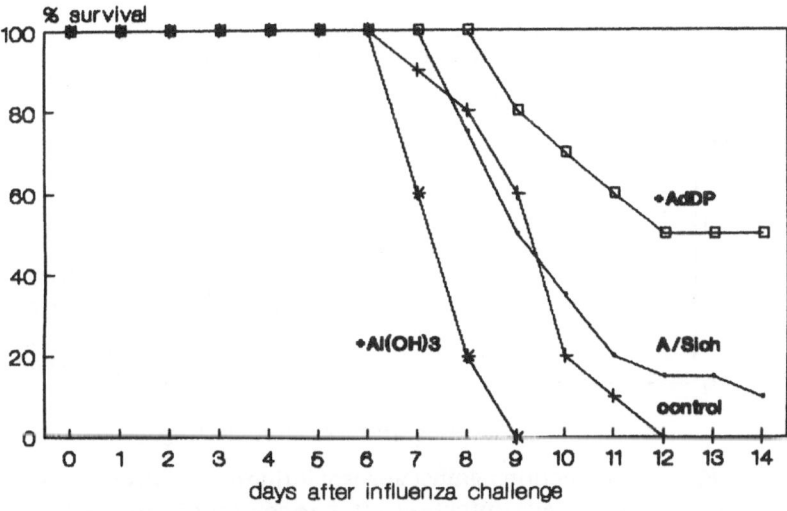

Fig. 5. Protection against aerosol influenza A/PR/8/34 (H1N1) challenge infection after primary immunization with influenza A/Sichuan/2/87 (H3N2) subunits alone or mixed with immunomodulators.

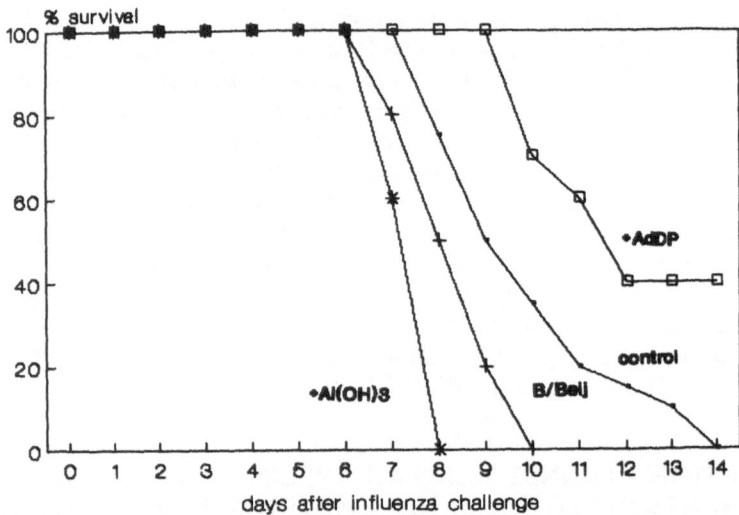

Fig. 6. Protection against aerosol influenza A/PR/8/34 (H1N1) challenge infection after primary immunization with influenza B/Beijing/1/87 subunits alone or mixed with immunomodulators.

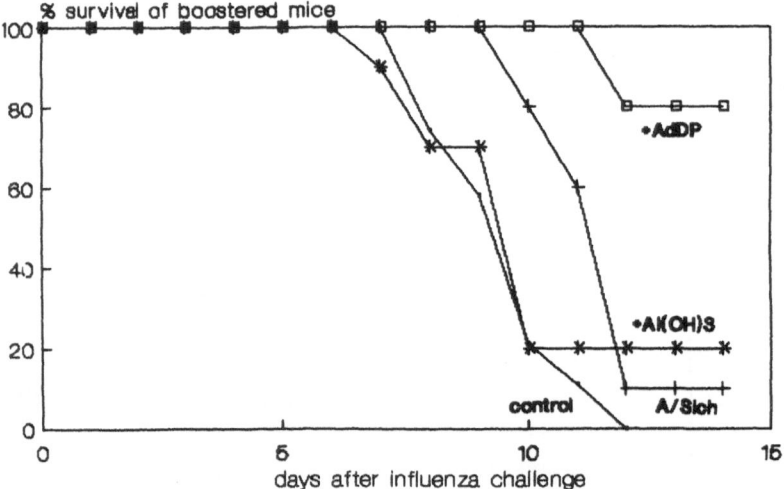

Fig. 7. Protection against aerosol influenza A/PR/8/34 (H1N1) challenge infection after secondary immunization with influenza A/Sichuan/2/87 (H3N2) subunits alone or mixed with immunomodulators.

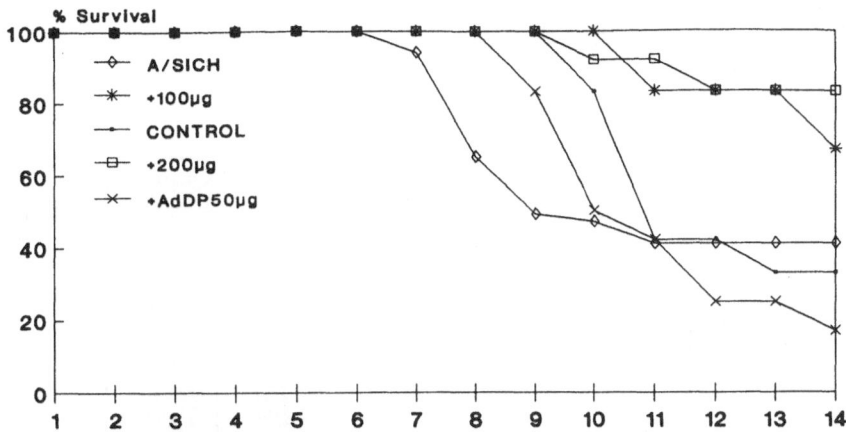

Fig. 8. Protection against aerosol influenza A/PR/8/34 (H1N1) challenge infection after primary immunization with influenza A/Sichuan/2/87 (H3N2) subunits alone or mixed with different doses of the immunomodulator AdDP.

secondary immunization. The results show that the inclusion of AdDP to heterologous A/Sichuan/2/87 (H3N2) subunit (Figure 5) and even to B/Beijing/1/87 subunit vaccine (Figure 6) conferred significant protection against H1N1 influenza A/PR/8/34. Animals receiving these vaccines mixed with Al(OH)$_3$ succumbed to challenge infection (17).

No protection was observed following the secondary immunization in groups receiving A/Sichuan subunits alone or mixed with Al(OH)$_3$. In contrast, mice immunized with A/Sichuan containing AdDP were significantly protected (Figure 7). Inclusion of immunomodulators to B/Beijing vaccine afforded only marginal resistance after secondary immunization (17).

The effect of varying the amount of AdDP in the subunit vaccine was also investigated. Groups of 20 mice were administered influenza A/Sichuan/2/87 (H3N2) influenza virus subunits alone or mixed with 50 µg, 100 µg and 200 µg of AdDP. One month after the immunization, animals were given a booster injection with the same dose of the subunit vaccine alone. Mice were given an aerosol of mouse-pathogenic influenza A/PR/8/34 (H1N1) virus infection 2 weeks after the secondary immunization. The results presented in Figure 8 show that the protection conferred against influenza A/PR/8/34 (H1N1) virus infection by influenza A/Sichuan/2/87 (H3N2) influenza virus subunits alone or adjuvantized with AdDP was dose-dependent. Significant protection was conferred by A/Sichuan/2/87 (H3N2) adjuvantized with 100 µg of AdDP and could be further improved by 200 µg of AdDP.

ASSESSMENT OF CELL-MEDIATED IMMUNE REACTIONS ELICITED BY INFLUENZA SUBUNIT VACCINE ALONE OR WITH IMMUNOMODULATOR AdDP

Chemiluminescence

Mice were immunized with 2.5 µg HA of A/Sichuan/2/87 (H3N2) subunit vaccine adjuvantized with 1, 25, 50 or 200 µg of AdDP. Four weeks after immunization, splenic cells were assayed for CL generation in response to particulate trigger zymosan. The

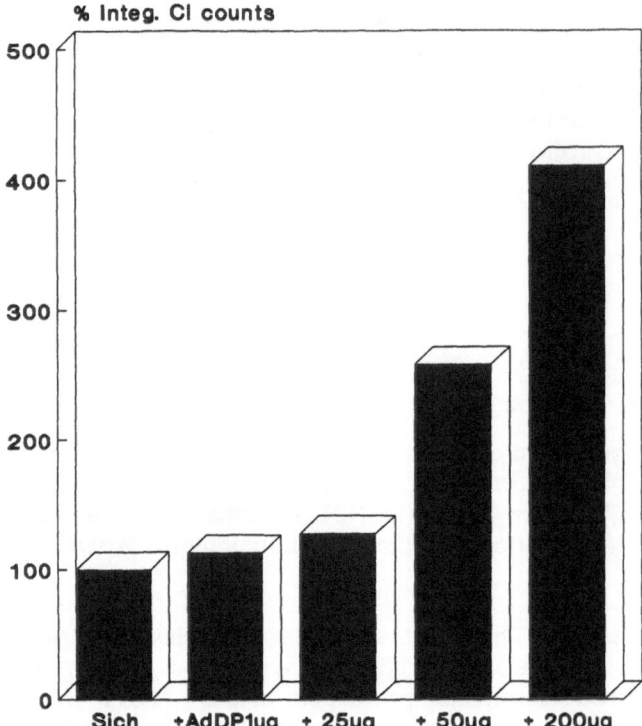

% Integ. Cl counts

Fig. 9. Zymosan-induced chemiluminescence response of spleen cells from mice immunized with influenza A/Sichuan/2/87 (H3N2) subunits alone or mixed with different doses of the immunomodulator AdDP.

results presented in Figure 9 show that CL responses induced by A/Sichuan/2/87 (H3N2) subunit vaccine alone or containing 1 μg and 25 μg of AdDP were essentially similar. In contrast, addition of 50 μg and 200 μg AdDP to A/Sichuan/2/87 (H3N2) subunit vaccine significantly elevated the CL responses compared with A/Sichuan/2/87 (H3N2) subunit vaccine alone. Data of the CL experiments extend the observation of dose-related efficacy of A/Sichuan/2/87 (H3N2) subunit vaccine adjuvantized with higher doses of AdDP and, in addition, demonstrate that the phagocytic cell activity can be increased by AdDP.

Delayed-Type Hypersensitivity

Twenty mice were injected with 2.5 μg HA of A/Sichuan/2/87 (H3N2) subunit vaccine containing 100 μg of AdDP. Further priming of animals was carried out at 4 weeks with inactivated A/Sichuan/2/87 (H3N2) subunits and at 6 weeks with infectious influenza A/PR/8/34 (H1N1) virus. The delayed type hypersensitivity (DTH) reaction was elicited by footpad injection of 1000 HAU of infectious influenza A/PR/8/34 (H1N1) virus to a group of 10 mice or infectious A/Sichuan/2/87 virus to the second group of 10 mice. The results presented in Figure 10 show that the swelling in mice receiving A/Sichuan/2/87 in the footpad was similar irrespective of whether AdDP was

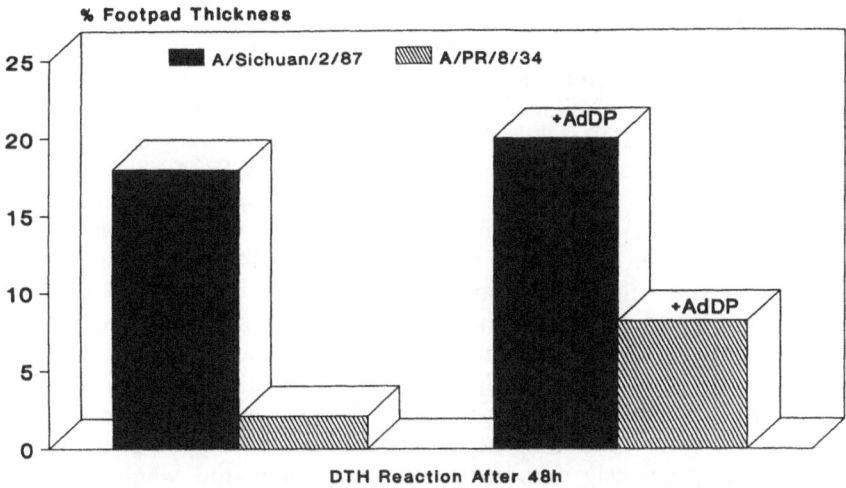

Fig. 10. Effect on DTH of mice treated with A/Sichuan/2/87 (H3N2) subunits alone or mixed with immunomodulator AdDP.

present in the vaccine or not. In contrast, the DTH reaction in response to A/PR/8/34 (H1N1) injection into the footpads was significantly increased in mice receiving A/Sichuan/2/87 (H3N2) subunit vaccine adjuvantized with AdDP. This stimulation of enhanced DTH reaction in mice receiving A/Sichuan/2/87 (H3N2) subunit vaccine plus AdDP against heterologous A/PR/8/34 (H1N1) strain suggest that cellular immune responses are responsible, in part, for the observed heterologous protection.

DISCUSSION

A major obstacle in effective control of influenza epidemics is the inherent variation of the influenza virus surface hemagglutinin and neuraminidase antigens. This requires an annual review of the composition of influenza vaccines. The appearance of completely new subtypes possessing differing surface antigens than are present in the currently circulating influenza viruses cannot be predicted reliably. In an event of antigenic shift, the benefits conferred by existing vaccines can be significantly curtailed. Modalities capable of generating protective immune responses that extend to different influenza A subtypes and to other influenza viruses are, therefore, of particular interest.

The results of preclinical studies with immunomodulator AdDP confirm and demonstrate that the homotypic immunity induced by influenza subunit vaccines can be broadened to a heterologous immune response. The exact roles played by the antiviral amantadine and the immunomodulatory dipeptide in the hybrid molecule AdDP remain speculative. A recent study, however, showed that the entire AdDP molecule is essential for inducing protection in this murine aerosol influenza model (18). Animals treated with either amantadine or dipeptide were not protected against sublethal aerosol influenza infection whereas significant survival of animals immunized with influenza subunit vaccine containing AdDP was obtained demonstrating that AdDP as a complete substance constitutes an essential pre-requisite for the observed protection (18).

The adjuvant effect of Al(OH)$_3$ on antibody response to influenza A/Sichuan subunit vaccines after primary immunization and to B/Beijing vaccine following the secondary immunization was confirmed. Subunit vaccine containing A/Sichuan (H3N2) and Al(OH)$_3$ stimulated high antibody levels. Despite the presence of high circulating antibody, animals were not protected against heterologous H1N1 influenza A/PR/8/34 infection. Mice immunized with A/Sichuan vaccine containing AdDP induced lower levels of antibody than vaccine with $Al(OH)_3$ but were significantly protected against A/PR/8/34 challenge. Similar immunization with influenza B/Beijing vaccines containing $Al(OH)_3$ or AdDP induced barely detectable antibody on day 28 but animals receiving AdDP were partially protected against A/PR/8/34 challenge. Secondary immunization greatly boosted the antibody response to A/Sichuan in animals receiving the subunits with $Al(OH)_3$, but not in the AdDP group, when compared with subunit vaccine alone. However, protection against A/PR/8/34 reached 80% in the AdDP group whereas, despite high the HI antibody, it did not exceed 10% in the other two groups. In another study (19), mice injected with inactivated H3N1 influenza REC31, a recombinant between A/England and A/PR/8, together with $Al(OH)_3$ and saponin produced high levels of HI antibody but poor protection against A/Victoria/3/75 (H3N2) challenge. Animals receiving live REC31 had, in contrast, lower HI titers than inactivated preparations containing adjuvants yet showed a good protection (19).

It has been previously demonstrated that inactivated whole virus or subunit vaccines can induce high levels of antibody but do not sufficiently prime for cytotoxic T cell activity (20). T lymphocyte responses, in particular that of cytotoxic T cells, are cross-reactive for a wide spectrum of influenza A (21). Viral titers can be reduced by adoptive transfer of T cells with cytotoxic activity or by specific cloned cytotoxic T cells (22, 23). Resistance to influenza has not shown consistent correlation with antibody (24) and there is an increasing awareness that inactivated influenza vaccines should carry components which trigger cell-mediated responses (23).

The results of the present study show that the DTH reaction in response to A/PR/8/34 (H1N1) injection into the footpads was significantly increased in mice receiving A/Sichuan/2/87 (H3N2) subunit vaccine adjuvantized with AdDP. This stimulation of enhanced DTH reaction in mice receiving A/Sichuan/2/87 (H3N2) subunit vaccine plus AdDP against heterologous A/PR/8/34 (H1N1) strain suggests that cellular immune responses are generated by AdDP and may be responsible, in part, for the observed heterologous protection. In a recent study, pretreatment with AdDP was shown significantly to increase DTH to sheep red blood cells and to stimulate graft-versus-host reaction (GVHR) (25). AdDP has been also shown to possess immunoadjuvant effects in a model of delayed-type hypersensitivity and have capacity to enhance DNA biosynthesis in thymus, spleen and liver (15). In other studies, macrophages exposed to AdDP have been shown to exhibit enhanced cytotoxic activity against P 815 mastocytoma target cells and to stimulate elevated levels of tumor necrosis factor (26). Both GVHR and DTH are considered to be *in vivo* tests of cell-mediated immunity.

Antigenic variation is a refined strategy evolved by pathogens such as influenza and human immunodeficiency viruses to evade the host immune response. Immuno-modulator AdDP can broaden the limited homotypic immunity induced by influenza subunit vaccine to a heterologous immune response. Immune system oriented interventions capable of influencing the spectrum of activity against different antigenic variants may offer a rationally more salubrious approach towards control of certain infections.

REFERENCES

1. T. A. Bektimirov, Current status of amantadine and rimantadine as anti-influenza-A agents, Bull. Wld. Hlth. Org. 63: 51 (1985).
2. D. B. Calne, The role of various forms of treatment in the management of Parkinson's disease, Clin. Neuropharmacol. 5 (suppl.1):38 (1982).

3. D. A. Fisher, Recurrent herpes simplex sciatica and its treatment with amantadine hydrochloride, Cutis 29:467 (1982).

4. A. W. Galbraith, Prevention of post-herpetic neuralgia by amantadine hydrochloride, Br. J. Clin. Pract. 37:304 (1983).

5. W. C. Koff, J. L. Jr. Elm, and S. B. Halstead, Inhibition of dengue virus replication by amantadine hydrochloride, Antimicrob. Agents Chemother. 18:125 (1980).

6. A. Helenius, M. Marsh, and J. White, Inhibition of Semliki Forest virus penetration by lysosmotropic weak bases, J. Gen. Virol. 58:47 (1982).

7. R. Schlegel, R. B. Dickson, M. C. Willingham, and I. H. Pastan, Amantadine and dansylcadaverine inhibit vesicular stomatitis virus uptake and receptor-mediated endocytosis of alpha 2-macroglobulin, Proc. Natl. Acad. Sci. USA 79:2291 (1982).

8. A. Widell, B. G. Hansson, B. Oberg, and E. Nordenfelt, Influence of twenty potentially antiviral substances on *in vitro* multiplication of hepatitis A virus, Antiviral Res. 6:103 (1986).

9. K. N. Masihi, W. Brehmer, W. Lange, and E. Ribi, Effects of mycobacterial fractions and muramyl dipeptide on the resistance of mice to aerogenic influenza virus infection, Int. J. Immunopharmac. 5:403 (1983).

10. F. X. Dietrich, H. K. Hochkeppel, and B. Lukas, Enhancement of host resistance against virus infections by MTP-PE, a synthetic lipophilic muramyl peptide. 1.Increased survival in mice and guinea pigs after single drug administration prior to infection, and the effect of MTP-PE on interferon levels in sera and lungs, Int. J. Immunopharm. 8:931 (1986).

11. K. N. Masihi, W. Lange, and B. Rohde-Schulz, Stimulation of antiviral activity by immunomodulators, in: "Antiviral Drugs, Basic and Therapeutic Aspects," R. Calio and G. Nistico, eds., Pythagora Press, Rome (1989).

12. S. Ikeda, T. Negishi, and C. Nishimura, Enhancement of nonspecific resistance to viral infection by muramyl dipeptide and its analogs, Antiviral Res. 5:207 (1985).

13. I. Azuma, J. Iida, K. Nishimura, C. Ishihara, S. Tokura, and Y. Yamamura, Prevention of Sendai virus infection with synthetic MDP and chitin derivatives in mice, in: "Immunomodulators and Nonspecific Host Defence Mechanisms Against Microbial Infections," K. N. Masihi and W. Lange, eds., Pergamon Press, Oxford (1988).

14. C. X. George, R. K. Jain, C. M. Gupta, and N. Anand, Enhancement in anti-Semliki Forest virus activity of ds RNA by muramyl dipeptide, FEBS. 200:37 (1986).

15. K. Masek, J. Seifert, M. Flegel, M. Krojidlo, and J. Kolinsky, The immunomodulatory property of a novel synthetic compound adamantylamide dipeptide, Meth. Find. Exptl. Clin. Pharmacol. 6:667 (1984).

16. K. N. Masihi, W. Lange, B. Rohde-Schulz, and K. Masek, Antiviral activity of immunomodulator adamantylamide dipeptide, Int. J. Immunotherapy III:89 (1987).

17. K. N. Masihi, W. Lange, S. Schwenke, G. Gast, P. Huchshorn, A. Palache, and K. Masek, Effect of immunomodulator adamantylamide dipeptide on antibody response to influenza subunit vaccines and protection against aerosol influenza infection, Vaccine 8:159 (1990).

18. A. Borecki and K. N. Masihi, Immunomodulatory effects of influenza subunit vaccine adjuvantized with adamantylamide dipeptide (AdDP) or its constituent components, in: "Immunotherapeutic Prospects of Infectious Diseases," K. N. Masihi and W. Lange, eds., Springer Verlag, Berlin (1990).

19. F. Y. Liew, S. M. Russel, G. Appleyard, C. M. Brand, and J. Beale, Cross-protection in mice infected with influenza A virus by the respiratory route is correlated with local IgA antibody rather than serum antibody or cytotoxic T cell reactivity, Eur. J. Immunol. 14:350 (1984).

20. R. G. Webster and B. A. Askonas, Cross-protection and cross-reactive cytotoxic T cells induced by influenza virus vaccines in mice, Eur. J. Immunol. 10:396 (1980).

21. K. L. Yap, G. L. Ada, and I. F. C. Mc Kenzie, Transfer of specific cytotoxic T-lymphocytes protects mice inoculated with influenza virus, Nature 273:238 (1978).

22. A. E. Lukacher, V. L. Braciale, and T. J. Braciale, *In vivo* effector function of influenza virus-specific cytotoxic T lymphocyte clones is highly specific, J. Exp. Med. 160:814 (1984).

23. P. M. Taylor and B. A. Askonas, Influenza nucleoprotein-specific cytotoxic T-cell clones are protective *in vivo*, Immunology 58:417 (1986).

24. R. M. Kris, R. Asofsky, C. B. Evans, and P. A. Small, Protection and recovery in influenza virus-infected mice immunsuppressed with anti-IgM, J. Immunol. 134:1230 (1985).

25. K. Sula, Z. Zidek, and K. Nouza, Effects of adamantylamide dipeptide (AdDP) on cell mediated immunity *in vivo*, in: "Immunotherapeutic Prospects of Infectious Diseases," K. N. Masihi and W. Lange, eds., Springer Verlag, Berlin (1990).

26. J. Müller, K. Nouza, A. Macakova, K. Sula, and Z. Zidek, *Ex vivo* and *in vivo* effects of adamantylamide dipeptide (AdDP) on macrophages and lymphocytes, in: "Immunotherapeutic Prospects of Infectious Diseases," K. N. Masihi and W. Lange, eds., Springer Verlag, Berlin (1990).

IMMUNOMODULATION BY MEDICINAL PLANTS

Hideyo Yamaguchi

Department of Microbiology, Teikyo
University School of Medicine
Tokyo, Japan

INTRODUCTION

Kampo medicines originally developed in ancient China are crude drugs containing extracts from 5 to 10 or more different species of herbs or plants in a specific combination and proportion. These were the mainstream of medicine in Japan for over 1,000 years until replaced by western synthetic medicines in the late 19th Century. However, in the last few decades, Kampo medicines have again become increasingly used for the treatment of a variety of disorders or diseases affecting the liver, kidney, and circulatory and/or respiratory systems, such as hepatitis, nephrosis, heart insufficiency and chronic bronchitis. Out of 120 Kampo prescriptions currently available in Japan, some are more specifically indicated for the management of those several chronic diseases, including autoimmune, allergic or chronic inflammatory disorders, which may not be adequately treated with existing western drugs(1).

Our special interest has been directed to those Kampo medicines which are recognized to improve the general condition of physiologically debilitated patients with weakness caused by some illness or surgical treatment, and immunocompromised patients, particularly cancer patients suffering from severe untoward side effects of intensive anticancer chemotherapy or radiation therapy, because of the possibility that these prescriptions may act as beneficial immunomodulators(2). Among them is Juzen-taiho-to, coded as TJ-48 and whose immunobiological activities have been most extensively studied.

In this review I will focus on the studies thus far made with TJ-48 concerning its effects on functions of lymphocytes and macrophages using several *in vitro*, *ex vivo* and *in vivo* murine systems, as well as its *in vivo* efficacy in the murine model of systemic candidiasis.

TJ-48: Prescription and Preparation

TJ-48 is made from spray-dried aqueous extracts of 10 species of herbs in exact proportions as shown in Table 1. The test sample of TJ-48 was prepared by dissolving it in saline. Administration of this sample to mice was carried out by the oral route using a gastric gavage. An equal amount of saline was given to a control group of mice.

Microbial Infections, Edited by H. Friedman *et al.*
Plenum Press, New York, 1992

Table 1. Composition of Juzen-taiho-to (TJ-48) Based On the Weight of
Spray-dried Aqueous Extract (from Takemoto et al. (5))

Herb	Ratio
Angelicae (radix)	1
Astragali (radix)	1
Atractylodis lauceae (rhizoma)	1
Cinnamimi (cortex)	1
Cnidii (rhizoma)	1
Ginseng (radix)	1
Hoelen	1
Paeoniae (radix)	1
Rehmanniae (radix)	1
Glycyrrhizae (radix)	0.5

Table 2. Effect of TJ-48 on Induction of Cytotoxic T Lymphocyte As
Measured by ^{51}Cr-release Assay[a]

Responder cells from:	Stimulator cells	% Cytotoxicity[b]
Untreated control mice	-	4.3 ± 0 58
	+	66.6 ± 3.67
TJ-48-treated mice	-	3.8 ± 0.40
	+	83.6 ± 1.64*

[a] Responder cells (5 x 10^6 BALB/c spleen cells) were cultured with or without
stimulator cells (5 x 10^6 C3H/He spleen cells treated with mitomycin C), and then
incubated with target cells (1 x 10^5 ^{51}Cr-preloaded L929 cells).

[b] Mean ± SE of four samples. *: $p < 0.01$

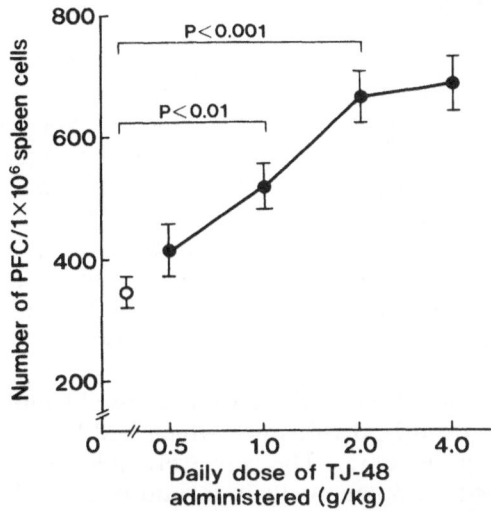

Fig. 1. Anti-SRBC response in untreated mice (●) and TJ-48-treated mice (O) as measured in terms of the relative number of plaque forming cells (PFC) in the spleen. (from Komatsu et al.(3))

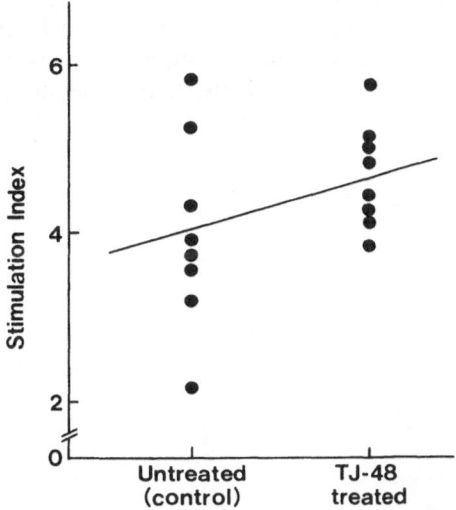

Fig. 2. Effect of treatment of mice with TJ-48 on mixed lymphocyte culture response. (from Takemoto et al., (5)). Results are expressed as a simulation index (SI) obtained from two separate experiments in which spleen cells from 4 each untreated and TJ-48-treated mice were individually assayed.

Effect on Lymphocyte Function In *Ex vivo* Assays

It was studied using the following three assays: (i) plaque forming cell assay; (ii) mixed lymphocyte culture response; and (iii) cytotoxic T lymphocyte response.

Komatsu et al., (3) reported the effect of TJ-48 on the anti-sheep red blood cell (SRBC) response in mice. Seven to 10 wk-old BDF_1 mice were administered with TJ-48 at daily dosages of 0.5 to 4 g/kg for 7 to 21 consecutive days, and then were injected intravenously with SRBC. Four days later, the number of plaque forming cells relative to that of whole spleen cells was counted after the method of Cunningham and Szenberg (4). As shown in Fig. 1, the value increased linearly with increasing doses of the drug up to 2 g/kg/day. When mice were treated with this daily dose for 7 consecutive days, the potentiated anti-SRBC response lasted for 3 days after the last dosing and then gradually lowered (3).

Takemoto et al., (5) studied the effect of TJ-48 on the mixed lymphocyte culture response using spleen cells taken from BALB/c mice which had been treated with 2 g/kg/day of the drug for 7 consecutive days. The cells were cultured for 4 days with irradiated BALB/c or C57BL/6 mouse spleen cells in a ratio of 3 to 10, and then the mixed cultures were pulsed with ^{3}H-thymidine for 16 hrs before harvesting cells for radioactive assay. As illustrated in Fig. 2, spleen cells from TJ-48-treated mice showed higher, although not significantly, mixed lymphocyte culture response than that from spleen cells from untreated control mice.

Takemoto and his coworkers (5) also reported the results of experiments in which the effect of TJ-48 on cytotoxic T lymphocyte response was tested using two assay methods: the ^{51}Cr-release assay and the plaque reduction assay (6). In the first test, spleen cells taken from BALB/c mice which had been treated with 2 g/kg/day of TJ-48 for 7 consecutive days before testing were used as responder cells. They were cultured for 5 days with spleen cells from mitomycin C-treated C3H/He mice used as stimulator cells, and then the cultures were incubated with ^{51}Cr-preloaded L929 cells as target cells. After 5 hrs, the released radioactivity in the supernatant was measured. As shown in Table 2, in the absence of stimulator cells, there was no difference in the cytotoxic activity between the cells from treated mice and those from untreated control

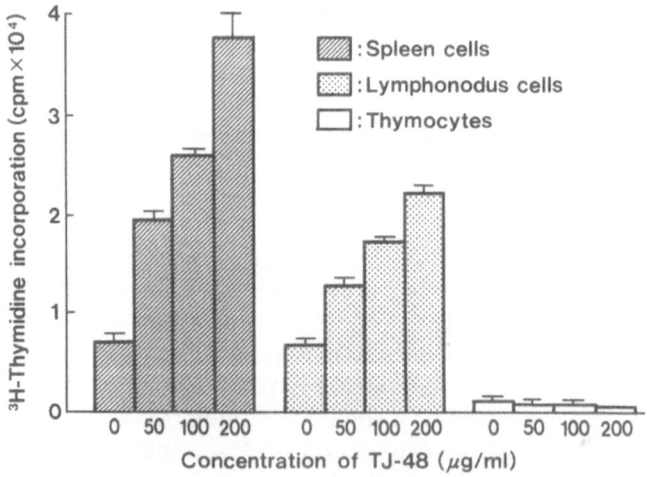

Fig. 3.   Mitogenic activity of TJ-48 against spleen cells, lymph node cells and thymocytes from BALB/c mice. (from Takemoto et al., (8))

mice. In contrast, in the presence of stimulator cells, the spleen cells from treated mice showed significantly greater cytotoxic activity than that from the spleen cells from untreated mice. Virtually the same results were obtained when measured by the hemolytic plaque reduction technique using BALB/c mice-derived anti-SRBC hybridoma cells as target (5). The effectiveness of orally administered TJ-48 in reducing the immunosuppression induced by *cis*-diaminedichloro-platinum(II) in C3H/He mice was also demonstrated by means of plaque forming cell assay and ^{51}Cr-releasing cytotoxic assay (7).

Effect on Lymphocyte Function *In vitro* with Special Reference to Mitogenic Activity

The mitogenic activity of TJ-48 in murine lymphoid cells was studied by Takemoto and his coworkers (8). Spleen cells, lymph node cells and thymocytes were taken from BALB/c mice and cultured for 72 hrs in RPMI 1640 medium with various concentrations of TJ-48. Then the cultures were pulsed with ^{3}H-thymidine for 4 hrs before harvesting for radioactive assay. As given in Fig. 3, TJ-48 at concentrations up to 200 μg/ml showed dose-dependent mitogenic activity on spleen cells and lymph node cells, while it had no effect on thymocytes. Spleen cells taken from BALB/c(nu/nu) mice also responded to the mitogenic activity of TJ-48 to virtually the same extent as did for spleen cells from BALB/c mice (8).

Fig. 4 shows the results of experiments in which spleen cells from BALB/c mice were pretreated with graded concentrations of anti-Thy 1.2 (left side) or anti-Ig antibody (right side) plus complement, and then the pretreated cells were cultured with or without 200 ug/ml of TJ-48 for 72 hrs for proliferation assay. The proliferative effect of TJ-48 on spleen cells was scarcely influenced by the pretreatment with anti-Thy 1.2 antibody, while it was almost completely abolished by the pretreatment with sufficiently high concentrations of anti-Ig antibody.

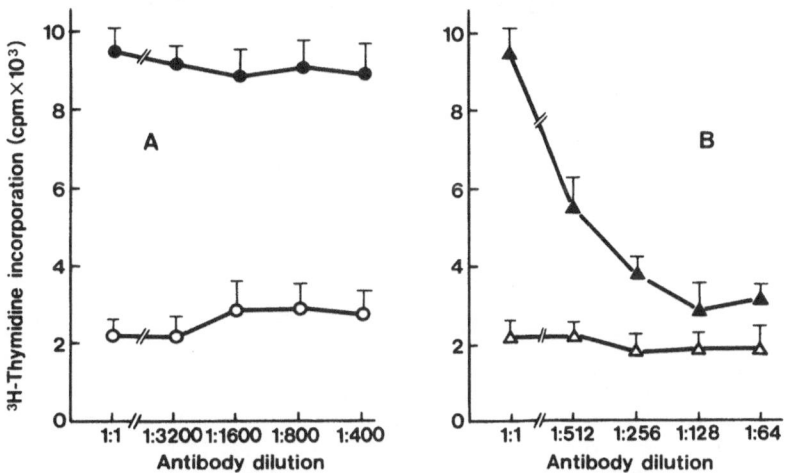

Fig. 4.

Mitogenic activity of TJ-48 against anti-Thy 1.2 antibody-treated (A) and anti-Ig antibody-treated (B) spleen cells from BALB/c mice. (from Takemoto et al.(8))

Spleen cells treated with various concentrations of anti-Thy 1.2 and anti-Ig antibody plus complement were cultured with (solid symbols) or without (open symbols) TJ-48 at 200 μg/ml.

Takemoto et al., (8) also performed an experiment designed to determine a role of spleen macrophage in the TJ-48-induced proliferation of splenocytes. When whole spleen cells were employed for proliferation assay, the substantial mitogenic activity of TJ-48 was observed. This was not the case, however, if non-adherent splenocytes fractionated from whole spleen cells were used. TJ-48 again became effective in inducing mitosis in splenocytes when adherent cells were added to them in an appropriate proportion (8).

All these data strongly suggest that TJ-48 is a B cell mitogen and that this mitogenic activity is T cell-independent and macrophage-dependent.

Effect on Lymphocyte/Macrophage Functions *In vivo*

The possibility that TJ-48 may potentiate the immune response of lymphocytes and macrophage *in vivo* was tested using plaque forming assay in SRBC-immunized BDF_1 mice(3) and delayed-type hypersensitivity (DTH) reaction in BALB/c mice sensitized with Type B influenza virus vaccine (5). When BDF_1 mice were treated with 2 g/kg/day of TJ-48 for 7 consecutive days before and 4 consecutive days after the immunization, the relative number of plaque forming cells in the spleen increased 2-fold as compared with the value for untreated mice (Fig. 5).

The DTH reaction to viral vaccine was induced by intradermal injection of 1,000 HA units of vaccine into the back of mice and, 6 days later, 500 HA units of vaccine was injected into the footpad of the hind leg. After 24 hrs, the thickness of the footpad was measured. An experimental group of mice was treated with 2 g/kg/day of TJ-48 for 13 consecutive days before sensitization. As seen in Fig. 6, the DTH reaction was significantly enhanced in these mice.

Effect on Macrophage Function in *Ex vivo* and *In vitro* Assays

The possible immunopotentiating activity of TJ-48 against macrophage was studied using *ex vivo* assay systems testing the phagocytic activity and generation of oxygen burst

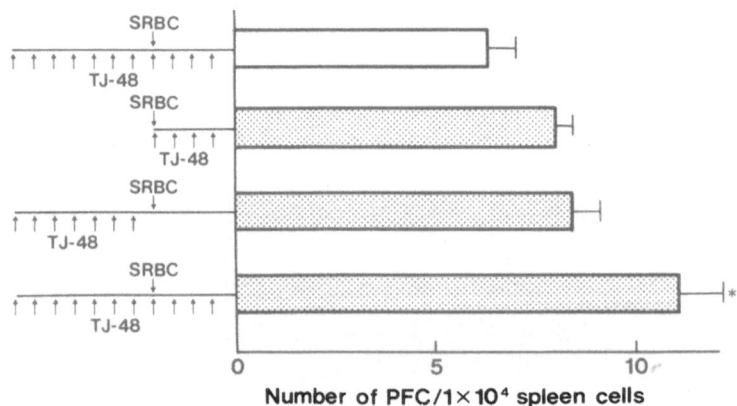

Fig. 5. Effect of TJ-48-treatment of mice on *in vitro* anti-SRBC response in their spleen cells. (from Komatsu et al.(3))*: $p < 0.05$

(9). BALB/c mice were treated with TJ-48 at various daily dosages for 7 consecutive days and peritoneal exudate cells (PEC) and bone marrow cells were harvested from these treated mice and from untreated control mice. The harvested cells were mixed with 5% fresh BALB/c mouse serum and viable *Candida parapsilosis* cells, and then the mixtures were incubated for 3 hrs. Phagocytic activity was determined by measuring viable counts of Candida remaining in the supernatant of the incubation mixture. As shown in Fig. 7, the ability of both PEC and bone marrow cells harvested from mice that had been treated with 1 g/kg/day or higher dosages of TJ-48 to phagocytize Candida cells was significantly enhanced as compared with corresponding cells from untreated mice. The TJ-48-stimulated activity of PEC gradually decreased, but was still maintained at a significantly higher level than that for PEC from control mice for 5 days after the termination of dosing (9). Chemiluminescence analysis performed with PEC from control and TJ-48-primed mice also demonstrated the potentiating effect of TJ-48 on PEC in terms of the generation of oxidative burst in response to zymosan (9).

The stimulation by TJ-48 of macrophage function was also observed in *in vitro* experimental systems used (9). As shown in Fig. 8, the ability of PEC to phagocytize Candida cells was significantly enhanced in the presence of TJ-48 at a concentration of 3 μg/ml or above.

The macrophage-or reticuloendothelial system-stimulating activity of TJ-48 and some constituents of this prescription has been proved by Haranaka and his coworkers (10), who demonstrated that these herbal drugs given orally were active in enhancing

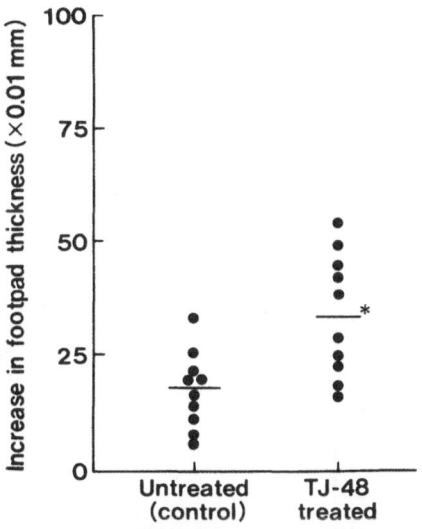

Fig. 6. Enhancing effect of TJ-48 on delayed-type hypersensitivity in mice sensitized and subsequently challenged with viral hemagglutinin. (from Takemoto et al.(5))*: $p < 0.05$

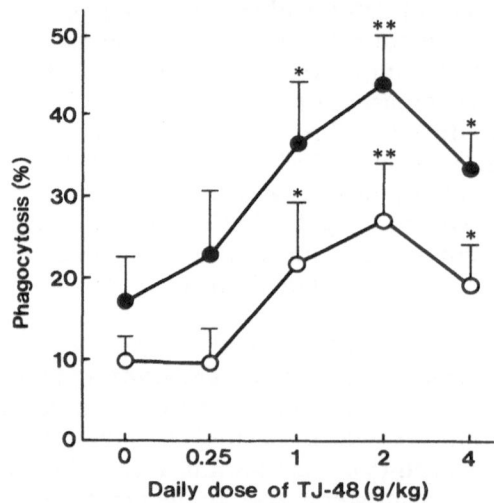

Fig. 7.　Ability of peritoneal exudate cells (●) and bone marrow cells (○) from mice treated with varying dosages of TJ-48 to phagocytize viable *C. parapsilosis* cells in *in vitro* assay. (from Maruyama et al.(9))

The ratio of effector cells to target cells was 10:1.　*: $p < 0.05$; **: $p < 0.01$

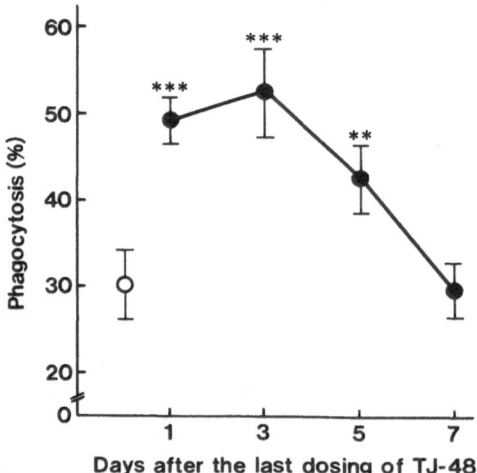

Fig. 8.　Phagocytizing ability of murine peritoneal exudate cells treated *in vitro* with (●) or without (○) 100 μg/ml of TJ-48. (from Maruyama et al., (9)).　*: $p < 0.05$.

294

as priming agent the production of tumor necrosis factor (TNF) induced by subsequent LPS administration in DDY mice. Sakagami et al., (11) reported that the production of interferon-γ and interleukin-2 from the PHA-stimulated human peripheral blood mononuclear cells was significantly enhanced when TJ-48 was present in the reaction mixture.

Effect on Microbial Infections

Satomi et al., (12) reported that pretreatment with TJ-48 and several other Kampo prescriptions of DDY mice had a protective effect against the lethal infection of *Pseudomonas aeruginosa.* We evaluated *in vivo* efficacy of orally administered TJ-48,

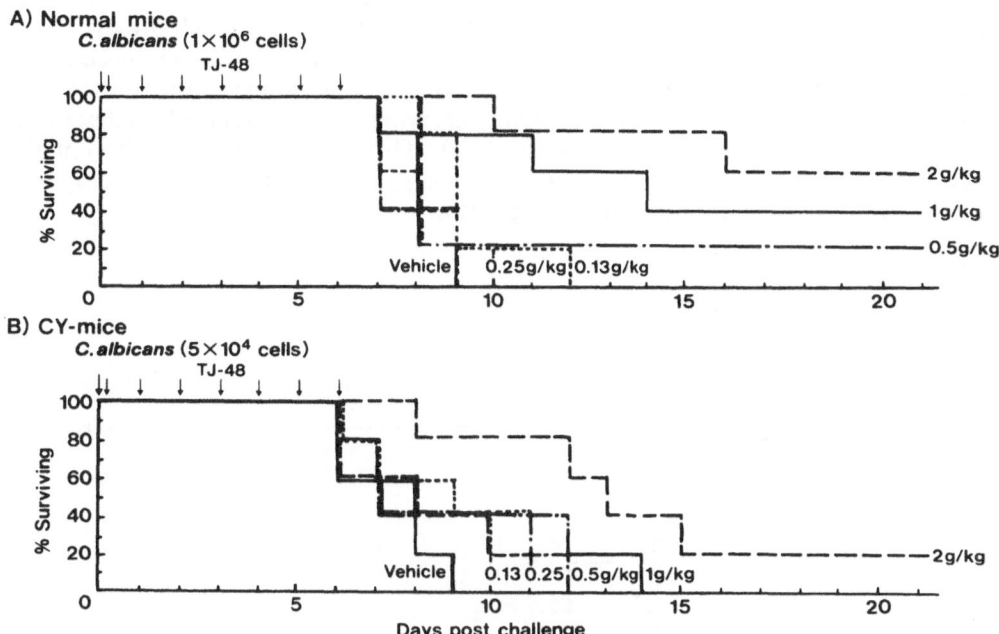

Fig. 9. Survival of normal mice (A) and cyclophosphamide-induced immunocompromised mice (B) challenged intravenously with *C. albicans* and treated from day 0 through day 6 with TJ-48 at various daily dosages.

alone and in combination with a triazole antifungal agent fluconazole, against systemic candidiasis in normal mice and cyclophosphamide-induced immunosuppressed ICR mice.

First, both normal immunosuppressed mice were infected intravenously with a lethal dose of *C. albicans* and administered TJ-48 at oral dosages of 0.13 to 2 g/kg/day for 6 consecutive days starting on the day of infection. As illustrated in Fig. 9, TJ-48-treatment was effective in prolonging life span and increasing the survival rate, although to a rather slight extent, of both normal and immunosuppressed mice with disseminated candidiasis.

Next, the combination effect of TJ-48 and fluconazole was tested using the same animal model of systemic candidiasis. Immunosuppressed mice were orally administered fluconazole at daily dosages of 0.01 to 10 mg/kg for 4 consecutive days, alone and in combination with 2 g/kg/day of TJ-48, starting on the day of infection. Mice treated with fluconazole plus TJ-48 survived longer than did mice treated with fluconazole alone (Fig. 10).

DISCUSSION AND CONCLUSIONS

All of the above-mentioned data strongly suggest that TJ-48 is active in modulating functions of both lymphocyte and macrophage, and that it has a potential usefulness in the adjunctive immunotherapy for some microbial infections. One of the most important and challenging problems to be solved with Kampo medicines and all other crude herbal drugs is to characterize the major biologically active principles contained in the prescriptions. It is also necessary to standardize such prescriptions in a rational way. Obviously, standardization of a complex mixture is much more difficult than standardization of a single compound. Despite that, with regard to TJ-48, Yamada and his coworkers (13) have been intensively carrying out biochemical analysis of this prescription to characterize its active principle(s). This approach to evaluation of TJ-48 on the basis of modern medical science will be supported by an understanding of the details of its immunopharmacological properties.

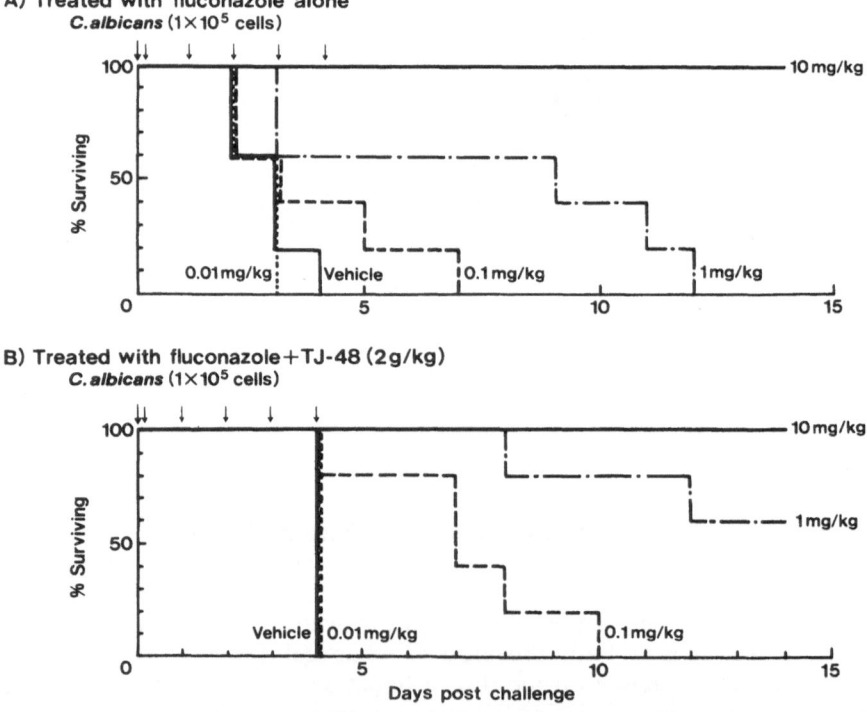

Fig. 10. Survival of cyclophosphamide-induced immunocompromised mice challenged intravenously with C. albicans and treated from day 0 through day 4 with oral fluconazole, alone or in combination with TJ-48 at a daily dosage of 2 g/kg.

REFERENCES

1. E. Hosoya and Y. Yamamura(eds.), Recent Advances in the Pharmacology of Kampo Medicine, Excerpta Medica, Amsterdam/Princeton/Geneva/Tokyo (1989).
2. T. Yamamoto, H. Okada and A. Kuroiwa.: Effects of Oriental herbal medicines on immunity in the elderly. In: Oda,T.(ed.) Recent Advances in Traditional Medicine in East Asia, p.96-105, Excerpta Medica, Amsterdam/Princeton/-Geneva/Tokyo (1985).
3. Y. Komatsu, N. Takemoto, H. Maruyama, H. Tsuchiya, M. Aburada, E. Hosoya, S. Shinohara and H. Hamada, Effect of Juzen-taihoto on the anti-SRBC response in mice, Jpn. J. Inflam.,6: 405-408 (1986).
4. A. J. Cunningham and A. Szenberg: Further improvement in the plaque forming technique for detecting single antibody-forming cells, Immunol., 14: 599-600 (1968).
5. N. Takemoto, H. Kawamura, H. Maruyama, Y. Komatsu, M. Aburada and E. Hosoya: Effect of TJ-48 on murine cellular immunity, Jpn. J. Inflam., 9: 49-52 (1989).
6. T. Ezaki and J. Marbrook: the haemolytic plaque reduction technique to measure cytotoxicity with hybridoma cells as a target. J. Immunol. Methods. 54: 281-290 (1982).
7. S. Ebisuno, A. Hirano, T. Ohkawa, H. Maruyama, H. Kawamura and E. Hosoya: Immunomodulating effects of Juzen-taiho-to in immunosuppressive mice, Biotherapy, 4: 112-116 (1990).
8. N. Takemoto, H. Maruyama, H. Kawamura, Y. Komatsu, M. Aburada and E. Hosoya: Mitogenic activity of Juzentaihoto (TJ-48) on murine lymphoid cells, Jpn. J. Inflam., 9: 137-140 (1989).
9. H. Maruyama, H. Kawamura, N. Takemoto, Y. Komatsu, M. Aburada and E. Hosoya: Effect of Juzentaihoto on phagocytes. Jpn. J. Inflam., 8: 461-465 (1988).
10. K. Haranaka, N. Satomi, A. Sakurai, R. Haranaka, N. Okada and M. Kobayashi: Antitumor activities and tumor necrosis factor producibility of traditional Chinese medicines and crude drugs, Cancer Immunol. Immunother., 20: 1-5 (1985).
11. Y. Sakagami, Y. Mizoguchi, K. Miyajima, H. Kuboi, K. Kobayashi, K. Kioka, H. Takeda, T. Shin, S. Morisawa and S. Yamamoto: Antitumor activity of Shi-Quan-Da-Bu-Tang and its effects on interferon-γ and interleukin 2 production, Allergy(Tokyo), 37: 57-60 (1988).
12. N. Satomi, A. Sakurai, R. Haranaka and K. Haranaka: Traditional Chinese medicines and drugs in relation to the host-defense mechanism. In: G. Gillissen, W. Opferkuch, G. Peters and G. Pulverer (eds.) The Influence of Antibiotics on the Host-Parasite Relationship III, p.77-86, Springer-Verlag, Berlin/Heidelberg (1989).
13. H. Yamada, H. Kiyohara, J. -C. Cyong, N. Takemoto, Y. Komatsu, H. Kawamura, M. Aburada and E. Hosoya: Fractions and characterization of mitogenic and anti-complementary active fractions from Kampo (Japanese herbal) medicine Juzen-taiho-to. Planta Med., 56: 386-391 (1990).

IMMUNOMODULATING EFFECTS OF POLYSACCHARIDES FROM

MEDICINAL PLANTS

J. Kraus and G. Franz

Department of Pharmaceutical Sciences
University of Regensburg
W-8400 Regensburg, Germany

INTRODUCTION

Several hundred natural polysaccharides are currently known and provide one of the richest and oldest reservoirs of structurally and functionally diverse biopolymers. Water soluble polysaccharide hydrocolloids are abundant in nature, present in many plant sources but only a restricted number is of therapeutical interest. For medical purpose many uses were proposed and have been established ever since, and new areas of pharmaceutical/medical uses are constantly being discussed (1).

In the past 30 years various biological activities, e.g. antitumoral, immunomodulating, anticoagulatory, antiviral, antiinflammatory, and hypoglycemic effects of different polysaccharides have been reported (2-5). Polysaccharides exhibiting antitumoral activities have been isolated from different biological origin like bacteria, fungi, algae, lichen as well as higher plants. Among the great variety of the polysaccharides investigated, ß-1.3/1.6-glucans from fungi have shown to be the most active

compounds, being effective against allogeneic, syngeneic, and even autochtonous tumors (6-8).

In contrast to these fungal polysaccharides, polymers isolated from higher plants often show complex structural features. In order to investigate structure-activity relationships, the elucidation of the detailed polysaccharide structure gains more and more importance. In this context the primary structure represented by the type of sugar linkages as well as the overall architecture (conformation) has to be determined.

Immunomodulating polysaccharides have been reported from higher plants and from microbial origin. Neutral as well as acidic polysaccharides having different structural features were shown to have immunomodulating activities, e.g. enhancement of phagocytosis, mitogenic effects, induction of cytotoxic macrophages and natural killer cells or an increased release of cytokines like TNF, IL-1, IL-2, or interferons (9-12).

One of the most severe points of error with the consequence of false interpreted results is the contamination of polysaccharides with bacterial endotoxins (lipopolysaccharides). Other than polysaccharides from microbial origin, which can be cultivated under controlled conditions, higher plants are bound to be infected by microorganisms from which, after a complex extraction and fractionation procedure,

Microbial Infections, Edited by H. Friedman *et al.*
Plenum Press, New York, 1992

may result a more or less severe contamination with endotoxin. These lipopolysacch-arides will drastically alter especially the *in vitro* results but not so much the *in vivo* studies carried out with tumor bearing mice (13). For this reason it is essential to determine, and if necessary, to reduce the amount of these immunologically highly active endotoxins. Removal of lipopolysaccharides by treatment with Polymyxine-B can be achieved for neutral polysaccharides (14). However, this method does not succeed in the case of acidic polysaccharides (15). Consequently, the immunological experi-ments of endotoxin contaminated polysaccharides have to be carried out using C3H/HeJ LPS-low responder mice, which are less sensitive to endotoxins due to a genetic defect (10).

Earlier results encouraged us to start with a broad screening program for the search of new polysaccharides with improved immunological as well as antitumoral properties. Since most of the results reported had been obtained with crude plant extracts or partially purified polymers, we also put emphasize on the structural investigations of the isolated polysaccharides.

NATURAL POLYSACCHARIDES

Structural Characteristics

The search for immunomodulating and antitumoral polysaccharides was started with a series of in Europe well-known medicinal plants. The structure of the polysac-charides was determined using various analytical methods, e.g., partial acid hydrolysis, methylation analysis, or NMR spectroscopy (17). The biological origin, the structural features, the average molecular weight, and the endotoxin content of the polysaccharides investigated are summarized in Table 1. In addition, the cell-wall glucans from in vitro cultivated fungi belonging to the genus Phytophthora were also included in our screening program.

The neutral polysaccharides were represented by a ß-2.1 fructan from Solidago canadensis and a ß-1.3/1.6 glucan from Phythophthora parasitica. The polysaccharides from Cassia angustifolia as well as Hibiscus sabdariffa represent the structural characteristics of a rhamnogalacturonan backbone with arabinogalactan side chains, which are known to be common for plant polysaccharides. These polysaccharides are differing in the uronic acid content. While the polysaccharide fraction GF3 from Ginkgo biloba is structurally similar to the former mentioned polymers, GF2a is characterized by an α-1.2-linked mannan backbone, which is highly branched via C-4 with single rhamnose residues or short chains of α-1.4-linked glucuronic acid. This polymer seems to be unique for Ginkgo biloba (18).

Antitumor Activity

The antitumoral activity of the various polysaccharides was evaluated against the allogeneic Sarcoma-180 in CD1-mice. This tumor model is considered to be highly sensitive to immunomodulating compounds (19). With the exception of SOL-F, which was used at 25 mg/kg, all substances were tested at a dose of 5 mg/kg by i.p. application daily from day 1 to 10.

As demonstrated in Table 2, the acidic polysaccharides HIB, SEN, SEN-A, GF2a, and GF3 showed no significant inhibition of tumor growth. On the contrary, the ß-1.3/1.6 glucans PHYT-G and SPG exhibited significantly high antitumor activities with inhibition rates of almost 100%. The ß-2.1 fructan SOL-F also showed a marked tumor inhibition of 80%. These results indicate that a relatively high uronic acid content is disadvantageous for exhibiting antitumor activities.

300

Table 1. Characterization of The Tested Compounds

origin	substance	structural features	average MW(kd)	LPS (ng/mg)
Phytophthora parasitica	PHYT-G	ß-1.3/1.6-glucan	10	< 0.04
Schizophyllum comune	SPG	ß-1.3/1.6-glucan	450	< 0.04
Solidago canadensis	SOL-F	ß-2.1-fructosan	4	0.08-0.8
Ginkgo biloba	GF2a	α-1.2-mannan + rhamnose	500	< 0.04
	GF3	rhamnogalacturonan + arabinogalactan	40	4
Hibiscus sabdariffa	HIB	rhamnogalacturonan (90 % gal-A)	100	1200
Cassia angustifolia	SEN	rhamnogalacturonan + arabinogalactan	1000	640
	SEN-A	rhamnogalacturonan (= backbone of SEN)	12	30

Table 2. Antitumor Activity Against Sarcoma-180

substance	dose[a] (mg/kg)	average tumor weight (g)	inhibition rate[b] (%)
Control	--	4.52	--
PHYT-G	5	0.01	99[c]
SPG	5	0.01	99[c]
SOL-F	25	0.81	82[c]
Control	--	4.05	--
GF2a	5	3.71	9[d]
GF3	5	3.89	4[d]
Control	--	8.80	--
HIB	5	5.28	40[d]
SEN	5	4.80	44[d]
SEN-A	5	6.68	22[d]

[a] Treatment was performed daily from day 1-10, i.p.
[b] % = (C-T/C)x100; T = average tumor weight of treated group,
C = average tumor weight of control group
[c] Difference from control significant at P < 0.01
[d] Not significant

Immunological Activities

In order to evaluate the immunological properties of the polysaccharides we investigated the mitogenic activity, the induction of cytotoxic macrophages as well as the enhancement of phagocytosis. Due to the endotoxin content of some polysaccharides the immunological tests were performed with cells from LPS-low responder C3H/HeJ mice.

As shown in Table 3 the acidic and antitumor inactive polysaccharides SEN, HIB, and GF3 expressed a strong enhancement of lymphocyte proliferation with stimulation indices of 8.9, 5.65, and 3.18, respectively. SEN-A, the rhamnogalacturonan backbone of SEN as well as the acidic branched mannan GF2a, both in vivo antitumor inactive, showed no mitogenic effect. On the other hand, the antitumor active polysaccharides SPG and SOL-F caused no significant increase in lymphocyte proliferation. Only the ß-1.3/1.6 glucan PHYT-G expressed a moderate mitogenic effect at the highest concentration of 200 µg/ml with a stimulation index of 2.5.

Table 3. Proliferation assay with C3H/HeJ Spleen Lymphocytes

substance	200[a]	20[a]	2[a]
PHYT-G	2.58[b]	1.37	1.36
SPG	1.59	1.11	0.87
SOL-F	1.88	1.27	0.80
GF2a	0.26	1.12	0.92
GF3	3.18	3.69	2.07
HIB	5.65	3.88	1.44
SEN	8.90	4.51	1.34
SEN-A	1.26	1.03	1.04
	10[a]	1[a]	0.1[a]
LPS	3.7[b]	3.58	1.81

[a] concentrations in µg/mL
[b] mitogenic activity is expressed as stimulation index (cpm test/ cpm control)

Due to the fact that cytotoxic macrophages play an important role in host's defence mechanism against cancer (20), the ability of polysaccharides to render macrophages cytotoxic against tumor cells was investigated. The macrophages were coincubated with 2 units/ml gamma-interferon, which served as a T-cell derived signal. The results are summarized in Fig. 1.

Antitumor inactive polysaccharides showed a strong inhibition of tumor cell proliferation at concentrations of 10 and 100 µg/ml, indicating induction of cytotoxic macrophages. Again, the antitumor active polysaccharides SPG, PHYT-G, and SOL-F were unable to render macrophages cytotoxic against P-815 tumor cells. SEN-A as well as the Ginkgo polysaccharides GF2a and GF3 did not influence the activity of macrophages. In this context it has to be mentioned that all polysaccharides tested showed no direct cytotoxic effects against P-815 mastocytoma cells.

Although phagocytosis plays a secondary role in host's defense against cancer, we investigated the influence of the polysaccharides on the phagocytic activity of macrophages. Cultivated bone marrow cells of LPS-low responder C3H/HeJ mice were used as a macrophage source. The phagocytic activity was determined after 15 hr preincubation indirectly via chemiluminescence measurement using zymosan as phagocytotic agents.

As shown in Fig. 2, only the acidic polysaccharides SEN and HIB were able to enhance markedly phagocytosis of zymosan. SEN was the most active compound,

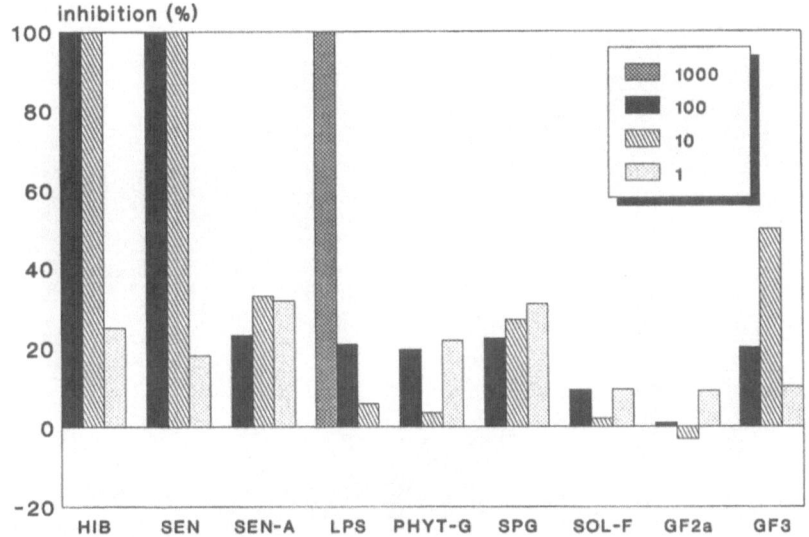

Fig. 1.   Induction of cytotoxic macrophages against P-815 mastocytoma cells. Activity is expressed in percent inhibition of tumor cell growth. Concentrations are given as µg/ml (polysaccharides) and ng/ml (LPS).

causing an increase in phagocytosis of 270% at a concentration of 20 µg/ml, followed by HIB with an increase of 180% at a concentration of 200 µg/ml. SEN-A caused only a moderate enhancement at 200 µg/ml. All other compounds, including the acidic Ginkgo polysaccharides had no significant influence on macrophages.

SYNTHETICALLY MODIFIED POLYSACCHARIDE DERIVATIVES

In order to get more information about structure-activity relationships, mainly in the case of ß-1.3-glucans, we chemically modified curdlan, a linear ß-1.3-glucan, and lichenan, a linear ß-1.3/1.4-glucan by addition of various glycosidic as well as non-glycosidic side chains.

The aim of these studies was to evaluate the requirement of a ß-1.3-glucan backbone for the expression of antitumor and/or immunomodulatory activity. Additionally, the influence of chemically different substituents had to be examined.

Structural Characteristics of Chemically Modified Polysaccharides

Curdlan as well as lichenan were regiospecifically branched at C-6 with glucose, arabinose, rhamnose, and in the case of curdlan also with gentiobiose. Ethylsulfonyl, propylsulfonyl, and butylsulfonyl acid were used for the preparation of the non-glycosidi-

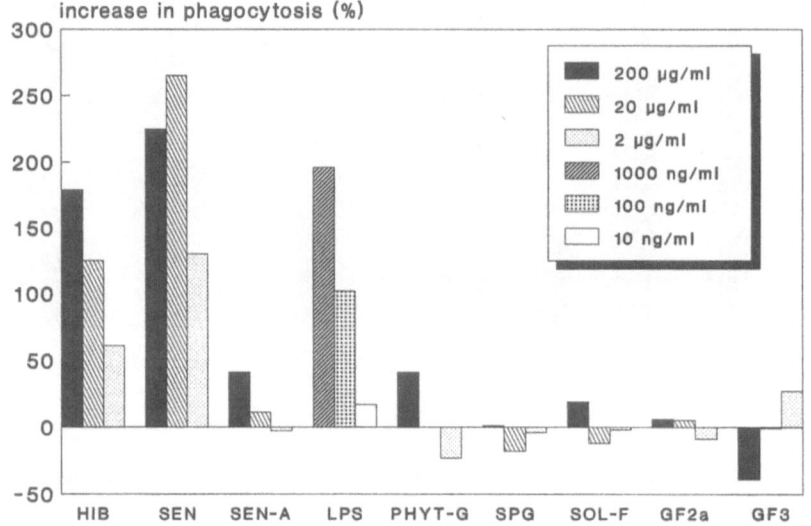

Fig. 2.   Increase (%) in phagocytosis of zymosan by C3H/HeJ bone marrow macrophages.

cally substituted glucan derivatives. While the sugar residues could be regiospecifically introduced in position C-6 (21-22), the sulfoalkyl groups were preferentially linked to C-6, and secondary to position C-2 (23).

The molecular weight, determined by GPC using pullulan standards, was >300.000d for the sulfoalkyl curdlan and lichenan derivatives, and ¯25.000d for the glycosidically branched derivatives, respectively. The structural characteristics of the polysaccharide derivatives are shown in Table 4.

Table 4.   Structural Characteristics of Curdlan and Lichenan Derivatives

starting polymer	substituent	branch frequency (%)	degree of substitution	LPS (ng/mg)
Lichenan	glucose	8		4
	rhamnose	10		40
	arabinose	10		40
	ethylsulfonic acid		0.27	8
	propylsulfonic acid		0.75	8
	butylsulfonic acid		0.40	4
Curdlan	glucose	29		4
	gentiobiose	29		40
	rhamnose	33		<0.04
	arabinose	33		<0.04
	ethylsulfonic acid		0.43	<0.04
	propylsulfonic acid		0.45	<0.04
	butylsulfonic acid		0.27	40

Antitumor Activity

The antitumor activity of the various curdlan and lichenan derivatives was evaluated against the allogeneic Sarcoma-180. The polysaccharide derivatives were tested at 25 mg/kg, intraperitoneally given for 10 consecutive days, starting 24 hr after tumor inoculation. The inhibition of tumor growth was determined at day 30.

As shown in Table 5, the glycosidically as well as the non-glycosidically substituted curdlan derivatives caused a more or less strong inhibition of tumor growth with inhibition ratios of 60 to 99%. The rhamnose and arabinose as well as the sulfoethyl and sulfobutyl derivatives turned out to be the most active compounds.

In contrast to the active curdlan derivatives, the lichenan derivatives did not show a significant tumor inhibition, with the exception of the glucose branched derivative, which expressed an inhibition rate of 88%.

These results indicate that the ß-1.3 linked glucan backbone of curdlan is essential for the expression of a strong antitumor effect. On the other hand, a high proportion of ß-1.4 linked glucose residues in the backbone seems to be disadvantagous. Surprisingly, the structure of the substituents has only a slight influence on the antitumor activity of the polysaccharide derivatives.

Immunological Activities of Polysaccharide Derivatives

The immunological effects of curdlan and lichenan derivatives were evaluated by testing the mitogenic activity, the induction of cytotoxic macrophages, and the enhancement of phagocytosis. The immunological assays were performed with cells from LPS-low responder C3H/HeJ mice. Table 6 provides an overview about the activities of the different derivatives exhibited in the immunological test systems.

Table 5. Antitumor Activity of the Curdlan and Lichenan Derivatives against Sarcoma-180

polysaccharide derivative	dose[a] (mg/kg)	average tumor weight (g)	tumor area[b] (mm^2)	inhibition[c] ratio (%)
control		4.03		
L-glucose[d]	25	0.48		88[e]
control		5.40		
L-rhamnose	25	5.98		-11[f]
L-arabinose	25	4.95		8[f]
control			164	
L-sulfoethyl	25		133	19[f]
L-sulfopropyl	25		108	34[f]
L-sulfobutyl	25		133	19[f]
control		5.40		
C-glucose[g]	25	2.20		59[f]
C-gentiobiose	25	1.26		77[e]
C-rhamnose	25	0.04		99[e]
C-arabinose	25	0.04		99[e]
C-sulfoethyl	25	0.01		99[e]
C-sulfobutyl	25	0.58		89[e]
control			554	
C-sulfopropyl	25		139	75[e]

[a] Treatment was performed daily from day 1-10, i.p.
[b] Tumor area at day 20 was used for evaluation of inhibition ratio
[c] % = (C-T/C)x100; T = average tumor weight (tumor area) of treated group, C = average tumor weight of control group
[d] L = lichenan derivative
[e] Difference from control significant at p < 0.01
[f] Not significant
[g] C = curdlan derivative

The antitumor inactive lichenan derivatives showed high immunostimulating effects in the *in vitro* assays. The moderately antitumor active glucose branched lichenan derivative had no influence on the immune cells. On the contrary, the antitumor active curdlan derivatives did not exhibit any immunostimulating properties. Furthermore, in the phagocytosis assay the antitumor active curdlan derivatives caused an inhibition of phagocytosis of zymosan, probably due to a hindering of the macrophage ß-1.3-glucan receptor (24).

CONCLUDING REMARKS

A possible treatment of tumor cells with non-toxic polysaccharides merits continued exploration. These immunostimulants are no longer in its infancy but they are still not mature enough for an indication of precise guidelines or therapeutically strict rules.

Table 6. Immunological Effects of the Curdlan and Lichenan Derivatives

starting polymer	substituent	mitogenic activity	induction of cytotoxic M0	enhancement of phagocytosis
Lichenan				
	glucose	-	-	-
	rhamnose	+	+ +	+
	arabinose	+	+ +	+
	sulfopropyl	+ +	-	+ +
	sulfobutyl	+	-	+ +
Curdlan				
	glucose	+	-	I
	gentiobiose	+	-	-
	rhamnose	-	-	I
	arabinose	-	-	I
	sulfoethyl	-	-	I
	sulfopropyl	-	-	I
	sulfobutyl	-	-	I

+ + = highly active; + = active; - = no activity; I = inhibition

At the moment such polysaccharide BRM's of fungal origin alone or in combination with chemotherapy are suitable for the adjuvant treatment of cancer. The search for an ideal type of macromolecule is not finished but has to be continued through a collaboration of natural product chemistry, pharmaceutical and medical sciences.

REFERENCES

1. G. Franz, Struktur und biologische Funktion von Polysacchariden, in: "Polysaccharide - Eigenschaften und Nutzung", W.Burchard, ed., Springer Verlag, Berlin, Heidelberg, New York, Tokyo (1985).
2. R. L.Whistler, A. Bushway, P. O. Singh, W. Nakahara, and R. Tokuzen, Noncytotoxic, antitumor polysaccharides, Adv. Carbohydr. Chem. Biochem. 32:235 (1976).
3. G. Chihara, Immunopharmacology of Lentinan and the glucans, Rev. Immunol. Immunofarmacol. 4:85 (1984).
4. J. Kraus, Biopolymere mit antitumoraler und immunmodulierender Wirkung, Pharmazie in unserer Zeit 19:157 (1990).
5. E. De Clerq, Chemotherapeutic approaches to the treatment of acquired immune deficiency syndrome (AIDS), J. Med. Chem. 29:1561 (1986).
6. G. Chihara, J. Hamuro, Y. Maeda, T. Shiio, T. Suga, N. Takasuka, and T. Sasaki, Antitumor and metastasis-inhibitory activities of Lentinan as an immunomodulator: an overview, Cancer Detection and Prevention Suppl. 1:423 (1987).
7. H. Furue, Biological characteristics and clinical effect of sizofilan (SPG), Drugs of Today 23:335 (1987).

8. J. Kraus and G. Franz, ß(1-3)glucans: anti-tumor activity and immunostimulation, in: "Fungal Cell Wall and Immune Response", J.-P. Latgé and D. Boucias, eds., NATO ASI Series H53, Springer Verlag, Berlin, Heidelberg, New York, London, Paris, Tokyo, Hong Kong, Barcelona, Budapest (1991).

9. H. Wagner, A. Proksch, I. Riess-Maurer, A. Vollmar, S. Odenthal, H. Stuppner, K. Jurcic, M. Le Turdu, and J. N. Fang, Immunstimulierend wirkende Polysaccharide (Heteroglykane) aus höheren Pflanzen, Arzneim. Forsch./Drug Res. 35:1069 (1985).

10. H. Wagner, H. Stuppner, W. Schäfer, and M. Zenk, Immunologically active polysaccharides of Echinacea purpurea cell cultures, Phytochemistry 27:119 (1988).

11. B. Luettig, C. Steinmüller, G. E. Gifford, H. Wagner, and M.-L. Lohmann--Matthes, Macrophage activation by the polysaccharide arabinogalactan isolated from plant cell cultures of Echinacea purpurea, J. Natl. Cancer Inst. 81:669 (1989).

12. K. Othani, K. Mizutani, S. Hatono, R. Kasai, R. Sumino, T. Shiota, M. Ushijima, J. Zhou, T. Fuwa, and O. Tanaka, Sanchinan-A, a reticuloendothelial system activating arabinogalactan from Sanchi-Ginseng (Roots of Panax notoginseng), Planta Med. 53:166 (1987).

13. A. Haslberger, J. Hildebrandt, C. Lam, E. Liehl, H. Loibner, I. Macher, B. Rosenwirth, E. Schütze, H. Vyplel, and F. M. Unger, Immunopharmacology of lipopolysaccharides (endotoxins) from gram-negative bacteria, Triangle 26:33 (1987).

14. A. C. Issekutz, Removal of gram-negative endotoxin from solutions by affinity chromatography, J. Immunol. Methods 61:275 (1983).

15. J. Kraus and F. Roßkopf, Relationship between immunological activity in vitro and antitumor effect in vivo of various polysaccharides, PPL: Pharm. Pharmacol. Lett. 1:11 (1991).

16. L. P. Ruco and M. S: Meltzer, Macrophage activation for tumor cytotoxicity: tumoricidal activity by macrophages from C3H/HeJ mice requires at least two activation stimuli, Cellular Immunology 41:35 (1978).

17. K. Gomaa, J. Kraus, and G. Franz, Structural investigations of glucans from cultures of Glomerella cingulata Spaulding & von Schrenck, Carbohydr. Res. 217:153 (1991).

18. J. Kraus, The water-soluble polysaccharides from Ginkgo biloba leaves, Phytochemistry 30:3017 (1991).

19. G. S. Tarnowski, I. M. Mountain, and C. C. Stock, Influence of genotype of host on regression of solid and ascites forms of Sarcoma-180 and effect of chemotherapy on the solid form, Cancer Res. 33:1885 (1973).

20. R. B. Johnston, Monocytes and macrophages. N. Engl. J. Med. 318:747 (1988).

21. S. Demleitner, J. Kraus, and G. Franz, Synthesis and antitumor activity of C-6 branched curdlan and lichenan derivatives, Carbohydr. Res. in press (1991).

22. S. Demleitner, J. Kraus, and G. Franz, Synthesis and antitumor activity of sulfoalkyl derivatives of curdlan and lichenan, Carbohydr. Res. in press (1991).

23. N. K. Kochetkov, A. B. Bochkov, T. A. Sokolovskaya, and V. J. Snyatkova, Modifications of the orthoester method of glycosylation, Carbohydr. Res. 16:17 (1971).

24. J. K. Czop, The role of ß-glucan receptors on blood and tissue leukocytes in phagocytosis and metabolic activation, Pathol. Immunopathol. Res. 5:286 (1986).

EFFECTS OF SHO-SAIKO-TO ON CYTOKINE CASCADE AND

ARACHIDONIC ACID CASCADE

Yasuhiro Mizoguchi, Yasuhiro Komatsu*, Yasufumi Ohkura*

The Third Department of Internal Medicine
Osaka City University Medical School and
*Tsumura Research Institute for Pharmacology
Osaka, Japan

The traditional Chinese medicine Sho-saiko-to (Xiao-Chai-Hu-Tang) is prepared from the herbs *Bupleuri radix*, *Pinelliae tuber*, *Scutellariae radix*, *Zizyphi fractus*, *Ginseng radix*, *Glycyrrhizae radix* and *Zingiberis rhizoma*, and has been used for the treatment of chronic hepatitis. There is accumulating evidence suggesting that Sho-saiko-to has a protective effect on biological membranes in addition to having an anti-inflammatory and anti-allergic effect (1, 2). However, most of the active substances contained in this medicine and their mechanisms of action await elucidation, except for certain saponin compounds and glycyrrhizin, the pharmacological effects of which have been studied in some detail. We previously demonstrated that Sho-saiko-to protects liver cells against experimentally-induced cell damage and suggested that this may be due at least in part to the action of saikosaponin on the liver cell membrane (3). In this report, we present the results of a multicenter open study of Sho-saiko-to on 80 patients with HBe-antigen (Ag)-positive chronic hepatitis. In order to assess the effects of the medicine on these hepatitis patients, we examined the HBe virus-related antigen and antibody levels and laboratory parameters of hepatic function. We also studied the effects of Sho-saiko-to on immune responses and arachidonic acid cascade.

EFFECTS OF SHO-SAIKO-TO ON PATIENTS WITH HBeAg-POSITIVE CHRONIC HEPATITIS

A clinical study was performed at 10 centers by Oda et al. (4), who treated 80 patients with HBeAg-positive chronic hepatitis with Sho-saiko-to (7.5 g/day) for 6 months.

Effects on HBeAg and Anti-HBe Antibody (Ab) Levels

Of the patients treated with Sho-saiko-to, 8 (19%) seroconverted from HBeAg to anti-HBeAb; 15 (18.8%) became seronegative; and in 11 (13.8%) HBeAg level decreased by more than 50%. A significant fall in the level of HBeAg was seen at 2 months, and this decline continued throughout the 6 month study period (Figure 1). Similarly, anti-HBeAb level rose significantly at 2 months and continued to rise

Microbial Infections, Edited by H. Friedman *et al.*
Plenum Press, New York, 1992

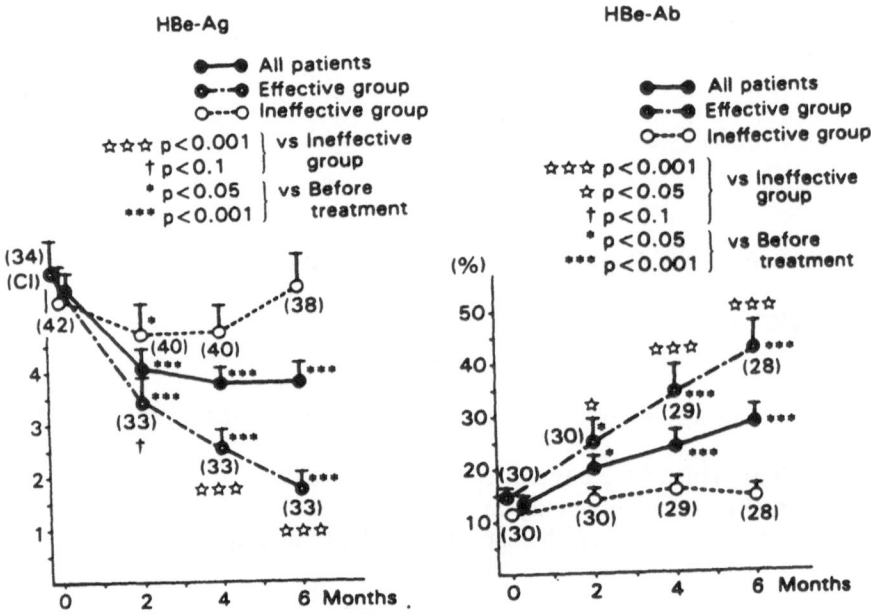

Fig. 1. HBe antigen and anti-HBe antibody levels in 80 patients with HBe antigen-positive chronic hepatitis treated with Sho-saiko-to. Cases were divided into an effective group and an ineffective group based on an HBe antigen cut-off index of 50% reduction.

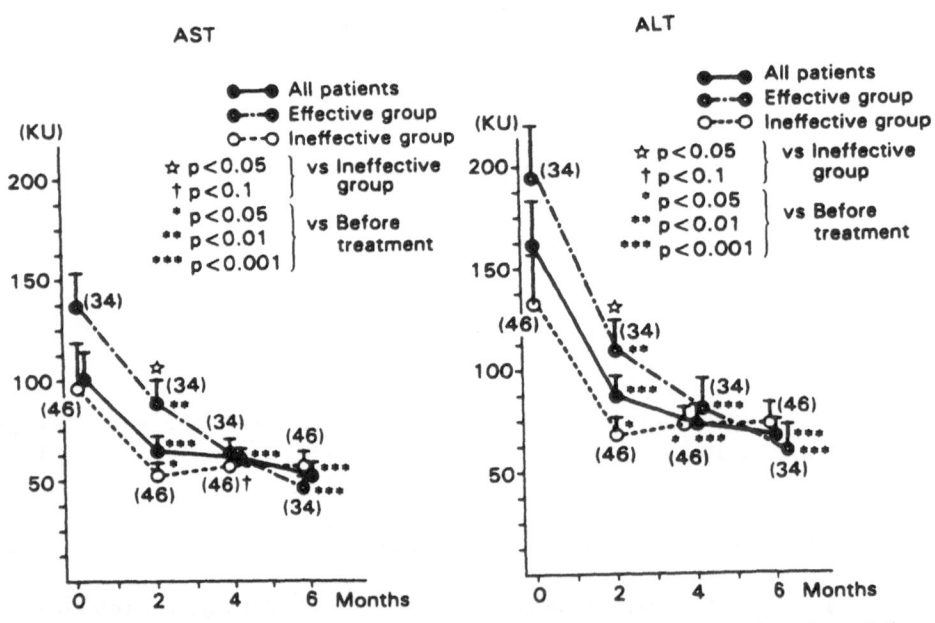

Fig. 2. ALT and AST levels in 80 patients with HBe antigen-positive chronic hepatitis treated with Sho-saiko-to.

throughout the treatment. These results suggested that Sho-saiko-to has an immuno-enhancing effect.

Effects on Serum AST and ALT Levels

For the analysis of serum AST and ALT levels, the patients were divided into effective and ineffective cases. In both groups, serum AST and ALT levels fell significantly after 2 months and continued to fall throughout the period of treatment (Figure 2), although immunoenhancing drugs usually only improve these levels in cases where virological improvement is also seen. These results indicated that Sho-saiko-to also has an anti-inflammatory effect. We therefore concluded from the results of the multicenter open study that Sho-saiko-to may be effective for the treatment of HBeAg-positive chronic hepatitis because of its immunoenhancing effect as well as its anti-inflammatory effect.

EFFECTS OF SHO-SAIKO-TO ON IMMUNE RESPONSES

The immunoenhancing effect of Sho-saiko-to was next studied by examining its effects on the cytokine cascade.

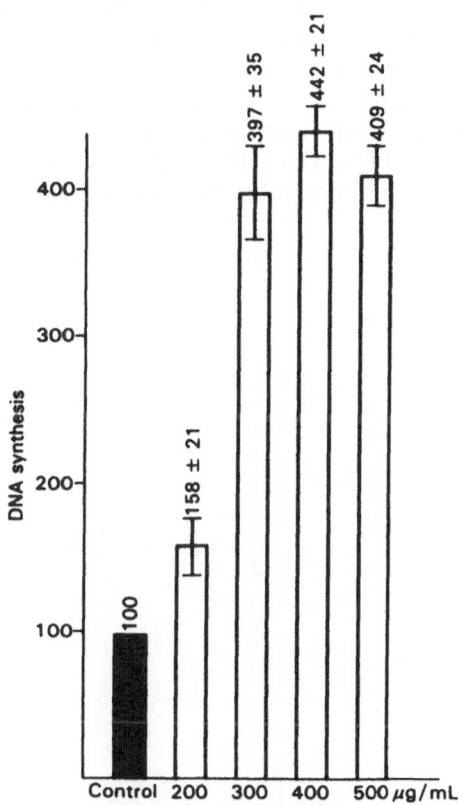

Fig. 3. Effects of Sho-saiko-to on production of interleukin 1. Interleukin 1 production was measured by thymocyte proliferation assay.

Interleukin (IL) 1 Production by Peripheral Blood Monocytes

Peripheral blood monocytes from healthy adults were incubated with Sho-saiko-to at 37°C for 48 hr, and IL1 activity in the culture supernatant was assayed by estimating DNA synthesis in phytohemagglutinin-stimulated C_3H/HeJ mouse thymocytes. As a result, IL1 activity in the culture supernatant from Sho-saiko-to-treated monocytes increased significantly (Figure 3). In addition, IL2 and IL4 production by human peripheral blood mononuclear cells and IL3 production by mouse splenocytes were observed to be enhanced by Sho-saiko-to (data now shown).

IL6 Production by Peripheral Blood Mononuclear Cells

Peripheral blood mononuclear cells from healthy adults were incubated with Con A and Sho-saiko-to at 37°C for 48 hr, and IL6 activity in the culture supernatant was determined by B9 assay. IL6 activity in the culture supernatant from Sho-saiko-to-treated mononuclear cells increased significantly (Figure 4).

Humoral Immune Response in Peripheral Blood Mononuclear Cells

Polyclonal antibody response was induced when peripheral blood mononuclear cells from healthy adults were stimulated with pokeweed mitogen (PWM). The humoral immune response was estimated by counting the hemolytic plaque forming cells incubated with protein A coated SRBC and complement according to the method of Gronowicz et al. (5). The number of plaque forming cells induced by stimulation with PWM was about twofold that of the control in which the mononuclear cells were cultured without PWM. The induction of antibody forming cells by PWM was enhanced by the simultaneous addition of Sho-saiko-to, and the number of plaque forming cells was significantly increased by adding Sho-saiko-to (Figure 5). The increased production of cytokines by Sho-saiko-to may account for this augmentation of antibody production.

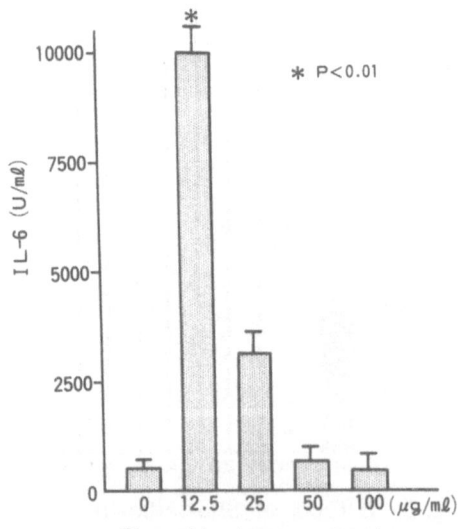

Fig. 4. Effects of Sho-saiko-to on production of interleukin 6.

312

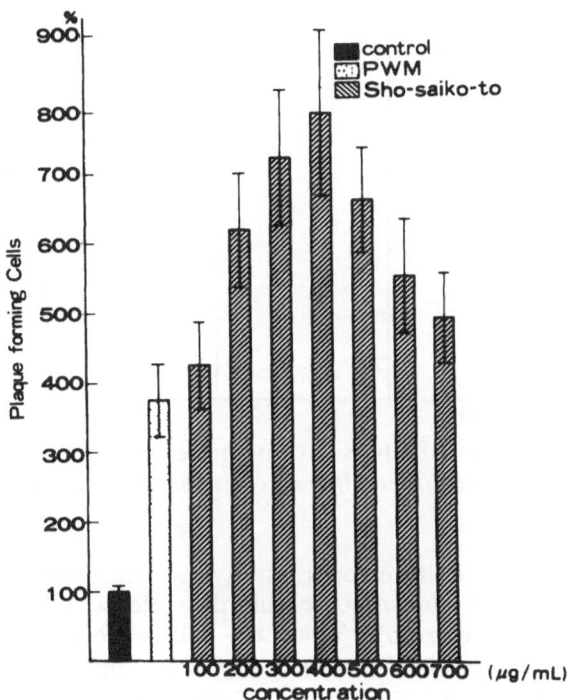

Fig. 5. Effects of Sho-saiko-to on PWM-
induced plaque forming cells.

EFFECTS OF SHO-SAIKO-TO ON ARACHIDONIC ACID CASCADE

To clarify the anti-inflammatory effect of Sho-saiko-to, its effects on the release of arachidonic acid from macrophages and on the production of lipocortin or a lipocortinoid substance were studied. The effects of Sho-saiko-to on the production of leukotriene from macrophages were also evaluated.

Glucocorticoids are known to bind with receptors of the cytoplasm in the cell nucleus and to stimulate the production of a certain protein and activate its physiological activities. This protein is called "lipocortin". Lipocortin is known to inhibit the metabolism of phospholipids contained in the cell membrane with a double layer structure. The inert membrane phospholipids have a hydrophilic part in the outer layer and a hydrophobic part in the inner layer. When the membrane is stimulated, phospholipids near the stimulated receptors are easily affected by phospholipase A_2 and then arachidonic acid is released. This released arachidonic acid is then metabolized via the lipooxygenase pathway to form leukotrienes, which are well-known physiologically active substances. Therefore, lipocortin is thought to be a substance that suppresses the release of arachidonic acid by interfering with phospholipase A_2 in Sho-saiko-to treatment.

Effects on Release of Arachidonic Acid

Peritoneal exudate cells were collected from Hartley guinea pigs treated with Marcol 52 4 days earlier. These peritoneal exudate cells were incubated with Sho-saiko-to at 37°C for 6 to 48 hr and labeled with 14C-arachidonic acid. The labeled cells were stimulated with 10 mM of formyl methinyl leucyl phenylalanine for 10 min,

313

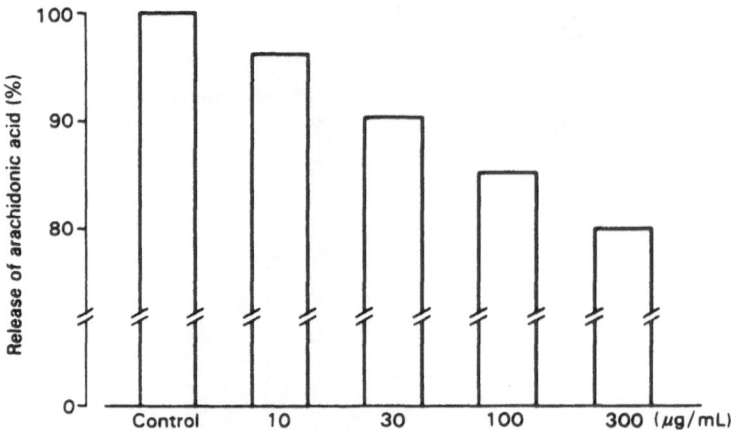

Fig. 6. Effects of Sho-saiko-to on release of arachidonic acid by guinea pig peritoneal exudate macrophages.

and the radioactivity in the culture supernatant was measured by a liquid scintillation counter. Sho-saiko-to was shown to inhibit the release of arachidonic acid from peritoneal exudate cells (Figure 6). When the cells were treated with 100 µg/ml of Sho-saiko-to, arachidonic acid release was especially significantly suppressed. The suppression of its release was seen from 12 hr after the addition of Sho-saiko-to, and maximum suppression was seen at 36 hr.

Effects on Phospholipase A_2 Activity

The peritoneal exudate cells were incubated with 100 µg/ml of Sho-saiko-to at 37°C for 36 hr, and then frozen and thawed. Anti-phospholipase A_2 activity in the culture supernatant was assayed with alpha-palmitoyl-β-14C-arachidonyl phosphatidyl-choline as a substrate, and the released 14C-arachidonic acid was measured by a liquid scintillation counter. The soluble fraction of macrophages treated with Sho-saiko-to was found to contain active factors which inhibited the activity of phospholipase A_2 (Figure 7). In the next experiment, 100 µg/ml of Sho-saiko-to was added to the same macrophage suspension, which was incubated at 37°C for 36 hr. After being frozen and thawed, this was centrifuged and fractionated by Sephadex G-100 column chromatography. The protein concentration and the active factors that inhibited phospholipase

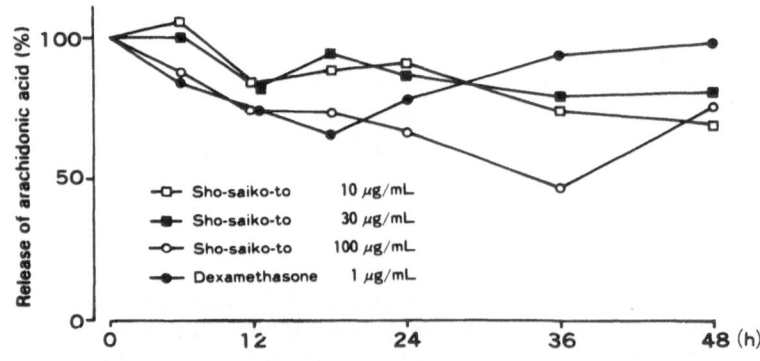

Fig. 7. Effects of Sho-saiko-to on phospholipase A2 activity in suspension of guinea pig peritoneal exudate macrophages.

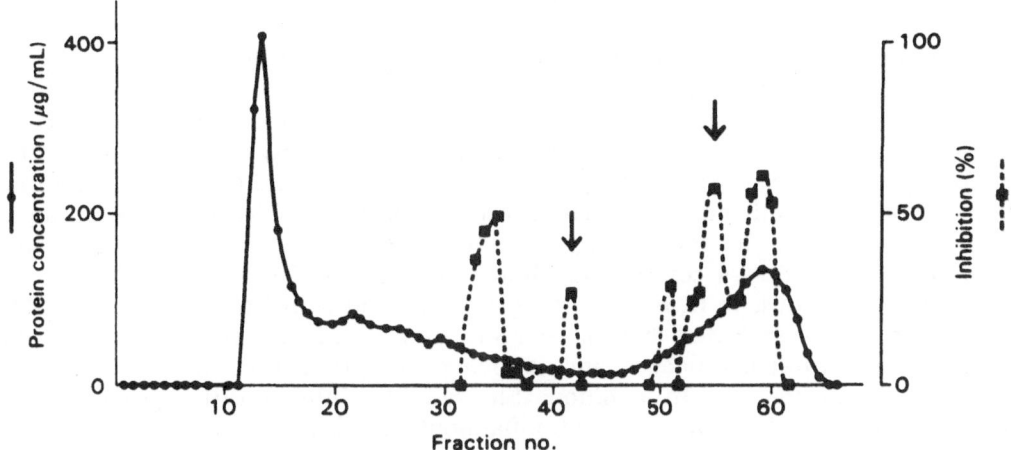

Fig. 8. Sephadex G-100 column chromatogram of phospholipase A2 activity-inhibiting proteins in suspension of guinea pig peritoneal exudate macrophages treated with Sho-saiko-to or dexamethasone. Active proteins were revealed in 5 fractions from the Sho-saiko-to-treated suspension. Two of these coincide exactly with the peaks obtained from the suspension treated with dexamethasone (arrows).

A_2 activity were measured in each eluted fraction. As shown in Figure 8, active phospholipase A_2-inhibiting factors were found in 5 fractions. Furthermore, the fractions isolated after treatment of macrophages with dexamethasone were identical with the two indicated by the arrows in Figure 8. The molecular weights of the two fractions were 32000 and 60000 daltons. These results indicated that when macrophages were treated with Sho-saiko-to, lipocortin or a lipocortinoid substance was produced.

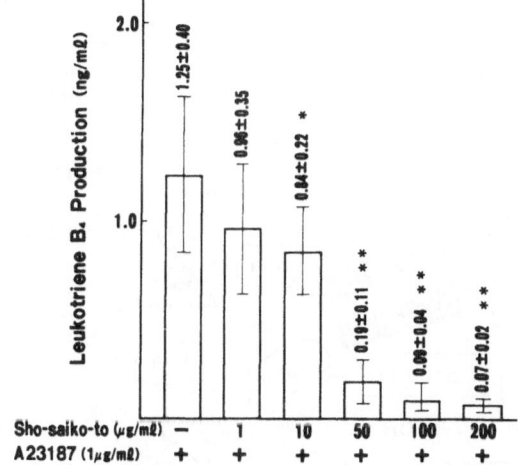

Fig. 9. Effects of Sho-saiko-to on production of leukotriene B4 by guinea pig peritoneal exudate macrophages in response to stimulation with Ca ionophore A23187.

Effects on Production of Leukotriene B$_4$

Sho-saiko-to was added to the same macrophage suspension at 5×10^6 cells/ml, which was then incubated at 37°C for 24 hr. After incubation, this was diluted twofold by adding saline. Ca ionophore A23187 (1 µg/ml) was then added, and the mixture was incubated for another 15 min. Four milliliters of ethanol was also added, and the culture supernatant was separated by centrifugation, frozen and dried. After addition of 0.5 ml of 50% methanol, fractionation was performed by high performance liquid chromatography and leukotriene B$_4$ was quantified. An untreated macrophage suspension was used as a control. As shown in Figure 9, Sho-saiko-to inhibited the production of leukotriene B$_4$ in macrophages stimulated with Ca ionophore A23187.

These results indicated that Sho-saiko-to suppressed the release of arachidonic acid by acting on macrophages and inducing the production of lipocortin or a lipocortinoid substance. In addition, Sho-saiko-to inhibited the production of leukotriene B$_4$, thereby showing its anti-inflammatory effect.

DISCUSSION

A multicenter open study was carried out on 80 patients with HBeAg-positive chronic hepatitis to determine the effects of Sho-saiko-to on hepatitis B virus-related antigen-antibody and on laboratory parameters of hepatic function. Our results showed a 50% improvement in hepatic function tests, 10% virological improvement in seroconverted patients and 18.8% in seronegative patients, with a decrease in HBeAg level to less than half of initial levels in 13.8%. These results suggested that the clinical effects of Sho-saiko-to are mediated by immunoenhancement and anti-inflammation. Based on these background factors, we studied the effects of the medicine on the production of cytokines and the arachidonic acid cascade. Our results clearly indicated that Sho-saiko-to acted on macrophage-monocytes which can produce and secrete several kinds of cytokines. It stimulated not only the production of IL1, which is one of the cytokines, but also the production of lipocortin or a lipocortinoid substance

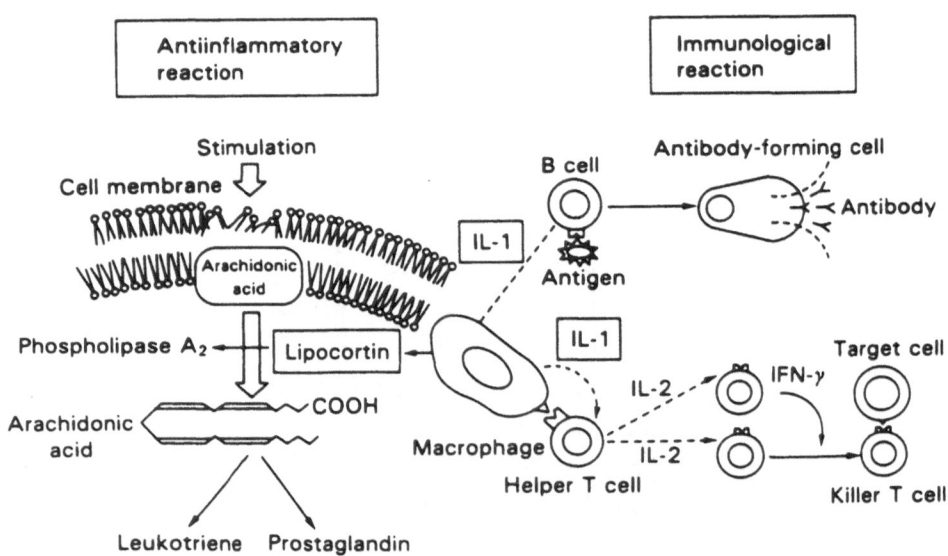

Fig. 10. Modes of action of Sho-saiko-to.

(Figure 10). By its direct effects on macrophage-monocytes, it augmented cellular immunity and antibody production through IL1 production, while at the same time showing an anti-inflammatory effect through the production of lipocortin or a lipocortinoid substance. These two different types of effects of Sho-saiko-to on the immune system and inflammatory reaction therefore appear to be beneficial for chronic hepatitis with the hepatitis B virus.

REFERENCES

1. K. Shimizu, S. Amagaya, and Y. Ogihara, Combined effect of Shosaikoto (Chinese traditional medicine) on the anti-inflammatory action of steroids, J. Pharmacobiodyn. 7:891 (1984).
2. K. Fujiwara, The effects of Sho-saiko-to (TJ-9) on the liver in experimental models, in: "Recent Advances in the Pharmacology of Kampo (Japanese Herbal) Medicines," Eikichi Hosoya and Yuichi Yamamura, ed., Excerpta Medica, Tokyo (1988).
3. H. Abe, M. Sakaguchi, H. Konishi, T. Tani, and S. Arichi, The effects of Saikosaponins on biological membranes. I. The relationship between the structures of Saikosaponins and hemolytic activity, Planta Medica 34:160 (1978).
4. K. Fujiwara and H. Oka, Kampo Medicines, Med. Clin. Jpn. 12:1150 (1986) (in Japanese).
5. E. Gronowicz, A. Coutinho, and F. Melchers, A plaque assay for all cells secreting Ig of a given type or class, Eur. J. Immunol. 6:588 (1976).

EFFECTS OF MEDICINAL PLANTS ON HEMOPOIETIC CELLS

Susumu Ikehara[1], Hideki Kawamura[2], Yasuhiro Komatsu[2],
Haruki Yamada[3], Hiroko Hisha[1], Ryoji Yasumizu[1],
Yoko Ohnishi-Inoue[1], Hiroaki Kiyohara[3], Masumi Hirano[3],
Masaki Aburada[2], and Robert A. Good[4]

[1]1st Department of Pathology, Kansai Medical University
Moriguchi, Osaka 570, Japan
[2]Tsumura Research Institute for Pharmacology, Ibaraki 300-11, Japan
[3]Oriental Medicine Research Center of the Kitasato Institute
Minato-ku, Tokyo 108, Japan
[4]All Children's Hospital, University of South Florida
St. Petersburg, FL 33701, U.S.A.

SUMMARY

It has been found that Juzentaihoto (TJ-48), one of the traditional Japanese kampo (herbal) medicines, improves the general condition of cancer patients receiving chemotherapy and radiation therapy. We analyze how TJ-48 elicits such effect, and show that oral administration of TJ-48 accelerates recovery from hemopoietic injury induced by radiation and the anti-cancer drug mitomycin C. The effects are found to be due to its stimulation of spleen colony-forming units. Based on the present findings, we propose that the administration of TJ-48 should be of benefit to patients receiving chemotherapy, radiation therapy or bone marrow transplantation.

INTRODUCTION

Juzentaihoto (TJ-48), one of the Japanese kampo (herbal) medicines (originating from traditional Chinese medicines), is comprised of ten medicinal plants: Astragalus root, Cinnamomum cortex, Rehmannia root, Paeonia root, Cnidium rhizome, Atractylodes lancea rhizome, Angelica root, Panax (ginseng) root, Hoelen, and Glycyrrhiza root. Because TJ-48 facilitates recovery from the physical debilitation of surgical patients or those suffering chronic diseases, it has traditionally been administered to patients with anemia, anorexia, or fatigue. It has also recently been found to improve the general condition of cancer patients receiving chemotherapy and/or radiation therapy (1).

In mice, we have found that TJ-48 is immunopotentiating (2) and enhances phagocytosis (3). It has also been shown to be radioprotective: when given to mice in drinking water after whole-body irradiation (5.5 Gy), the 30-day survival rate was 77.5%, whereas that of the controls was 32.5% (4).

Microbial Infections, Edited by H. Friedman *et al.*
Plenum Press, New York, 1992

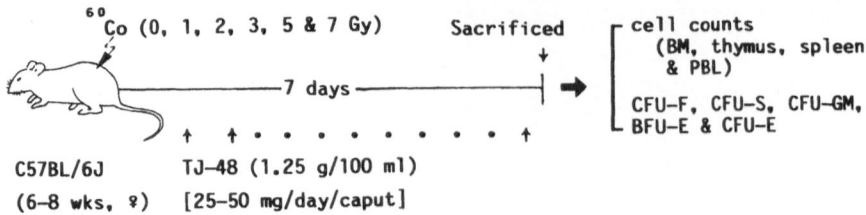

Fig. 1. Experimental protocol in the first experiment.

Radiation and anti-neoplastic drugs have toxic effects on the hemopoietic system (5, 6). There have been many reports with respect to various compounds that enhance hemopoiesis (7-12). During the clinical application of TJ-48, we found that peripheral blood counts were maintained at a higher level in patients who took TJ-48 orally in combination with anti-cancer drugs (1). This finding prompted us to examine the effects of TJ-48 on reversing hemopoietic suppression after treatment with radiation or mitomycin C (MMC). In the present study, we analyze how TJ-48 elicits recovery, and show that it increases spleen colony-forming unit (CFU-S) counts. In addition, we show that a hexane-soluble fraction extracted from TJ-48 has the capacity to enhance GM-CFC activity *in vitro*.

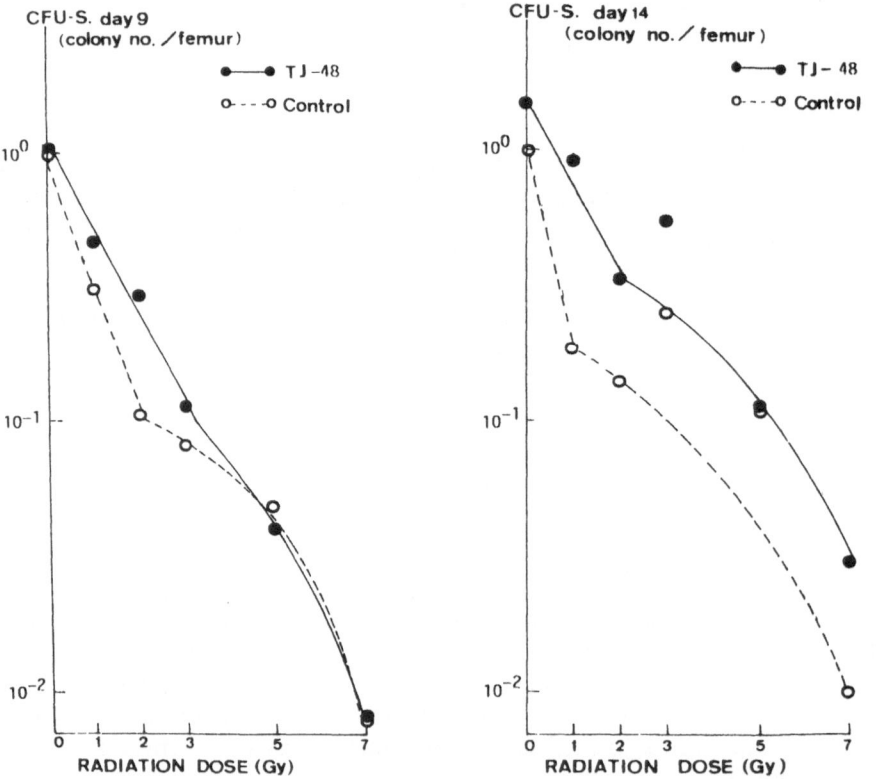

Fig. 2. Hemopoietic stem cell counts on day 7 after irradiation (0-7 Gy). Each point represents the mean of day-9 CFU-S or day-14 CFU-S counts of five mice reconstituted with 5×10^4 BMCs. o-o, control groups; •-•, TJ-48-treated groups.

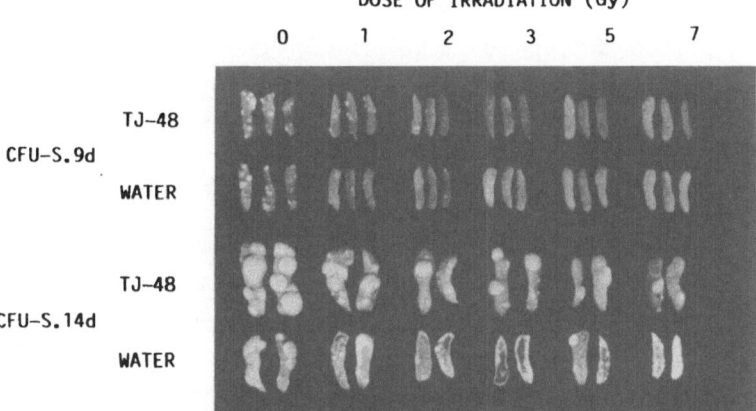

DOSE OF IRRADIATION (Gy)

Fig. 3. Day-9 and day-14 CFU-S. Mice were given TJ-48
solution 7 days after irradiation (0-7 Gy). They were
sacrificed, and BMCs were used for CFU-S assays. It
should be noted that the number of colonies in the assay
of day-14 CFU-S is greater in TJ-48-treated groups than
in the control (see Fig. 1).

STIMULATION OF CFU-S BY TJ-48

Mice were irradiated at 1, 2, 3, 5 and 7 Gy, then fed the TJ-48 solution for 7 days
(Figure 1). The peripheral blood cells were counted 7 days after irradiation and the
bone marrow cells were used for CFU assays.

The viable cells of bone marrow, thymus, spleen, peripheral white blood and
platelets decreased in a radiation dose-dependent manner similarly in both experimental
and control groups. Granulocyte-macrophage colony-forming unit (CFU-GM) and
erythroid colony-forming unit (CFU-E) assays showed no statistically significant
differences between the two groups. Fibroblast colony-forming unit (CFU-F) and
erythroid burst colony-forming unit (BFU-E) assays also revealed no significant
difference between the two groups. However, CFU-S assays on day 14 revealed that the
number of colonies in the TJ-48-treated group is greater than that in the control group
(Figure 2). The TJ-48-treated group showed an obviously good general condition
(including body weight) and heavier spleens with more and larger colonies than in the
control group (Figure 3).

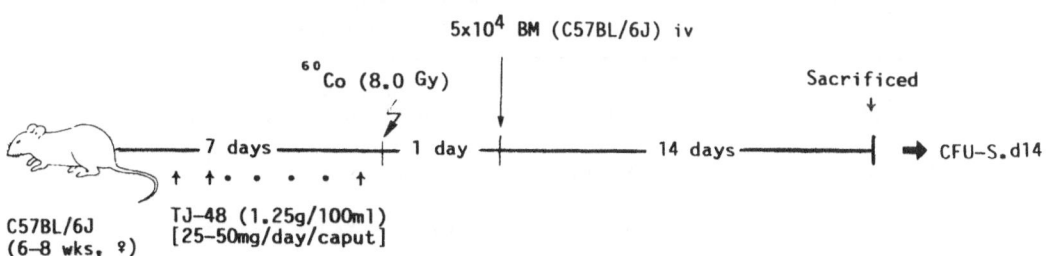

Fig. 4. Experimental protocol in the second experiment.

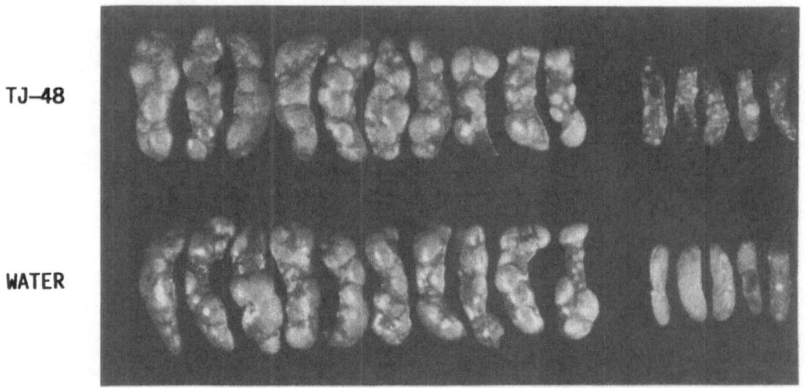

BMC(+)   BMC(−)

TJ–48

WATER

Fig. 5. Day 14 CFU-S in TJ-48-treated and non-treated mice. Mice were given TJ-48 solution for 7 days and then irradiated with 8 Gy. The mice were injected with 5x10⁴ BMCs from non-treated syngeneic mice, sacrificed on day 14 after cell transfer, and the number of colonies in the spleen counted (see Fig. 4).

EFFECT OF TJ-48 ON HEMOPOIETIC INDUCTIVE MICRO-ENVIRONMENT (HIM)

The next step was to examine whether the increased day-14 CFU-S counts were due to TJ-48 having an indirect effect via the HIM. Because it is known that the HIM is involved in the regulation of hemopoiesis (13-15), we attempted to examine the possibility that TJ-48 enhances day-14 CFU-S counts by activating the HIM. C57BL/6J (B6) mice were given TJ-48 solution for 7 days (Figure 4). The mice were irradiated at 8 Gy, since it is thought that the HIM (but not hemopoietic cells) is resistant to 8 Gy (16-18). The mice were reconstituted with 5×10^4 bone marrow cells from nontreated B6 donors 1 day after irradiation, sacrificed on day 14 after cell transfer, and the day-14 CFU-S counts were assessed. There was no difference between the TJ-48-treated and control groups in day-14 CFU-S count (Figure 5), body weight, spleen weight, or thymic weight (Table 1). These results suggest that the radioprotective effect of TJ-48 is due to stimulation of day 14 CFU-S.

Table 1. Day-14 CFU-S Counts and the Weight of Body, Spleen, and Thymus in the Second Experiment

Gr	n	TJ–48	8Gy	BMC	CFU–S.d14/10⁵	Weight		
						Body (g)	Spleen (mg)	Thymus (mg)
						Mean ± S.E.		
I	10	+	+	+	19.8 ±3.8	19.2 ±0.7	167.9 ±38.3	44.2 ± 6.1
II	10	−	+	+	18.6 ±3.4	18.8 ±0.6	180.0 ±24.3	43.6 ± 2.5
III	5	+	+	−	1.2 ±1.1	17.3 ±2.4	34.6 ± 5.9	32.4 ±13.9
IV	5	−	+	−	0.0 ±0.0	16.5 ±2.5	34.6 ± 6.1	30.6 ±17.4

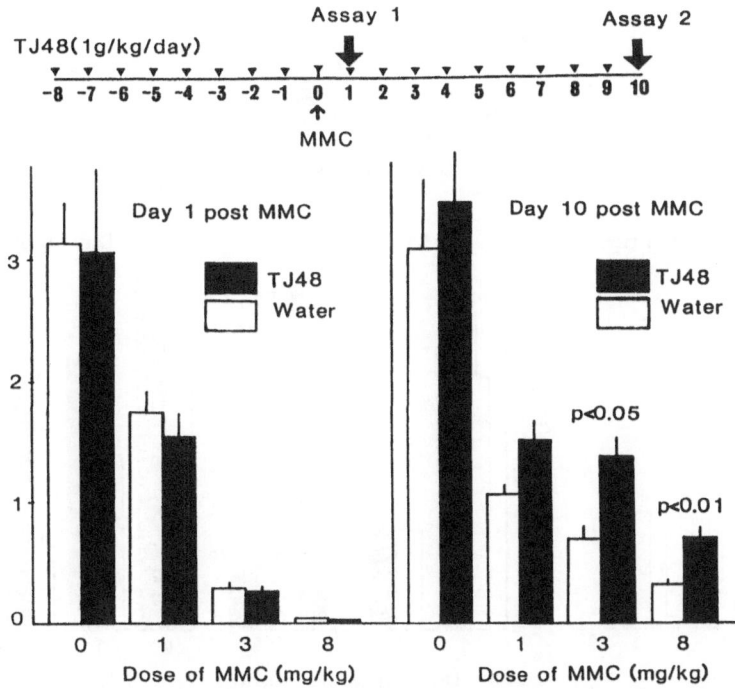

Fig. 6. Effect on CFU-S of TJ-48 administration before and after MMC injection in C57BL/6J mice. Mice were injected i.p. with MMC at 1, 3, or 8 mg/kg. One g/kg of TJ-48 or water alone was orally administered once a day from 8 days before MMC injection. Values represent the means ± S.E. of 5-6 determinations.

EFFECTS ON CFU-S OF TJ-48 ADMINISTRATION BEFORE AND AFTER MMC INJECTION

As shown in Figure 6, administration of TJ-48 (1 g/kg/day) was begun 8 days before MMC injection and continued throughout the experiment. When B6 mice were injected with 1, 3, or 8 mg/kg MMC, reductions of 44%, 91%, and 98% in CFU-S counts were observed on day 1 after the injection. Administration of TJ-48 for 8 days before MMC injection had no effect on these MMC-induced reductions in CFU-S counts. Ten days after the MMC injection, however, a marked increase in colony counts (about 40%-100%) over the control (MMC alone) was seen in the mice that had been treated with TJ-48, particularly in the groups that received high dosages (3 or 8 mg) of MMC.

EFFECT OF TJ-48 ON MMC-DOSE-DEPENDENT SUPPRESSION OF CFU-S

As shown in Figure 7, CFU-S counts were suppressed in proportion to the MMC dose. In contrast, the counts were enhanced by the administration of TJ-48; the greater the suppression, the greater the rate of recovery, and the longer the TJ-48 administration was continued, also the greater the rate of recovery.

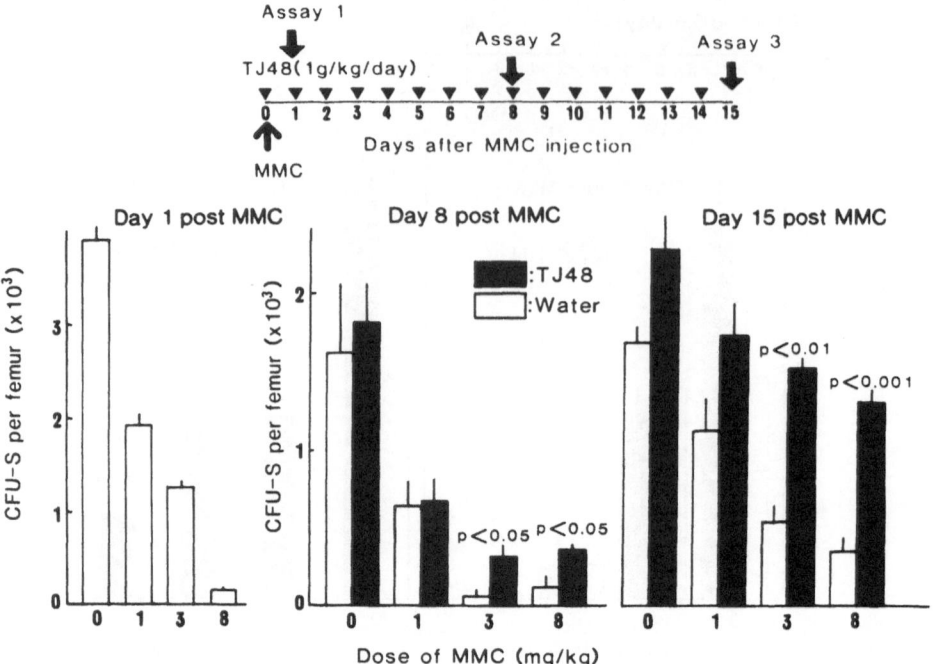

Fig. 7. Effects on CFU-S of TJ-48 administration after MMC injection.

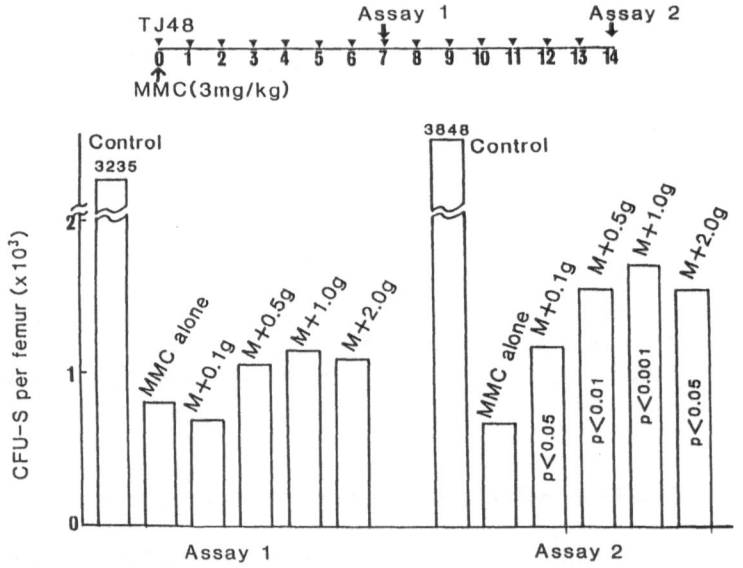

Fig. 8. Effects of TJ-48 concentrations on MMC-induced suppression in CFU-S.

DOSE-DEPENDENT EFFECT OF TJ-48 ON CFU-S AFTER MMC TREATMENT

After MMC (3 mg/kg) treatment, various doses of TJ-48 were administered i.p. to male B6 mice (7 to 9-week-old). There was no statistically significant difference in CFU-S counts in Assay 1 (7 days after TJ-48 administration), although 30% to 40% of CFU-S counts were increased by the administration of TJ-48 (> 0.5 g/kg/day) (Figure 8). After administration of TJ-48 for 14 days, however, significant increases in CFU-S counts were noted with TJ-48 doses of 0.1g to 2.0g/kg.

EFFECT OF TJ-48 EXTRACTS ON GM-CFC *IN VITRO*

Since we have demonstrated that TJ-48 enhances CFU-S counts after hematopoietic suppression induced by radiation and MMC treatment *in vivo*, we attempted to purify the active components in TJ-48 extracts.

As shown in Figure 9, TJ-48 was extracted by refluxing with MeOH (19); MeOH-soluble (Fr-1) and MeOH-insoluble fractions were obtained, and the latter was further fractionated into four fractions (Fr-2 to Fr-5) (19). Table 2 indicates that Fr-1 (MeOH-soluble fraction) and Fr-3 (dialyzable low-molecular weight fraction) enhance GM-CFC counts; only Fr-3 shows a dose-dependent enhancement. Further fractionation (Figure 10) revealed that Fr-1-1 (hexane-soluble fraction) obtained from Fr-1 has the capacity to enhance GM-CFC counts (Table 3). These results suggest that the active component is a lipophilic molecule with low molecular weight.

DISCUSSION

Although Japanese kampo medicines are known to be clinically effective in some cases, few studies have been carried out to elucidate the mechanisms of their actions. TJ-48 has been reported to induce various anti-tumor factors such as tumor necrosis

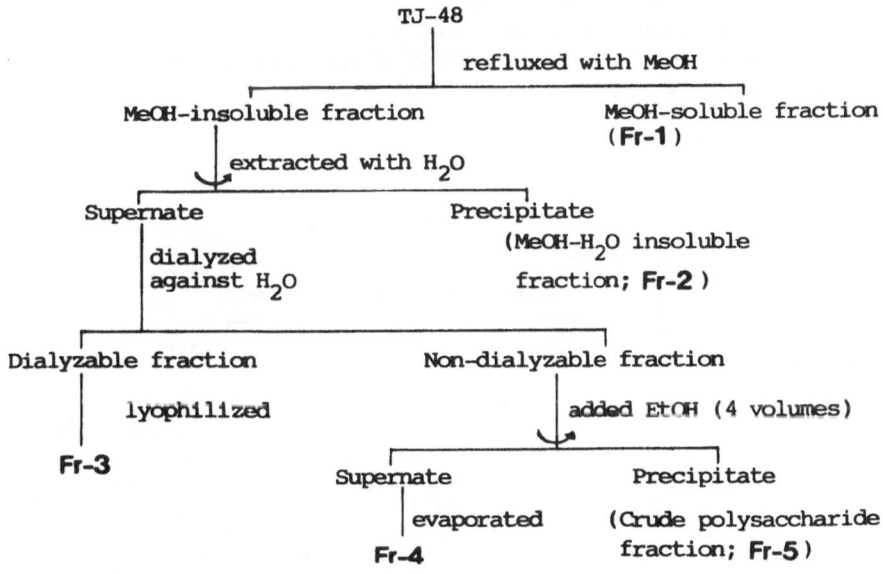

Fig. 9. Fractionation of TJ-48.

Table 2. GM-CFC Stimulatory Activity of Fractions Extracted from TJ-48

Fraction [‡]	Ratio of GM-CFC [‡‡]	
	$10\,\mu g/ml$	$100\,\mu g/ml$
Fr-1	2.11	1.84 [§]
Fr-2	n.d. [§]	n.d. [§]
Fr-3	1.26	2.18
Fr-4	1.51	0.93
Fr-5	1.19	1.58

[‡] All fractions were dissolved in PBS and used for GM-CFC assays.

[§] Fr-2 was insoluble in PBS and therefore not determined.

[‡‡]
$$ratio = \frac{number\ of\ GM\text{-}CFC\ on\ sample\ dish}{number\ of\ GM\text{-}CFC\ on\ control\ dish}$$

factor (20) and interferon (21), and to inhibit carcinogenesis (22). It has also been shown to prolong the survival of tumor-bearing mice when used in combination with an anti-cancer drug (23).

The present study was performed to determine why oral administration of TJ-48 accelerates the recovery from hemopoietic injury by radiation or the anti-cancer drug MMC. We have found that this action can be attributed to the capacity of TJ-48 to stimulate CFU-S. It has recently been noted that there are two distinct haemopoietic stem cell populations: one subset, which forms colonies detectable after 7 days, shows a relatively restricted capacity for self-replication and forms relatively small populations of differentiating progeny; the other subset, which is detectable after 14 days, perhaps more closely matches the behavior of repopulating cells as it has a considerably greater capacity for self-replication and generates much larger populations of differentiating progeny (24). Our data indicate that the administration of TJ-48 to irradiated mice for 7 days does not reduce the loss of mature hemopoietic cells (T cells, B cells, WBC, and platelets) or progenitor cells (CFU-GM, BFU-E, CFU-E, and CFU-F), but significantly activates CFU-S activity by day 14.

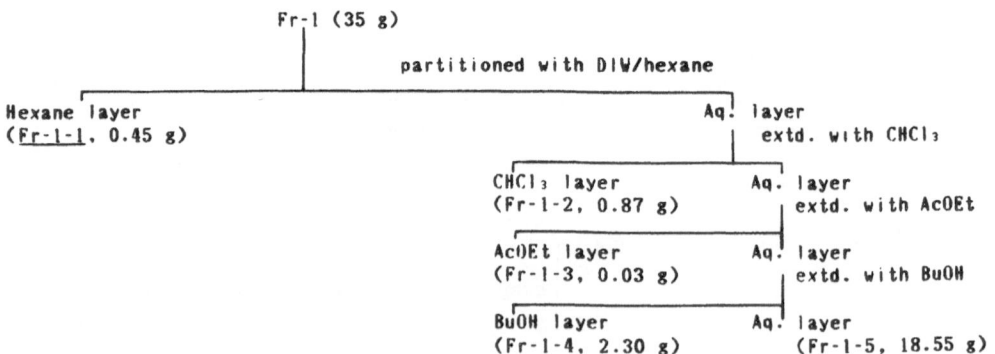

Fig. 10. Fractionation of GM-CFC-stimulating substances from Fr-1.

Table 3. GM-CFC Stimulatory Activity in Subfractions
of Fr-1

Fraction (10 μ g/ml)	Ratio of GM-CFC [*]
0.5% MeOH alone	1.08
Fr-1 (MeOH-soluble)	1.10
Fr-1-1 (Hexane-soluble)	2.77
Fr-1-2 ($CHCl_3$-soluble)	1.27
Fr-1-3 ($CH_3COOC_2H_5$-soluble)	1.41
Fr-1-4 (BuOH-soluble)	0.84
Fr-1-5 (H_2O-soluble)	1.05

Fractions other than Fr-1-5 were dissolved in MeOH
and used for GM-CFC assays, since they are insoluble
in PBS.
The final concentration of MeOH in each fraction was 0.5%.

[*] ratio = $\dfrac{\text{number of GM-CFC on sample dish}}{\text{number of GM-CFC on control dish}}$

Recently, HIM has attracted considerable attention because it plays a crucial role in supporting the proliferation of hemopoietic stem cells (13-15). The administration of TJ-48, however, did not lead to a statistically significant increase in the number of CFU-F. In addition, from the results of the second experiment, it seems that TJ-48 does not elicit this effect via HIM (Table 2). Thus, TJ-48 appears to promote recovery from radiation injury by stimulating the hemopoietic stem cells (CFU-S on day 14).

Pretreatment of mice with TJ-48 did not prevent MMC-induced suppression of CFU-S on day 1 after MMC injection, suggesting that TJ-48 cannot block the cytotoxic effect of MMC. The possibility that TJ-48 interrupts the activity of MMC by direct interaction can therefore be ruled out. TJ-48, however, markedly enhanced recovery from haematopoietic injury by MMC, and is capable of accelerating the recovery of CFU-S even when it is administered after MMC injection.

Various substances are known to have radioprotective effects (25-34). For example, carbon particles (25), thrombocyte-rich plasma (26), deuterated drinking water (27), and interleukin 1 (IL-1) (32) have been reported to stimulate hematological recovery in irradiated mice. Moor and Warren recently demonstrated that *in vivo* administration of IL-1 promotes hemopoietic regeneration after 5-FU treatment in mice, and acts synergistically with granulocyte colony-stimulating factors (G-CSF) in accelerating neutrophil regeneration (33). Of the numerous Chinese herbal medicines, ginseng (roots of Panax ginseng) is one in which the effective mechanism is known. It has recently been reported that ginseng increases the 30-day survival rate and accelerates the recovery of thrombocytes and erythrocytes as well as splenic weight in irradiated mice (28-30). It is therefore likely that traditional oriental medicines regulate homeostasis and correct abnormal states rather than making a selective attack on the target regions involved in the ailment (31).

In the present study, we have found that TJ-48 enhances the recovery from hemopoietic injury induced by radiation or the anti-cancer drug MMC, and that this is due to acceleration of the recovery of CFU-S. Thus, TJ-48 is a unique medicine that acts directly on the hemopoietic stem cells. We are now in the process of purifying its active components. Based on the present findings, we propose that the administration of TJ-48 should be of benefit to patients receiving chemotherapy, radiation therapy or bone marrow transplantation.

ACKNOWLEDGEMENTS

We thank Ms. K. Higuchi, Ms. Y. Shinno, Y. Umeda for expert technical assistance and Ms. S. Ohya, Ms. Y. Hatamoto and H. Eastwick-Field for help in manuscript preparation. This study was supported by the research funds from the "Traditional Oriental Medical Science Program" of the Public Health Bureau of the Tokyo Metropolitan Government.

REFERENCES

1. K. Nabeya and S. Ri, Effect of oriental medical herbs on the restoration of the human body before and after operation, Proc. Symp. Wakan-Yaku 16:201 (1983).

2. Y. Komatsu, N. Takemoto, H. Maruyama, H. Tsuchiya, M. Aburada, E. Hosoya, S. Shinohara, and H. Hamada, Effect of Juzentaihoto on the anti-SRBC response in mice, Jpn. J. Inflammat. 6:405 (1986).

3. H. Maruyama, H. Kawamura, N. Takemoto, Y. Komatsu, M. Aburada, and E. Hosoya, Effect of kampo medicines on phagocytes, Jpn. J. Inflammat. 8:65 (1988).

4. Y. Hosokawa, Radioprotective effects of oriental drugs on mice, Ther. Res. 2:31 (1985).

5. J. C. Marsh, The effects of cancer chemotherapeutic agents on normal hematopoietic precursor cells: a review, Cancer Res. 36:1953 (1976).

6. R. P. Gale, Antineoplastic chemotherapy myelosuppression: mechanisms and new approaches, Exp. Hematol. 13:3 (1985).

7. S. Ladish, D. G. Poplack, and J. M. Bull, Acceleration of myeloid recovery from cyclophosphamide-induced leukopenia by pretreatment with bacillus Calmette Guerin, Cancer Res. 38:1049 (1978).

8. Y. Maruyama, C. Magura, and J. Feola, Corynebacterium parvum-induced radiosensitivity and cycling changes of hematopoietic spleen colony-forming units, J. Natl. Cancer Inst. 59:173 (1979).

9. A. Hiraoka, M. Yamagishi, T. Ohkubo, T. Kamamoto, Y. Yoshida, and H. Uchino, Effect of a streptococcal preparation, OK432, on hematopoietic spleen colony formation in irradiated mice, Cancer Res. 41:2954 (1981).

10. O. Vos and W. S. O. Roos-Verhey, Protection against X-irradiation by some orally administered compounds, Int. J. Radiat. Biol. 45:479 (1984).

11. V. S. Gallicchio, Lithium and haematopoietic stem toxicity. 1. Recovery in vivo of murine haematopoietic stem cells (CFU-S and CFU-Mix) after single-dose administration of cyclophosphamide, Exp. Hematol. 14:395 (1986).

12. J. A. Laissue, H. J. Altermatt, E. Bally, J. O. Gebbers, Protection of mice from whole body gamma irradiation by deuteration of drinking water: hematologic findings, Exp. Hematol. 15:177 (1987).

13. A. Johnson and K. Dorshkind, Stromal cells in myeloid and lymphoid long-term bone marrow cultures can support multiple hemopoietic lineages and modulate their production of hemopoietic growth factors, Blood 68:1348 (1986).

14. V. C. Broudy, K. Kaushansky, J. M. Harlan, J. W. Adamson, Interleukin 1 stimulates human endothelial cells to produce granulocyte-macrophage colony-stimulating factor and granulocyte colony-stimulation factor, J. Immunol. 139:464 (1987).

15. P. Anklesaria, K. Kase, J. Glowacki, C. A. Holland, M. A. Sakakeeny, J. A. Wright, T. J. Fitzgerald, C-Y. Lee, and J. S. Greenberger. Engraftment of a clonal bone marrow stromal cell line in vivo stimulates hemopoietic recovery from total body irradiation, Proc. Natl. Acad. Sci. U.S.A. 84:7681 (1987).

16. W. H. Knospe, J. Blom, and W. H. Crosby, Regeneration of locally irradiated bone marrow. I. Dose dependent, long-term changes in the rat, with particular emphasis upon vascular and stromal reaction, Blood 28:398 (1966).

17. G. F. Rabotti, Bone marrow and spleen patterns in mice irradiated and protected with homologous cells, Ann. NY Acad. Sci. 114:468 (1964).

18. Y. Imai and I. Nakao, *In vivo* radiosensitivity and recovery pattern of the hematopoietic precursor cells and stem cells in mouse bone marrow, Exp. Hematol. 15:890 (1987).

19. H. Yamada, H. Kiyohara, J-C. Cyong, N. Takemoto, Y. Komatsu, H. Kawamura, M. Aburada, and E. Hosoya, Fractionation and characterization of mitogenic and anti-complementary active fractions from Kampo (Japanese herbal) medicine "Juzen-Taiho-To," Planta Med. 56:386 (1990).

20. K. Haranaka, N. Satomi, A. Sakurai, R. Haranaka, N. Okada, and M. Kobayashi, Antitumor activity and tumor necrosis factor producibility of traditional Chinese medicines and crude drugs, Cancer Immunol. Immunother. 20:1 (1985).

21. Y. Sakagami, Y. Mizoguchi, K. Miyajima, H. Kuboi, K. Kobayashi, K. Kioka, H. Takeda, T. Shin, and S. Morisawa, Antitumor activity of Shi-Quan-Da-Bu-Tang and its effects of interferon-r and interleukin-2 production, Jpn. J. Allergol. 37:57 (1988).

22. R. Haranaka, R. Hasegawa, S. Nakagawa, A. Sakurai, N. Satomi, and K. Haranaka, Antitumor activity of combination therapy with traditional Chinese medicine and OK432 or MMC, J. Biol. Response Mod. 7:77 (1988).

23. M. Aburada, S. Takeda, E. Ito, M. Nakamura, and E. Hosoya, Protective effects of Juzentaihoto, dried decoctum of 10 Chinese herbs mixture, upon the adverse effect of mitomycin C in mice, J. Pharm. Dyn. 6:1000 (1983).

24. D. Metcalf, "The Hemopoietic Colony Stimulating Factors," Elsevier, Amsterdam (1984).

25. S. Nakamura, Effect of carbon particles on hematological recovery of irradiated mice, Radiat. Res. 52:130 (1972).

26. W. Nakamura, Effect of thrombocyte-rich plasma on x-ray induced bone marrow death in mice, Radiat. Res. 55:118 (1973).

27. J. A. Laissue, H. J. Altermatt, E. Bally, and J-O, Gebbers, Protection of mice from whole body gamma irradiation by deuteration of drinking water: hematologic findings, Exp. Hematol. 15:177 (1987).

28. M. Yonezawa, Restoration of radiation injury by intraperitoneal injection of ginseng extract in mice, J. Radiat. Res. 17:111 (1976).

29. A. Takeda, M. Yonezawa, and N. Katoh, Restoration of radiation injury by ginseng. I. Responses of x-irradiated mice to ginseng extract, J. Radiat. Res. 22:323 (1981).

30. A. Takeda, N. Katoh, and M. Yonezawa, Restoration of radiation injury by ginseng. III. Radioprotective effect of thermostable fraction of ginseng extract on mice, rats and guinea pigs, J. Radiat. Res. 23:150 (1982).

31. K. Haranaka, N. Satomi, A. Sakurai, R. Haranaka, N. Okada, and M. Kobayashi, Antitumor activities and tumor necrosis factor producibility of traditional Chinese medicines and crude drug, Cancer Immunol. Immunother. 20:1 (1985).

32. R. Neta, S. Douches, and J. J. Oppenheim, Interleukin-1 is a radioprotector, J. Immunol. 136:2483 (1986).

33. M. A. S. Moore, and D. J. Warren, Synergy of interleukin 1 and granulocyte colony-stimulating factor: *in vivo* stimulating of stem-cell recovery and hematopoietic regeneration following 5-fluorouracil treatment of mice, Proc. Natl. Acad. Sci. U.S.A. 84:7134 (1987).

34. H. Kawamura, H. Maruyama, N. Takemoto, Y. Komatsu, M. Aburada, S. Ikehara, and E. Hosoya, Accelerating effect of Japanese kampo medicine on recovery of murine haematopoietic stem cells after administration of mitomycin C, <u>Int. J. Immunotherapy</u> V:35 (1989).

MULTIPLE IMMUNOLOGICAL FUNCTIONS OF EXTRACTS FROM THE

CONE OF JAPANESE WHITE PINE, *PINUS PARVIFLORA* SIEB. ET ZUCC

Hiroshi Sakagami[1], Kunio Konno[2], Yutaka Kawazoe[3], Patrick Lai[4], and Meihan Nonoyama[4]

Showa University School of Medicine[1], Tokyo, Japan, Jonan Hospital[2], Tokyo, Japan, Nagoya City University[3], Nagoya, Japan, Tampa Bay Research Institute[4], St. Petersburg, Florida, USA

INTRODUCTION

Japanese folklore suggests a therapeutic capacity of the extract of the pine cone of *Pinus parviflora* Sieb. et Zucc. for gastric cancer. We partially purified the antitumor substances (Fr. VI and Fr. VII) by acid- and ethanol- precipitation from the NaOH extract (1). It was unexpectedly found that the replication of human immunodeficiency virus (HIV) was potently inhibited by these fractions (2, 3). We report here their structural analysis and ability to induce several kinds of interesting biological activities.

STRUCTURAL ANALYSIS

Fr. VI, a brownish substance, has a MW of about 10 kD (1). Chemical analysis by UV, IR, NMR and ESR spectroscopy strongly suggested that this fraction consists of lignin structures complexed with polysaccharides (4). About 11% was neutral sugars, which were mainly fucose, mannose, galactose and glucose.

ANTIVIRAL SPECTRUM

Fr. VI inhibited HIV reverse transcriptase and stimulated the T-cell line (CEM) to produce a pepsin-sensitive HIV-1-inhibiting factor (HIF) (5). It inhibited the plaque formation of influenza virus in MDCK cells, and RNA-dependent RNA-polymerase activity (6). It inhibited the adsorption of herpes simplex virus (HSV-1, HSV-2) to African green monkey kidney cells and human adenocarcinoma cells (7).

Microbial Infections, Edited by H. Friedman *et al.*
Plenum Press, New York, 1992

IMMUNOMODULATION ACTIVITY

Antitumor Activity

Intraperitoneal administration of 10 mg/kg Fr. VI or Fr. VII significantly extended the survival time of sarcoma-180-bearing ICR mice (T/C = 227-271%), and suppressed the growth of solid tumor cells transplanted in mice, with occasional tumor regression and necrosis. These fractions had no direct cytotoxicity against cultured sarcoma-180 cells up to 500 µg/ml, suggesting host-mediated antitumor action (1).

Antimicrobial Activity

Pretreatment of mice with Fr. VI (i.p., s.c., i.m.) significantly prevented the lethal effect of *Staphylococcus aureus*, *Escherichia coli*, *Pseudomonas aeruginosa*, *Klebsiella pneumoniae* and *Candida albicans* (8). The administration of Fr. VI or Fr. VII (i.p.) induced significant accumulation of polymorphonuclear cells (PMN), and enhancement of luminol-dependent chemiluminescence (LDCL) by peritoneal exudate cells. The generation of LDCL by adherent cells was similarly enhanced by Fr. VI or Fr. VII treatment, but this was still only 10-20% of that by the peritoneal exudate cells (9).

Antiparasite Activity

Pretreatment with Fr. VI or Fr. VII significantly reduced the number of cysticercoids in the villi of small intestines after oral administration of *Hymenolepis nana* (Cestoda) eggs (10). Significant antiparasite effects were induced by subcutaneous administration of these fractions (10 mg/kg) to 1 week old mice, or by intraperitoneal or oral administration to 4 week old mice.

Hemolytic Plaque-Forming Cell Production

Various lignified products including pine cones extracts (Frs. V and VI) were potent stimulators of hemolytic plaque-forming cell production (11).

Cytokine Production

With an appropriate eliciting agent such as OK-432 (Picibanil), administration of Fr. VI (i.v.) transiently induced endogenous production of a cytotoxic factor (CF), possibly TNF, in normal mice (12). Production of CF elicited by OK-432 was significantly augmented by priming with Fr. VI. The production peaked 2 hr after OK-432 administration. The CF producibility depended greatly on both the dose and the interval between administration of Fr. VI and OK-432. It declined significantly during aging of the mice or after inoculation of mice with tumor cells (13).

EFFECTS OF Fr. VI AND Fr. VII ON MACROPHAGES/MONOCYTES, PMN AND SPLENOCYTES

They stimulated morphological maturation (enlargement and elongation) and the NBT-reducing activity of mouse peritoneal macrophages (14), and production of differentiation-inducing factor (DIF) by mouse macrophage-like J774.1 cells (1). They stimulated iodination (incorporation of radioactive iodine into an acid-insoluble fraction) of human peripheral blood monocytes, and the monocytes' production of CF and interleukin-1-like substance (unpublished data). They stimulated iodination of

PMN, time- and temperature- dependently. The stimulation effect was only observed in myeloperoxidase (MPO)-containing cells, and was significantly inhibited by MPO inhibitors (15). They stimulated the DNA synthesis of isolated mouse splenocytes (16).

TOXICOLOGY

Intravenously administered Fr. VI was moderately toxic on mice. The LD_{50} values of Fr. VI in male and female ICR mice were about 60 mg/kg. Macroscopic observation of dead animals suggested that Fr. VI caused pulmonary hemorrhage. However, Fr. VI administered orally to rats (male, SD strain) daily for 30 days at dose levels of 5 mg/kg to 405 mg/kg produced no toxic effects, such as weight gain or change in tissue weight/body weight ratio. The results suggest that Fr. VI is non-toxic when administered orally.

CONCLUSIONS AND PROSPECTS

Fr. VI and Fr. VII have been demonstrated to have various unique biological activities (such as antiviral activity, and stimulation of macrophages/monocytes and PMN), whereas most other biological response modifiers do not have these characteristics. This might be a result of its structure which is mainly lignin complexed with polysaccharides. Decomposition of the lignin structure with $NaClO_2$ led to significant loss of activity (14, 17, 18). The importance of the lignin structure was supported by our recent finding that synthetically polymerized phenylpropenoids had similar activity (19, 20). It is important to investigate whether these lignin-related products activate the MPO-halide-hydrogen peroxide system, which might lead to antimicrobial (21) or antiviral induction.

However, Fr. VI was more potent than its acid-hydrolyzed counterparts or synthesized phenylpropenoid polymers in inducing antimicrobial activity (unpublished data), and endogenous CF production in mice (12). This suggests the importance of complex bonding between the lignin and polysaccharide portions to augment the biological activity (22, 23).

ACKNOWLEDGEMENTS

We thank Dr. A. Simpson for editing the manuscript. This study was supported in part by a U. S. Public Health Services Grant U01 AI27280 from the National Institutes of Health.

REFERENCES

1. H. Sakagami, M. Ikeda, S. Unten, K. Takeda, J. Murayama, A. Hamada, K. Kimura, N. Komatsu, and K. Konno, Antitumor activity of polysaccharide fractions from pine cone extract of *Pinus parviflora* Sieb. et Zucc., <u>Anticancer Res.</u> 7:1153 (1987).
2. P. K. Lai, J. Donovan, H. Takayama, H. Sakagami, A. Tanaka, K. Konno, and M. Nonoyama, Modification of human immunodeficiency viral replication by pine conce extracts, <u>AIDS</u> <u>Res.</u> <u>Human</u> <u>Retroviruses</u> 6:205 (1990).

3. H. Takayama, G. Bradley, P. K. Lai, Y. Tamura, H. Sakagami, A. Tanaka, and M. Nonoyama, Inhibition of human immunodeficiency virus forward and reverse transcription by PC6, a natural product from cones of pine trees, AIDS Res. Hum. Retroviruses 7:349 (1991).

4. H. Sakagami, T. Oh-hara, T. Kaiya, Y. Kawazoe, M. Nonoyama, and K. Konno, Molecular species of the antitumor and antiviral fraction from pine cone extract, Anticancer Res. 9:1593 (1989).

5. Y. Tamura, P. K. Lai, W. G. Bradley, K. Konno, A. Tanaka, and M. Nonoyama, A soluble factor induced by an extract from *Pinus parviflora* Sieb et Zucc can inhibit the replication of human immunodeficiency virus *in vitro*, Proc. Natl. Acad. Sci. USA 88:2249 (1991).

6. K. Nagata, H. Sakagami, H. Harada, M. Nonoyama, A. Ishihama, and K. Konno, Inhibition of influenza virus infection by pine cone antitumor substances, Antiviral Res. 13:11 (1990).

7. K. Fukuchi, H. Sakagami, M. Ikeda, Y. Kawazoe, T. Oh-hara, K. Konno, S. Ichikawa, N. Hata, H. Kondo, and M. Nonoyama, Inhibition of herpes simplex virus infection by pine cone antitumor substances, Anticancer Res. 9:313 (1989).

8. T. Oh-hara, H. Sakagami, Y. Kawazoe, T. Kaiya, N. Komatsu, N. Ohsawa, M. Fujimaki, S. Tanuma, and K. Konno, Antimicrobial spectrum of lignin-related pine cone extracts of *Pinus parviflora* Sieb. et Zucc., In Vivo 4:7 (1990).

9. H. Harada, H. Sakagami, K. Konno, T. Sato, N. Osawa, M. Fujimaki, and N. Komatsu, Induction of antimicrobial activity by antitumor substances from pine cone extract of *Pinus parviflora* Sieb. et Zucc., Anticancer Res. 8:581 (1988).

10. M. Abe, K. Okamoto, K. Konno, and H. Sakagami, Induction of antiparasite activity by pine cone lignin-related substances, In Vivo 3:359 (1989).

11. T. Oh-hara, Y. Ikeda, H. Sakagami, K. Konno, T. Kaiya, K. Kohda, and Y. Kawazoe, Lignified natural products as potential medicinal resources, I. Potentiation of hemolytic plaque-forming cell production in mice, Chem. Pharm. Bull. 38:282 (1990).

12. H. Sakagami, S. Kohno, S. Tanuma, and Y. Kawazoe, Induction of cytotoxic factor in mice by lignified materials combined with OK-432 (Picibanil), In Vivo 4:371 (1990).

13. A. Hanaoka, H. Sakagami, and K. Konno, Pine cone antitumor substances stimulate cytotoxic factor production in young mice, but not in aged or tumor-bearing mice, Showa Univ. J. Med. Sci. 1:57 (1989).

14. K. Kikuchi, H. Sakagami, S. Fujinaga, Y. Kawazoe, T. Oh-hara, S. Ichikawa, Y. Kurakata, M. Takeda, and T. Sato, Stimulation of mouse peritoneal macrophages by lignin-related substances, Anticancer Res. 11:841 (1991).

15. S. Unten, H. Sakagami, and K. Konno, Stimulation of granulocytic cell iodination by pine cone antitumor substances, J. Leukocyte Biol. 45:168 (1989).

16. Y. Kurakata, H. Sakagami, M. Takeda, K. Konno, K. Kitajima, S. Ichikawa, N. Hata, and T. Sato, Mitogenic activity of pine cone extracts against cultured splenocytes from normal and tumor-bearing animals, Anticancer Res. 9:961 (1989).

17. H. Harada, H. Sakagami, K. Nagata, T. Oh-hara, Y. Kawazoe, A. Ishihama, N. Hata, Y. Misawa, H. Terada, and K. Konno, Possible involvement of lignin structure in anti-influenza virus activity, Antiviral Res. 15:41 (1991).

18. H. Sakagami, Y. Kawazoe, T. Oh-hara, K. Kitajima, Y. Inoue, S. Tanuma, S. Ichikawa, and K. Konno, Stimulation of human peripheral blood polymorphonuclear cell iodination by lignin-related substances, J. Leukocyte Biol. 49:277 (1991).

19. H. Sakagami, K. Nagata, A. Ishihama, T. Oh-hara, and Y. Kawazoe, Anti-influenza virus activity of synthetically polymerized phenylpropenoids, Biochem. Biophys. Res. Commun. 172:1267 (1990).

20. H. Sakagami, T. Oh-hara, K. Kohda, and Y. Kawazoe, Lignified materials as a potential medicinal resource. IV. Dehydrogenation polymers of some phenylpropenoids and their capacity to stimulate polymorphonuclear cell iodination, Chem. Pharm. Bull. 39:950 (1991).

21. S. J. Klebanoff, Myeloperoxidase-halide-hydrogen peroxide antibacterial system, J. Bact. 95:2131 (1968).

22. T. Oh-hara, Y. Kawazoe, and H. Sakagami, Lignified materials as potential medicinal resources. III. Diversity of biological activity and possible molecular species involved, Chem. Pharm. Bull. 38:3131 (1990).

23. H. Sakagami, Y. Kawazoe, N. Komatsu, A. Simpson, M. Nonoyama, K. Konno, T. Yoshida, Y. Kuroiwa, and S. Tanuma, Antitumor, antiviral and immunopotentiating activities of pine cone extracts: Potential medicinal efficacy of natural and synthetic lignin-related materials (Review), Anticancer Res. 11:881 (1991).

BIOLOGICAL RESPONSE MODIFIERS IN INFECTIOUS DISEASES:

PROSPECTIVES FOR THE FUTURE

Herman Friedman

Department of Medical Microbiology and Immunology
University of South Florida College of Medicine
Tampa, Florida 33612

It is less than 125 years that the etiologic cause of infectious diseases was described and widely accepted. Since then it has become well known that infections are due to microorganisms, either bacteria, fungi, rickettsia or viruses. Soon after discovery that microorganisms cause disease, early "biological" methods were developed to prevent many infections, i.e, vaccination with killed or attenuated microorganisms given to individuals to induce resistance to infection by specific humoral inhibitors (i.e. antibodies), cellular immunity, or both. Thus vaccines were really the first "biological response modifiers" (BRM). However, in the middle portion of this century antibiotics superseded many vaccine procedures to treat infections and became a staple for infectious disease control. It is now well known, however, that certain antibiotics may also affect the host immune response system, especially antibiotics which are antimetabolites and interfere with metabolic or synthetic pathways. Yet antibiotics are not considered biological response modifiers.

In the last two or three decades the field of BRM has emerged mainly for treatment or stimulation of host resistance against neoplasia. Nevertheless, many of the BRM systems developed for malignancy are now being considered for infectious disease therapy or even prevention. The symposium on which this proceeding is based dealt with various BRM models which are being developed or adapted to infectious disease areas. Among the first is the possible use of endotoxins or other bacterial lipopolysaccharides as immunomodulators which increase host resistance to microbial infection or have other physiological or beneficial effects on the host in regards to resistance to infectious organisms. In this symposium a wide variety of endotoxin models were described, including those which are being used in diverse model systems for either enhancing or even suppressing the immune response, including unwanted responses. Components of bacterial endotoxins, especially lipid A and synthetic preparations, are being utilized in a wide variety of experimental systems. Furthermore, antibodies to endotoxins, including monoclonal antibodies to endotoxins and their components, are also being utilized as biological response modifiers in microbial infection.

In terms of other bacterial products, a number of agents derived from bacteria have been suggested as nonspecific enhancers of immunity to infectious agents, including bacteria, fungi and viruses. The prototype substance has been muramyl dipeptide (MDP), which appears to be the smallest adjuvantic unit of a bacterial cell

Microbial Infections, Edited by H. Friedman *et al.*
Plenum Press, New York, 1992

wall, especially those derived from mycobacteria which nonspecifically can stimulate host resistance against infectious agents. Various derivatives of MDP are being utilized not only as immunoadjuvants for malignancy and even autoimmunity, but also against bacterial infections. MDP, as well as other BMRs derived from bacteria, induce a wide variety of cytokines, many of which are important in the immune response and result also in enhanced resistance. Coupling of MDP with cytokines chemically has been proposed, and studies are underway to show that these "hybrid" preparations have immunostimulatory properties that can protect against infectious disease, at least in experimental animal models.

MDP, as well as other bacterial products and certain cytokines including interferons and even the interleukins, activate macrophages important not only in cellular and humoral immune responses but also in resistance to infectious agents. Stimulation of macrophages with microorganisms or some of their products activate the cells and induce them to produce or release cytokines which, in turn, affect host resistance and immunity. Among the factors released by activated macrophages is tumor necrosis factor (TNF), first described as a soluble substance produced by lipopolysaccharide activated macrophages resulting in necrosis of certain tumors. TNF is now known to be a strong immunoregulator and immunomodulating cytokine, and also appears to have marked effects in enhancing resistance against certain infections, including those caused by opportunistic bacteria, fungi and viruses. Thus, there is much interest concerning cytokines in immune responses and resistance to microbial infectious agents. It appears that to date we have only scratched the surface of the potential use of cytokines in infectious diseases, especially therapeutic use of such cytokines in individuals who have already been infected with either pathogenic or opportunistic microorganism.

It is widely recognized that interferon may be useful in many infections by viruses. However, the interferons may also be of much value in infections by bacteria and fungi. Other cytokines such as interleukin 2, interleukin 8, or the many other interleukins being discovered almost on a daily basis, have effects not only against tumor cells but also against certain microorganisms, including bacteria and viruses, as well as fungi. Thus, there is a great deal of interest concerning cytokines and infectious diseases and whether cytokines activate cells directly or indirectly by as yet unknown mechanisms.

It is of great interest that BRMs are present and can be prepared not only from bacteria or synthesized *in vitro* to mimic bacterial components, but also can be found present in various botanical substances, including plants, herbs and other living components of the biosphere around us. A most interesting session in this symposium was that dealing with "Kampo Medicine" (or Oriental biological response modifiers). A number of exciting studies are being performed both in Europe as well as in Japan and China concerning use of plant extracts and their preparations to stimulate host resistance against infectious agents, as well as other diseases of man. For example, in studies in Germany, lentins and glycans prepared from various plants have been shown to have immunostimulatory activities and these activities are not due to contamination with endotoxins from bacteria. Similar distinct preparations in Japan as well as in China also have been shown to have immunostimulatory activity as well as activity against various diseases in man and experimental animals, but information concerning the nature and characteristics of these plant and herb derived BRMs, and whether or not they are contaminated with endotoxins, is not yet known. It can be assumed, however, that future studies concerning such "natural" immunostimulators from sources as diverse as plants and herbs may be of great excitement in the field of BRM and infectious diseases.

The BRM approach to infections highlights the prohost concept of treating an individual to resist or control infections by pathogenic as well as opportunistic microorganisms. The continuing increase in knowledge about the nature and characteristics of the immune response lends itself to developing immunologic concepts concerning control of infections which are not based on the specificity of a vaccine or the complete nonspecificity of antibiotics. The ability to recognize the importance of certain components of the immune response, whether T cells, NK cells, other cytolytic cells, or humoral antibodies, as well as complement, interferon, etc., to participate in host defenses against infections provides a conceptual basis for developing BRMs dependent upon certain cells or components of the host defense system which influence beneficial host response mechanisms rather than detrimental one. It can be anticipated that further investigations concerning immune interactions of a host with microbes, either bacteria, viruses or fungi, will permit not only an understanding but also development of better methods to control such microbes associated with infectious diseases. The publication of the proceedings of this symposium will hopefully stimulate many investigators to recognize the importance of BRMs in host resistance to infections and will stimulate even more studies in the future which should hold great promise in enhancing beneficial host/parasite interactions.

INDEX

Polysaccharide (*cont'd*)
  botanical, medicinal, 300
  capsular, streptococcal, 32, 125–135
  fungal, 299
  as immunomodulator, 31–38, 299–308
  natural, 300
    structure, 300
  streptococcal, capsular, 32, 125–135
  as tumor inhibitor, 300, 302
Polysaccharide polymerase, 32
Probit analysis method, statistical, 233, 239
Prohost approach, 13
Propionibacterium acnes, 157, 159, 253
Prostaglandin E2 (PGE2), 41–42, 45, 259
  assay, 41–42
Protein
  disulfide isomerase, 268
  isolation, 50
  39kDa, 49, 50, 53–60
  kinase (PKC), 39, 42, 43, 46, 47
  and lipopolysaccharide, 50
  tyrosine kinase (PTK), 46
Proteus mirabilis, 49, 50, 53–55, 60
Pseudomonas aeruginosa, 32, 93, 176–179, 190, 194, 203, 295
PTK, *see* Protein tyrosine kinase
Pyran copolymer (MVE-2), 15, 248

Quinilonamine, 247

Rat, lung infection, fungal, 186–189
Reed and Muench method, 277
Response modifer, biological (BRM) 145, 146, 243–251
  antibiotics as, 337
  antivirals as, 243–251
  bacteria as, 161
    a double-edged sword, 161
  cytokines as, 338
  diverse, 243
  future of, 337–339
  in host, immunosuppressed, 243–251
  immunomodulators as, 337
  interferons as, 338
  interleukins as, 338
  kampo medicines as, 338
  lipoteichoic acid, streptococcal as, 157–160
  macrophages as, 338
  muramyl dipeptide as, 175–184
    as prototype, 337
  plant polysaccharides as, 338
  tumor necrosis factor as, 338
  vaccines were the first, 337
13-*cis*-Retinoic acid (RA), 265, 266
Ribonucleic acid, *see* RNA
Rickettsia sp., 81
Rift Valley Fever virus, 247
RNA, 41, 99, 101, 111
RNase L, 248
Romurtide, 185–191, 260–261
  and *Aspergillus fumigatus* in rat, 186–189
  and infection, experimental, in animal, 185–191

Romurtide (*cont'd*)
  and *Klebsiella pneumoniae* in mouse, 185–186
  mode of action, 261
RU-*41740*, *see* Biostim

Saikosaponin, 309
Salmonella abortus equi, 50, 58
  S. enteritidis, 89, 158
  S. minnesota R*595*, 33–36
  S. typhi O-*901*, 40
  S. typhimurium, 34, 70, 89–94, 168, 194
Salmonellosis in mouse, 89–95
  cytokine protection against, 89–95
Saponin, 98, 309
Sarcoma-*180* in mouse, 201, 300, 301, 305
Schistosoma japonicum, 208
  S. mansoni, 81, 207–212
Semliki Forest virus, 247, 248, 275
Sendai virus, 254–259, 275
Serratia marcescens, 32
Spleen cell, 53–55, 126, 130, 131, 204, 205, 290–292, 302
Splenocyte, 332, 333
Staphylococcus aureus, 168, 137–143, 168, 190, 203, 235
Streptococcus faecalis ATCC9790, 158–160
  S. mutans, 32
  S. pneumoniae, 168, 176
  S. pyogenes, 157–160
Streptomyces olivoreticuli, 193
Superoxide anion, 79
Sho-saiko-to, 309–317
  and arachidonic acid cascade, 309–317
  and cytokine cascade, 309–317
  mixture of 7 herbs, 309
  mode of action, 316
Tamp cell, 226–229
T-cell, 2–3, 31, 45, 106, 113–118, 125–135, 166, 219
  activation, 31–32
  adult, *see* Adult T-cell
  and cytokines, 125–135
  and endotoxin, 32–36
  growth factor, *see* Interleukin-2
  line CEM, 331
  and *Listeria monocytogenes*, 113–123
  and monocytes, 113–123
  purification, 40
  subpopulations, 114, 116
  suppressor-, 2, 32, 33, 125–135
Therapy
  immunomodulation, 16
  immunorestoration, 16
7-Thia-8-oxoguanosine, 244, 247
Thioredoxin, 265–274
Thioredoxin reductase, 265–271
Thymidine
  assay, 267
  and DNA, 41
Thymocyte, 290, 291, 311, 312
Thymoma cell line EL-*4*, 82
Thymopentin, 18, 19
Thymopoietin, 14

Thymopoietin pentapeptide, 3
Thymosin, 3, 14
Thymulin, 14
Thymus, mammalian, 3
  and AIDS, 19
TIL, *see* Tumor-infiltrating lymphocyte
TJ-*48* (Juzen-taiho-to), 287–296, 319, 325
  and infection, microbial, 295–296
  and lymphocyte function in assays, 288–292
  mixture of 10 herbs, 288, 310
T-lymphotropic virus, human type I (HTLV),
      265
TNF, *see* Tumor necrosis factor
Toxoplasma gondii, 81, 83
Trehalose dimycholate (TDM), 253, 258
Tretinoin, 280
Tretinoin palmitate, 280
Trypanosoma cruzi, 83
Tuberculosis, 169
Tuftsin, 15
Tumor
  cell line
      WEHI-*164*-JD, 218
      WISH, 218
  and lymphocyte infiltration (TIL), 133

Tumor (*cont'd*)
  necrosis factor (TNF), 3, 25, 97, 107, 108, 141,
      157–159, 166, 179–182, 218–220, 228,
      229, 295, 332
    alpha, 70, 79, 81, 91–94
    beta, 91–93

Ubenimex (bestatin), 193–200
  and *Candida albicans* infection, 193–200
  and host resistance enhancement, 193–200
  in mouse, 193–200
  as response modifier, biological (BRM), 193–200
Uronic acid, 300

Vaccine adjuvant B*30*-MDP, 145
Vaccinia virus, 275
Vesicular stomatitis virus (VSV), 203, 204, 207, 218
Virosome, *see* Liposome, stable

Western blotting analysis, 266–269

Yeast glucan, 202–207, 212

Zidovudine, *see* AZT
Zymosan, 202, 293, 303–306